Statistical Methods

Springer
New York
Berlin
Heidelberg
Barcelona
Budapest
Hong Kong
London
Milan
Paris
Santa Clara
Singapore
Tokyo

David J. Saville Graham R. Wood

Statistical Methods
A Geometric Primer

With 88 figures

Springer

David J. Saville
Biometrics Unit
New Zealand Pastoral Agriculture Research Institute
P.O. Box 60, Lincoln
New Zealand

Graham R. Wood
Department of Mathematics and Computing
Central Queensland University
Rockhampton, Queensland 4702
Australia

Library of Congress Cataloging-in-Publication Data
Saville, David J.
Statistical methods : a geometric primer / David J. Saville, Graham R. Wood.
p. cm.
Includes bibliographical references and index.
ISBN 0-387-94705-1 (softcover : alk. paper)
1. Analysis of variance. 2. Regression analysis. 3. Geometry.
I. Wood, Graham R. II. Title.
QA279.S253 1996
519.5–dc20 96-10598

Printed on acid-free paper.

Production managed by Steven Pisano; manufacturing supervised by Jeffrey Taub.
Camera-ready copy prepared by the author.
Printed and bound by Edwards Brothers, Inc., Ann Arbor, MI.
Printed in the United States of America.

9 8 7 6 5 4 3 2 1

ISBN 0-387-94705-1 Springer-Verlag New York Berlin Heidelberg SPIN-10523864

To our wives, Sandra and Stephanie;
our children, Jessica and Jordan, Zoë and Verity;
and to our parents.

Preface

The purpose of this primer is to impart a real understanding of the most commonly used statistical methods in the simplest possible manner. To this end, four examples involving normally distributed data are analyzed using a novel geometric approach. These examples explain the mathematics underlying paired and independent sample t tests, analysis of variance and regression.

In essence, the approach reduces statistical analysis from a high-dimensional problem involving matrices to a two-dimensional problem involving only vectors. The approach makes it possible to learn statistics by drawing pictures on a piece of paper.

A sister text, *Statistical Methods: The Geometric Approach*, was published in 1991 by the same authors and publishers. This earlier text is aimed at second-year university students. It constitutes a more comprehensive treatment of the statistical methods most commonly used by research workers, and includes realistic, mainly agricultural examples, assumption checking, data transformations, presentation and interpretation of results, and computing. By way of contrast, this introductory primer also uses realistic, though mainly nonagricultural examples, but concentrates mostly on presenting the geometry in as simple a manner as possible. This simplification makes the underlying theory accessible to a much wider group of students and users of statistics.

The Authors

To introduce ourselves, David Saville is a consulting statistician working in agricultural research; Graham Wood is a university professor involved in teaching statistical methods. We have both worked for over twenty years in our current fields.

The Readers

We envisage four uses for the book. Firstly, we see it as a general interest book for statisticians, mathematicians, students of statistics, and users of statistics. These are people who are interested in answering the question: "What is the mathematics underlying methods such as the paired t test, analysis of variance, and regression?" Secondly, we see the book as an introduction to the geometric approach for lecturers of second-year applied statistics courses and their students. This primer and/or the more comprehensive sister text would be suitable as the text or texts for such a course. Thirdly, the book could serve as the basis for a first-year university course for students who plan to go on in either mathematics or statistics. Such a course could either teach linear algebra and statistics simultaneously from the start of the year, or could follow on from introductory linear algebra and statistics courses. Fourthly, the book could be used to enrich linear algebra courses by providing a concrete example of the application of linear algebra in another field of mathematics.

Origin of Material

The material presented in this book has its origin in a three-hour workshop which we ran at the Third International Conference on the Teaching of Statistics in Dunedin, New Zealand, in August, 1990. The enthusiasm of the seventy participants, about half high-school and half university teachers, inspired us to write up this workshop material as a simpler version of the sister text which we referred to above.

In expanding the workshop material to a small textbook, we have on occasions copied or modified material from the sister text. This is especially the case for class exercises and general exercises.

Acknowledgments

We are grateful to the many students who have contributed to the evolution of the geometric approach, and to the teachers whose enthusiasm for the approach inspired the writing of this lower-level text. Philip A. Carusi is especially thanked for stimulating the work on the alternative test statistic. Gilbert Wells and Murray Jorgensen are thanked for helpful discussions, Harold Henderson, Sandy Wright, Lesley Hunt, Garry Tee and Bill Baritompa for constructive criticism, and David Baird for computer support. Jordan Wood is thanked for calculating the percentiles of θ in Table T.2 using Maple. We have prepared the book ourselves using LATEX. Our employers, the NZ Pastoral Agriculture Research Institute Ltd., Lincoln, and the Mathematics Departments of the Universities of Canterbury, Christchurch and Central Queensland, Rockhampton are thanked for their support, with special thanks to John Keoghan. Martin Gilchrist, of Springer-Verlag, is also thanked for pointing out a simplification to Chapter 1.

David J. Saville
Christchurch, New Zealand

Graham R. Wood
Rockhampton, Australia

Contents

Chapter 1

Introduction

In §1.1 of this chapter we discuss the aim of this book. In §1.2 and §1.3 we pictorially introduce the geometric ideas upon which the most commonly used statistical methods are based. In §1.4 we mention an alternative to the usual t or F test statistics, and in §1.5 we mention estimation of confidence intervals. In §1.6 we describe the layout of the book and in §1.7 we have a word with teachers. Lastly, in §1.8 we mention other authors who have dealt with these geometric ideas.

1.1 Aim of This Book

The aim of this book is to present the mathematics underlying elementary statistical methods based on the normal distribution in as simple a manner as possible. The methods we refer to are paired and independent sample t tests, analysis of variance, and regression. The underlying mathematics is the theory of n-dimensional space.

To present this material in an elegant and succinct manner many statistical texts employ matrix methods. However, we prefer to use the simpler vector geometric methods since these reduce the dimensionality of the problem, provide more visual insight, and make the theory accessible to a much wider audience.

This primer has been written with two audiences in mind. Firstly, students of statistics, old or young, who wish to understand the mathematics underlying the statistical methods based on the normal distribution. This includes general interest readers as well as university students in their first or second year. Secondly, linear algebra or vector geometry students who desire the illumination provided by a concrete application of the theory.

For advancing students we see this text as a gentle lead-in to our more comprehensive text, *Statistical Methods: The Geometric Approach*, published in 1991 by the same authors and publishers.

1.2 The Basic Idea

To illustrate the basic idea behind our approach we turn to the simplest possible case, that of a sample of size two from a single population. We deal with such a small sample so that we can draw our pictures in two dimensions, as compared to the n dimensions required for a sample size of n.

Suppose a thermometer has been purchased and the question arises:"Is the thermometer accurate at freezing point?" One way of testing for bias at freezing point would be to take the reading on the thermometer after placing it in a beaker of finely chopped ice, on two different days. The data from such a study would be two temperatures, y_1 and y_2. How can we use these data to decide whether the thermometer is biased?

Imagine that you carry out the above study many times (on many independent pairs of days). Also imagine that the thermometer is perfectly accurate, so that the temperatures average to $\mu = 0$ in the long run. Then each pair of temperatures y_1, y_2, such as the pair shown in Figure 1.1(a), can be plotted as a single point ($\times$) on a two-dimensional graph as shown in Figure 1.1(b). Over many repetitions of the study these points will be dotted around the origin as shown in Figure 1.1(c).

Now imagine that the thermometer is biased, and that the temperatures average to $\mu = 1.2°$C in the long run. Then the corresponding pictures are Figures 1.1(d), (e) and (f). A typical sample point ($\times$) is now one in a scatter of points which is centered on the point $(\mu, \mu) = (1.2, 1.2)$.

With these pictures at our disposal, we are now able to translate the original question: "Is the thermometer biased at freezing point?" into a geometric question. In a real situation you will have only one set of experimental results, say $(y_1, y_2) = (1.3, 1.5)$. The equivalent geometric question is:

> Does the point belong to a scatter centered on the origin, or does it belong to one centered away from the origin?

More tersely, is $\mu = 0$ or is $\mu \neq 0$? What we need is a measure, termed a *test statistic*, which will distinguish the two cases. It must be small if $\mu = 0$ and large if $\mu \neq 0$.

Figures 1.1(c) and (f) provide the clues we require. When $\mu \neq 0$ the scatter of points is moved up the equiangular line, the line at 45° to both axes. For a typical sample point such as "$\times$," this has the effect of making the ratio of the distance "up the line" (A) to the distance "from the line" (B) greater when $\mu \neq 0$ than when $\mu = 0$, as shown in Figures 1.2(a) and (b). This suggests that we use the ratio

$$\frac{A}{B}$$

as our test statistic. This ratio is "small" when $\mu = 0$ and "large" when $\mu \neq 0$. Note that this test statistic works no matter what the spread of the

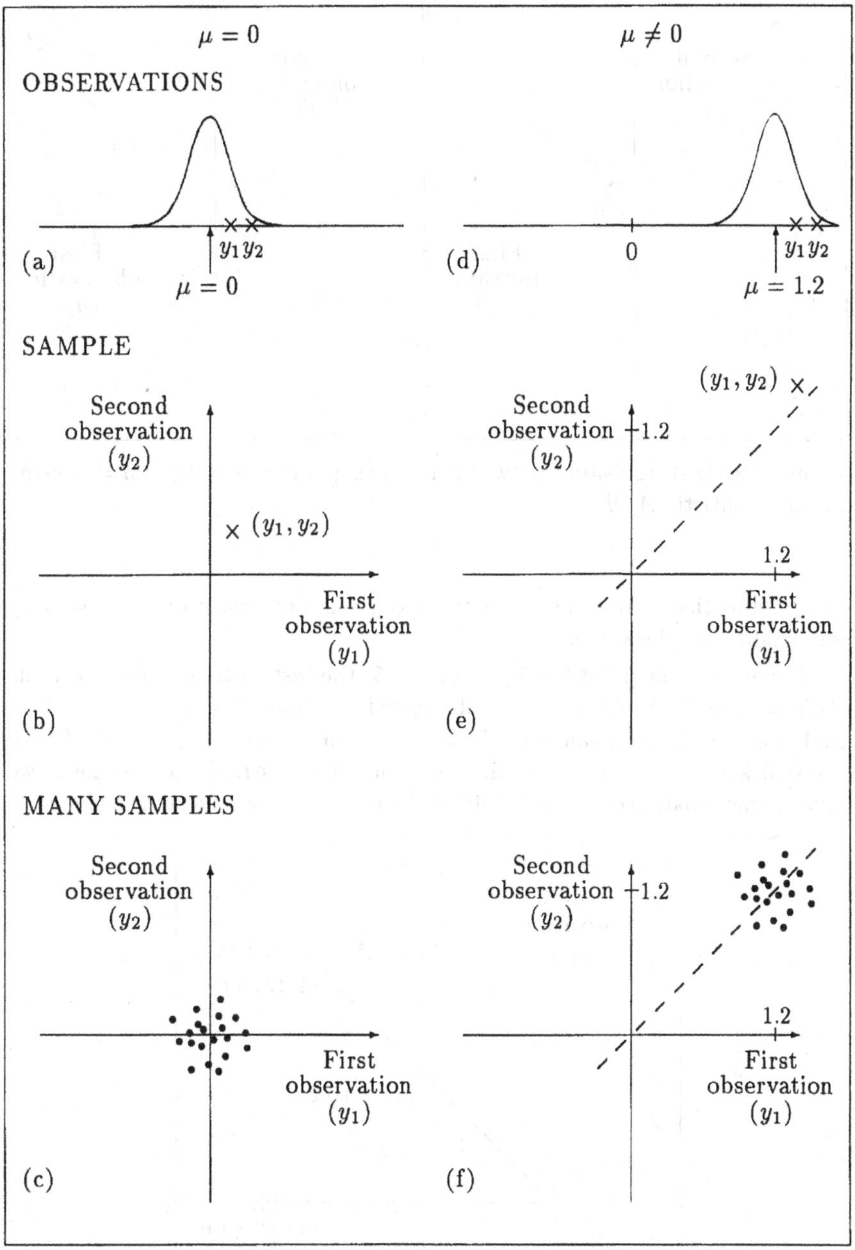

Figure 1.1: The correspondence between samples of size two and points (y_1, y_2) in 2-space. The scatter of points resulting from many repetitions of the thermometer study will be centered on the origin if $\mu = 0$, and will be centered on the point $(\mu, \mu) = (1.2, 1.2)$ if $\mu = 1.2$.

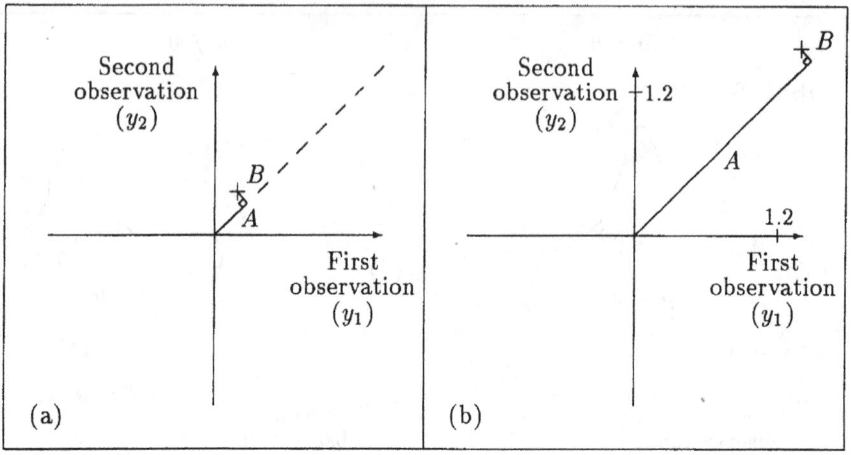

Figure 1.2: Distinguishing between the cases (a) $\mu = 0$ and (b) $\mu \neq 0$ using the test statistic A/B.

points, and that it does not depend on our units of measurement, two very important considerations.

For our sample point $(y_1, y_2) = (1.3, 1.5)$ the test statistic takes the value $A/B = 1.4\sqrt{2}/(0.1\sqrt{2}) = 14$, as illustrated in Figure 1.3. In Chapter 2 we shall see that if the mean μ really is zero, then our test statistic will follow the well-known "Student's t_1" distribution. (To be strictly correct here, we must regard distances A down-left of the origin as negative.) Knowledge of

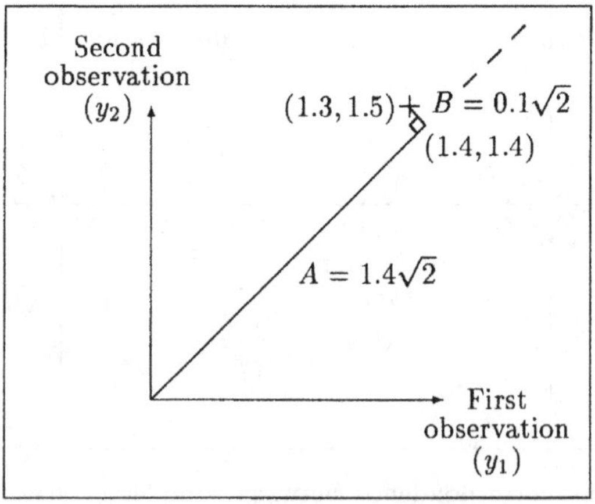

Figure 1.3: The sample point $(y_1, y_2) = (1.3, 1.5)$, the perpendicular onto the equiangular line, and the associated distances A and B.

this "reference" distribution enables us to be precise about a test statistic being "large." For example, the absolute value of the t_1 statistic is less than 12.7 in 95% of cases. Traditionally, statisticians declare these values to be "small," with values greater than 12.7 being declared "large." This means that our value of 14 would be declared large, so that we would have evidence to support the notion that the thermometer is biased.

1.3 The General Method

In order for us to more easily generalize the example in the last section to other situations it is helpful for us to rephrase what we have done in terms of vector geometry. In this section we immediately use vector geometric terminology without any formal introduction; when you come across an unfamiliar term or concept refer to Appendix A for elucidation.

As a first step we change from depicting our observations as a *point* in 2-space to depicting them as a *vector*, say $\mathbf{y} = [1.3, 1.5]^T$ as shown in Figure 1.4. (Note that we write vector labels such as $\mathbf{y}$ in **bold** type.)

Secondly, we choose an orthogonal coordinate system for 2-space which is tailored to our purpose. Recall from Figure 1.2 that the equiangular direction is a direction of special interest. We therefore choose this direction, $\mathbf{U_1} = [1, 1]^T/\sqrt{2}$, to be our first coordinate axis direction. This leaves only one choice for our second axis, the direction $\mathbf{U_2} = [-1, 1]^T/\sqrt{2}$ (or its negative to be strictly correct). These axes are also shown in Figure 1.4.

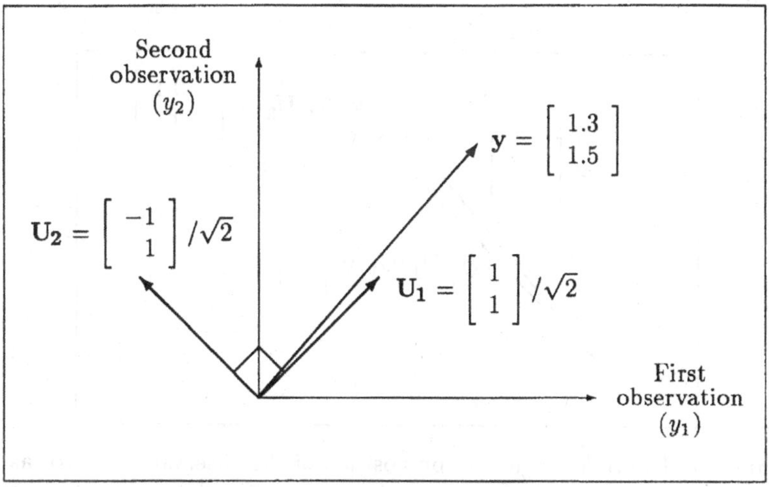

Figure 1.4: The observations displayed as a vector $\mathbf{y} = [1.3, 1.5]^T$. Also shown is an orthogonal coordinate system $\mathbf{U_1} = [1, 1]^T/\sqrt{2}$, $\mathbf{U_2} = [-1, 1]^T/\sqrt{2}$ which is tailored to our statistical problem.

Thirdly, we project our observation vector $\mathbf{y}$ onto each of our coordinate axes in turn and obtain the projection vectors

$$(\mathbf{y}\cdot\mathbf{U}_1)\mathbf{U}_1 = 1.4\sqrt{2}\mathbf{U}_1 = \begin{bmatrix} 1.4 \\ 1.4 \end{bmatrix} \text{ and } (\mathbf{y}\cdot\mathbf{U}_2)\mathbf{U}_2 = 0.1\sqrt{2}\mathbf{U}_2 = \begin{bmatrix} -0.1 \\ 0.1 \end{bmatrix}$$

as shown in Figure 1.5 (see Appendix A for details of projections and the definition of the "dot product" used in $\mathbf{y}\cdot\mathbf{U}_1$). This yields an orthogonal decomposition of the observation vector into a "model vector" and an "error vector,"

$$\begin{array}{ccccc} \mathbf{y} & = & (\mathbf{y}\cdot\mathbf{U}_1)\mathbf{U}_1 & + & (\mathbf{y}\cdot\mathbf{U}_2)\mathbf{U}_2 \\ \mathbf{y} & = & 1.4\sqrt{2}\mathbf{U}_1 & + & 0.1\sqrt{2}\mathbf{U}_2 \\ \begin{bmatrix} 1.3 \\ 1.5 \end{bmatrix} & = & \begin{bmatrix} 1.4 \\ 1.4 \end{bmatrix} & + & \begin{bmatrix} -0.1 \\ 0.1 \end{bmatrix} \\ \text{observation vector} & & \text{model vector} & & \text{error vector} \end{array}$$

This leads to the test statistic

$$\frac{\mathbf{y}\cdot\mathbf{U}_1}{|\,\mathbf{y}\cdot\mathbf{U}_2\,|} = \frac{1.4\sqrt{2}}{0.1\sqrt{2}} = 14$$

where $|\mathbf{y}\cdot\mathbf{U}_2|$ means the magnitude of $\mathbf{y}\cdot\mathbf{U}_2$. Note that this is the test statistic A/B obtained in the previous section (to within a sign).

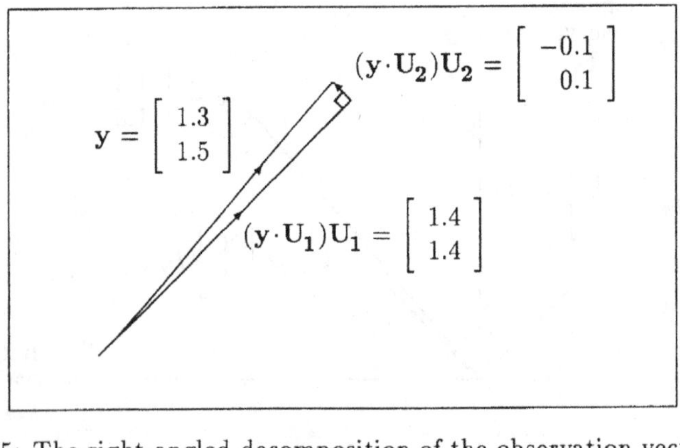

Figure 1.5: The right-angled decomposition of the observation vector as the sum of two projection vectors, one in the equiangular direction and the other in a direction perpendicular to this. Note that for illustrative purposes we have moved the vector $(\mathbf{y}\cdot\mathbf{U}_2)\mathbf{U}_2$ from the origin to the tip of the vector $(\mathbf{y}\cdot\mathbf{U}_1)\mathbf{U}_1$.

The associated Pythagorean breakup is

$$\|\mathbf{y}\|^2 = (\mathbf{y}\cdot\mathbf{U}_1)^2 + (\mathbf{y}\cdot\mathbf{U}_2)^2$$

or,
$$1.3^2 + 1.5^2 = (1.4\sqrt{2})^2 + (0.1\sqrt{2})^2$$

or,
$$3.94 = 3.92 + 0.02$$

as shown in Figure 1.6. (This breakup is the basis for the *analysis of variance* table often seen in statistics textbooks.)

This leads to another, equivalent test statistic, the ratio of the two squared projection lengths

$$\frac{(\mathbf{y}\cdot\mathbf{U}_1)^2}{(\mathbf{y}\cdot\mathbf{U}_2)^2} = \frac{3.92}{0.02} = 196$$

a form which is more convenient in advanced cases. Note that this is A^2/B^2 in the notation of the previous section. This value is compared with the percentiles of the $F_{1,1}$ distribution (e.g., 161 is the 95 percentile), and again leads us to conclude that we have evidence to support the notion that the thermometer is biased.

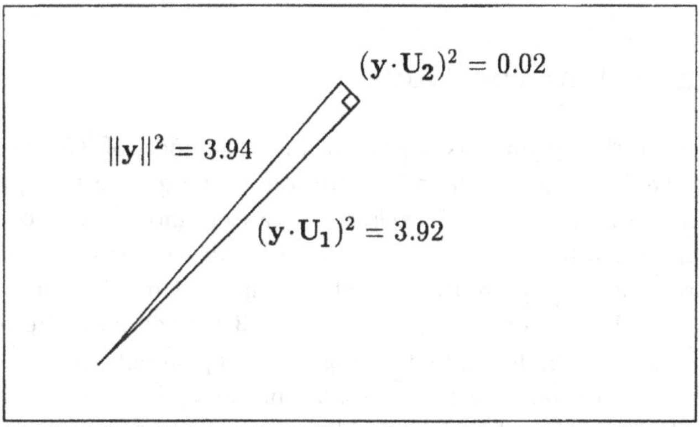

Figure 1.6: The Pythagorean breakup associated with the right-angled decomposition shown in Figure 1.5.

In Chapter 2 we apply the above general method to a paired samples data set with a sample size of two, then a sample size of three and, lastly, a sample size of nine. Then in Chapters 3–5 we apply the method in three other situations. In each example we first draw two scatters of points as in Figure 1.1, then use the picture to help us to set up a coordinate system which is tailored to the particular situation. From there on the method is quite automatic, following the pattern established above.

1.4 An Alternative Test Statistic

An alternative to using the ratio $t = A/B$ (or $F = A^2/B^2$) as our test statistic is to use the *angle* θ between the observation vector and the equiangular direction as our test statistic (this is the angle in the bottom left of Figure 1.5). Since A/B is the cotangent of the angle θ, either test statistic can be simply transformed into the other. Using the angle θ has some advantages over using the traditional t; for example, it has a very simple relationship with the probability value p. This intriguing test statistic is introduced and discussed in Appendix D.

1.5 Estimation

In general this primer is slanted toward hypothesis testing rather than estimation of confidence intervals. This is because the geometric approach leads naturally to fitted models (including estimation of effects) and Pythagorean breakups in the first instance. Estimation of confidence intervals is a further small step which we have not taken in the interests of brevity. Instead, in the summary section of each chapter we refer to the appropriate page of the sister text (Saville and Wood, 1991), which fully covers this topic.

1.6 Layout of the Book

The layout of this primer is based around the idea of "examples first, theory second." With this in mind, we apply the geometric approach to a new example at the start of each new chapter, moving on to deal with the general case later in the chapter. The topics covered are as follows. In Chapter 2 we compare the heights of male twins and female twins using the paired samples t test. In Chapter 3 we compare the heights of male students and female students using the independent samples t test. In Chapter 4 we move on to a medical example involving three independent samples; selenium levels for adults, full-term babies and premature babies (this topic is also referred to as "analysis of variance"). In Chapter 5 we use simple regression to examine the relationship between air pollution levels and a measure of the inversion effect. Chapter 6 is a summary chapter which further unifies the methods used in Chapters 2–5.

A basic familiarity with elementary statistical methods is assumed. A "geometric tool kit" is included in Appendix A to serve as an introduction or as a refresher course, and for the purpose of establishing our notation. Similarly, a small "statistical tool kit" is given in Appendix B. It is envisaged that these appendices will only be referred to when the reader requires clarification of some point while reading Chapters 2–5.

On the whole, readers are expected to do class exercises and general exercises without the aid of computers or hand-held calculators with statistical functions. However, to cater for later exercises which include significant amounts of data we include a brief Appendix C on computing.

To keep the story line clean we delay the introduction of the alternative test statistic θ until Appendix D. Each of the examples in the main body of the text is then reworked using the angle test statistic.

Written solutions to selected exercises are given in Appendix E. Tables of random numbers and percentiles of important reference distributions are given in Appendix T.

1.7 Word to Teachers

For teaching the material in this textbook we feel that the number one priority is to do the class exercises, since we feel it is essential for students of statistics to experience first hand the variation which is the stock and trade of a statistician's work.

For each chapter we therefore suggest that the teacher devote two or three 50-minute lectures to discussions of the material, with an additional 2-hour tutorial period devoted to the class exercise, scheduled after the first or second lecture.

1.8 References

We now list some of the other writers who have discussed the geometry underlying statistical methods. In their books, Scheffé (1959) discussed the topic on pages 10–13 and 42–45; Dyke (1988) devoted a chapter to the subject; Box, Hunter and Hunter (1978) discussed it on pages 178–182, 197–203, 212–215 and 500–501; and Box (1978) discussed it on pages 122–129. In addition, Corsten (1958) published a booklet describing his usage of vectors for teaching statistics at the Agricultural University of Wageningen in The Netherlands. Papers on the topic have also been published by Durbin and Kendall (1951), Margolis (1979), Herr (1980), Bryant (1984) and ourselves (Saville and Wood, 1986).

Chapter 2

Paired Samples

Paired samples occur when observations come in pairs, such as the heights of mixed-sex twins or the pulse rates of joggers before and after jogging. The interest is normally in the pairwise differences, such as the *difference* in height between the male and female in each twin pair, or the *change* in pulse rate of each jogger.

Provided our pairs have been chosen in a random manner from a defined population of study, we can treat our pairwise differences as a random sample from this larger population of differences, and proceed to test ideas about the mean of this population. More formally, in this chapter we shall assume that we have a random sample from a *normally distributed* population of differences with a certain mean (say μ) and a certain variance (say σ^2). Our main interest will usually be in estimating the mean and in testing whether it is zero or not. For example, in the case of the twins we would like to find out whether, on average, the males and females differ in height, and by how much. In the case of the joggers we would like to know whether, on average, pulse rate is increased by jogging, and by how much.

Essentially, we are dealing with the case of a *single* normally distributed population. The fact that our observations arise by calculating paired differences can be forgotten as soon as they have been calculated. From then on, we can simply think of our observed differences as a random sample of observations from a single population, as in §1.2.

In this chapter we shall summarize some height data for a sample of mixed-sex twins (§2.1), then analyze a subset of these data consisting of a sample of two paired differences in height (§2.2). The extension to a sample size of three is described in §2.3, and to a sample size of nine (the full data set) in §2.4. The chapter closes with a summary in §2.5, a class exercise, and general exercises.

2.1 Mixed-Sex Twins

To generate some data for this chapter, one of the authors sent out the following electronic mail message to the several hundred staff connected to the same personal computer network:

> *Data on mixed-sex twins*
>
> Hi! I'm currently writing a statistics textbook, and would like to include data on the heights of mixed-sex twins. Do you know any such twins? The idea is to find out whether the male twin is taller than the female twin (to compare with the more general population, where males are on average taller than females).
>
> If you could measure any such twins and send me the results I'd be very grateful. I'd preserve the anonymity of the twins, but it would be handy if you gave me their names or initials so I could delete duplicates (you might all know the same twins!).
>
> The data I'd like are:
> (1) height of female twin (state centimeters or inches);
> (2) height of male twin;
> (3) age of twins (nearest year approximately);
> (4) names or initials of twins (optional).
> You could send this by email or to the address below.
>
> Thanks in anticipation!
>
> (Name and address of author)

Thirteen replies were received. The resulting data were sorted in order of increasing age, then tabulated in Table 2.1 and graphed in Figure 2.1.

In considering the usefulness of these data, we must try to define the population to which the results may be extrapolated, and also consider whether the method of data collection has any inherent biases. Our study has serious defects in relation to the latter point, which we now discuss.

In Table 2.1 the first four twin pairs were children under 16 years of age. It was anticipated prior to the study that very young twins would not show as large a difference in height as more mature twins. (This appears to be confirmed by the data.) For the purposes of this chapter we would like to deal with a single study population, so we define this to be twins of 16 years of age or older living in New Zealand in 1992. We consider the last nine twin pairs from Table 2.1 to be a random sample from this population. The mean difference in height for these pairs is given at the bottom of the table.

Is it reasonable to regard the people who received the mail message as a random sample from the above population? No! is the clear answer. What we did was to sample an entire subpopulation of people living in New Zealand (plus people known to these people), and we also allowed individuals to

Age (years)	Height Male	Height Female	Difference in Height (M−F) Cm	Difference in Height (M−F) In.
1.5	77 cm	78.5 cm	−1.5	−0.6
3	94.2 cm	94.2 cm	0	0
8	125 cm	120 cm	5	2
8	123 cm	124 cm	−1	−0.4
16	187 cm	157 cm	30	11.8
17	185 cm	166 cm	19	7.5
17	5 ft 11 in.	5 ft 5 in.	15.2	6
21	183 cm	163 cm	20	7.9
23	5 ft 11 in.	168 cm	12.3	4.9
29	5 ft 6 in.	5 ft 2 in.	10.2	4
30	6 ft 1 in.	5 ft 10 in.	7.6	3
33	6 ft 4 in.	5 ft 6 in.	25.4	10
40	6 ft 0 in.	5 ft 3 in.	22.9	9
Mean	182.7 cm	164.7 cm	18.1	7.1

Table 2.1: Age, male and female height data as received, and difference in height (male minus female) expressed in both centimeters and inches. The last row gives the mean over the last nine twin pairs only (those aged 16 and above).

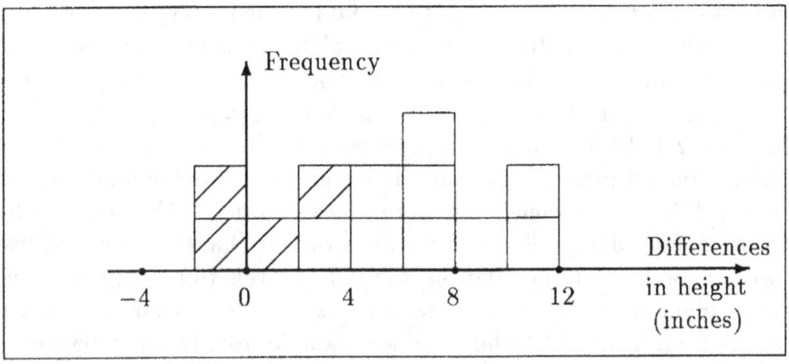

Figure 2.1: Histogram of differences in height between male and female twins. Classes are $-2 \le x < 0$, $0 \le x < 2$, $2 \le x < 4$, and so on. The differences for twins under 16 years of age are shaded, while those for twins of 16 years of age or older are left unshaded.

decide whether to respond or not. It would require a very vivid imagination for one to regard this as a random procedure. The people connected to the computer network were primarily employed by governmental bodies, with perhaps half involved in agricultural research or consultancy. Are such people typical of New Zealanders in terms of what we are trying to assess? Probably not. It is hard to point to any obvious difference, but there may be some subtlety of which we are unaware.

Were there any biases in terms of who responded on behalf of their family or friends? Yes, it is probable that there was a bias toward the more computer-literate staff. If so, it is hard to see how this would introduce a bias in terms of difference in height; however, again there may be some subtlety of which we are unaware. Another possibility is that the wording of the message may have biased the study. By spelling out the motivation behind the study (to find out whether the male twin is taller than the female twin), we may have inadvertently decreased the probability of response from people who knew twins where the female was taller than the male. In this respect it is comforting to see negative differences appearing in the child twin data.

All in all, our sampling procedure left a lot to be desired, and should not be taken as a prototype worthy of inclusion in any reputable statistics textbook! Nevertheless, to enable us to proceed we now set aside our misgivings, and treat our sample of nine twin pairs as a random sample from our defined study population.

2.2 Sample Size of Two

To consolidate the ideas introduced in §1.2 and §1.3, we first work with just two of our twin pairs. More precisely, we shall imagine that we have drawn a random sample consisting of the seventh and thirteenth pairs of twins in Table 2.1. The corresponding differences in height between male and female are 6 in. and 9 in., respectively. The resulting *observation vector* is

$$\mathbf{y} = \begin{bmatrix} 6 \\ 9 \end{bmatrix}$$

Objective

The study objective is as follows:

> To test whether there is, on average, a difference in height (μ) between male and female for mixed-sex twins who are 16 years of age or older and who were living in New Zealand in 1992.

The Basic Idea

At the risk of being repetitive, we restate the basic idea as given in §1.2, in relation to the key picture, Figure 1.2, which we reproduce as Figure 2.2.

The idea is that if the average difference in height μ is zero, our data point $(6,9)$ will be part of the scatter shown in Figure 2.2(a), while if μ is nonzero our data point will be part of the scatter shown in Figure 2.2(b). To decide between the possibilities $\mu = 0$ and $\mu \neq 0$ we therefore compare the distance of our data point $(6,9)$ *up* the line with its distance *from* the line.

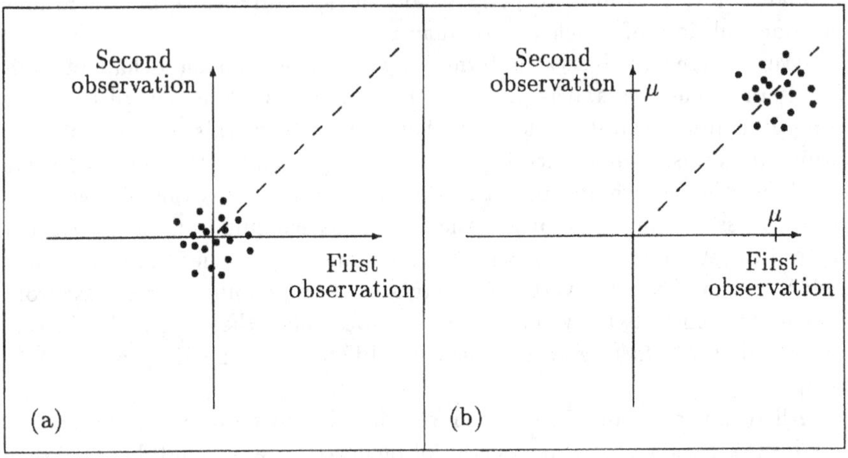

Figure 2.2: Scatters of data points representing many repetitions of our study. In the case (a) $\mu = 0$ the scatter is centered on the origin, while in the case (b) $\mu \neq 0$ the scatter is centered on the point (μ, μ).

An Appropriate Coordinate System

For this purpose an appropriate set of coordinate axes for 2-space is

$$\mathbf{U}_1 = \frac{1}{\sqrt{2}}\begin{bmatrix} 1 \\ 1 \end{bmatrix}, \quad \mathbf{U}_2 = \frac{1}{\sqrt{2}}\begin{bmatrix} -1 \\ 1 \end{bmatrix}$$

Here $\mathbf{U}_1 = [1,1]^T/\sqrt{2}$ is the *direction associated with the parameter of interest*, μ.

The Orthogonal Decomposition

We now project our observation vector $\mathbf{y} = [6,9]^{\mathrm{T}}$ onto each of our coordinate axes. The first projection vector is

$$(\mathbf{y} \cdot \mathbf{U}_1)\mathbf{U}_1 = \left(\begin{bmatrix} 6 \\ 9 \end{bmatrix} \cdot \frac{1}{\sqrt{2}}\begin{bmatrix} 1 \\ 1 \end{bmatrix}\right)\frac{1}{\sqrt{2}}\begin{bmatrix} 1 \\ 1 \end{bmatrix} = 7.5\begin{bmatrix} 1 \\ 1 \end{bmatrix}$$

The reader who is unfamiliar with vector geometry should refer to Appendix A for an introduction to projection.

The second projection vector is

$$(\mathbf{y}\cdot\mathbf{U}_2)\mathbf{U}_2 = \left(\begin{bmatrix}6\\9\end{bmatrix}\cdot\frac{1}{\sqrt{2}}\begin{bmatrix}-1\\1\end{bmatrix}\right)\frac{1}{\sqrt{2}}\begin{bmatrix}-1\\1\end{bmatrix} = \begin{bmatrix}-1.5\\1.5\end{bmatrix}$$

The resulting orthogonal decomposition of the observation vector is

$$\begin{array}{ccccc} \mathbf{y} & = & (\mathbf{y}\cdot\mathbf{U}_1)\mathbf{U}_1 & + & (\mathbf{y}\cdot\mathbf{U}_2)\mathbf{U}_2 \\ \begin{bmatrix}6\\9\end{bmatrix} & = & (15/\sqrt{2})\mathbf{U}_1 & + & (3/\sqrt{2})\mathbf{U}_2 \\ \begin{bmatrix}6\\9\end{bmatrix} & = & \begin{bmatrix}7.5\\7.5\end{bmatrix} & + & \begin{bmatrix}-1.5\\1.5\end{bmatrix} \\ \text{observation vector} & & \text{model vector} & & \text{error vector} \end{array}$$

That is, we have broken $\mathbf{y}$ into two orthogonal components, a *model vector* and an *error vector*, as shown in Figure 2.3.

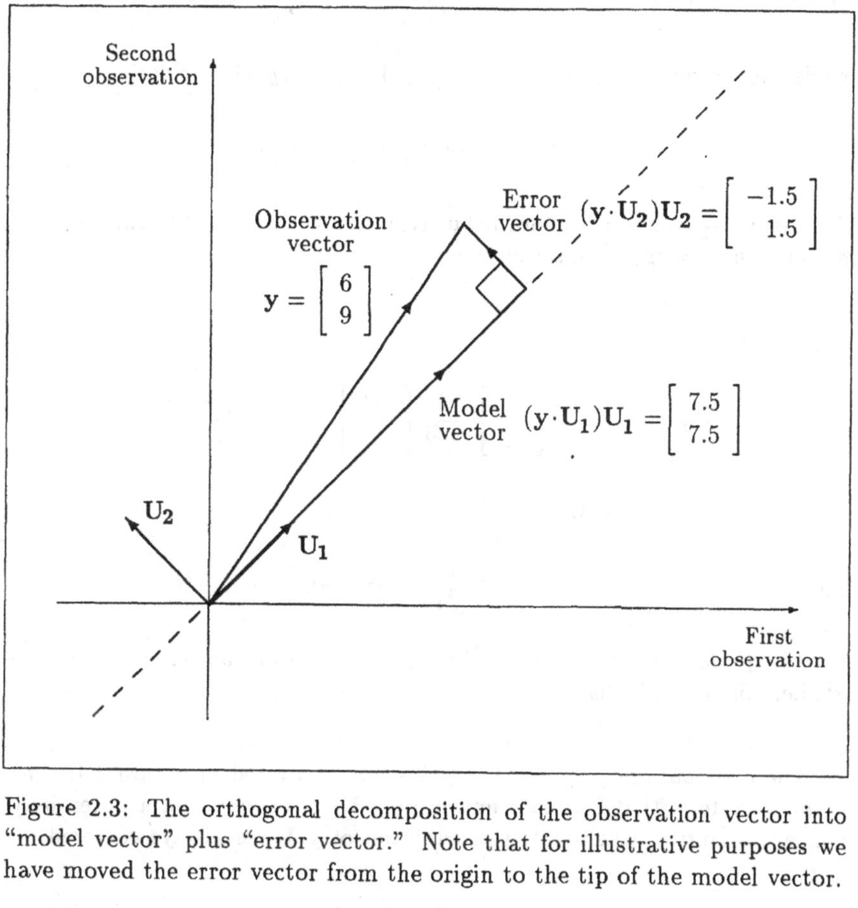

Figure 2.3: The orthogonal decomposition of the observation vector into "model vector" plus "error vector." Note that for illustrative purposes we have moved the error vector from the origin to the tip of the model vector.

Distributions of the Projection Lengths

To prepare for testing whether the true mean difference in height could plausibly be zero, we now work out the distributions of the projection lengths $\mathbf{y} \cdot \mathbf{U}_1$ and $\mathbf{y} \cdot \mathbf{U}_2$.

Firstly,

$$\mathbf{y} \cdot \mathbf{U}_1 = \begin{bmatrix} y_1 \\ y_2 \end{bmatrix} \cdot \frac{1}{\sqrt{2}} \begin{bmatrix} 1 \\ 1 \end{bmatrix} = \frac{y_1 + y_2}{\sqrt{2}}$$

where the symbols y_1 and y_2 refer to independent observations from a normal distribution with mean μ and variance σ^2 (as defined in Appendix B).

$$\text{Hence the mean of } \mathbf{y} \cdot \mathbf{U}_1 = \frac{\mu + \mu}{\sqrt{2}} = \frac{2\mu}{\sqrt{2}} = \sqrt{2}\mu$$

$$\text{while the variance of } \mathbf{y} \cdot \mathbf{U}_1 = \left(\frac{1}{\sqrt{2}}\right)^2 [\text{variance}(y_1) + \text{variance}(y_2)]$$

$$= \frac{1}{2}\left[\sigma^2 + \sigma^2\right] = \sigma^2$$

That is, $\mathbf{y} \cdot \mathbf{U}_1$ comes from a normal distribution with mean $\sqrt{2}\mu$ and variance σ^2, i.e., an $N(\sqrt{2}\mu, \sigma^2)$ distribution.

Secondly,

$$\mathbf{y} \cdot \mathbf{U}_2 = \begin{bmatrix} y_1 \\ y_2 \end{bmatrix} \cdot \frac{1}{\sqrt{2}} \begin{bmatrix} -1 \\ 1 \end{bmatrix} = \frac{y_2 - y_1}{\sqrt{2}}$$

$$\text{so the mean of } \mathbf{y} \cdot \mathbf{U}_2 = \frac{\mu - \mu}{\sqrt{2}} = 0$$

$$\text{and the variance of } \mathbf{y} \cdot \mathbf{U}_2 = \left(\frac{1}{\sqrt{2}}\right)^2 \left[\sigma^2 + \sigma^2\right] = \sigma^2$$

That is, $\mathbf{y} \cdot \mathbf{U}_2$ comes from a normal distribution with mean zero and variance σ^2, i.e., an $N(0, \sigma^2)$ distribution.

The distributions of $\mathbf{y} \cdot \mathbf{U}_1$ and $\mathbf{y} \cdot \mathbf{U}_2$ are illustrated in Figure 2.4. The point to note is that the distribution of $\mathbf{y} \cdot \mathbf{U}_1$ is centered on a potentially nonzero quantity, whereas the distribution of $\mathbf{y} \cdot \mathbf{U}_2$ is always centered on zero.

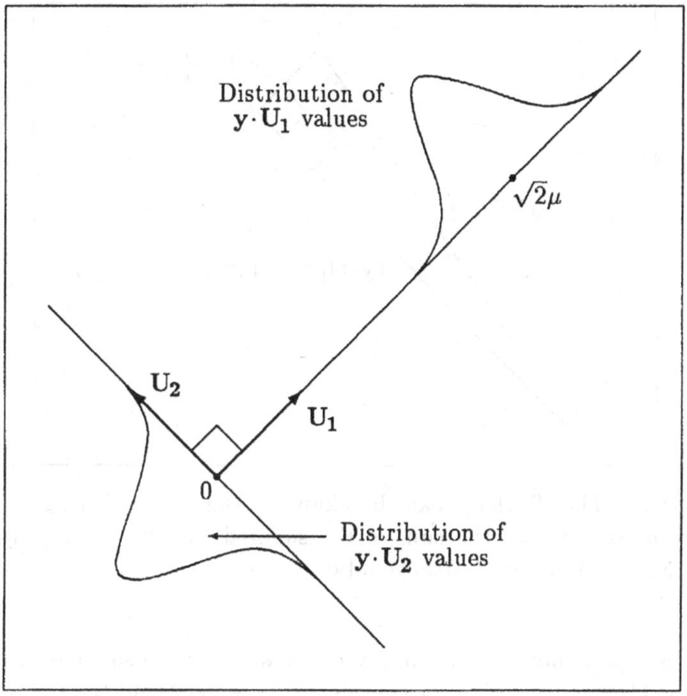

Figure 2.4: The distributions of the lengths of the projections, $\mathbf{y}\cdot\mathbf{U}_1$ and $\mathbf{y}\cdot\mathbf{U}_2$, of the observation vector, $\mathbf{y}$, onto an appropriate set of orthogonal directions, $\mathbf{U}_1$ and $\mathbf{U}_2$.

Testing the Hypothesis

We now proceed to investigate the study objective: Is the true mean difference in height between male and female twins zero? More formally, we wish to test the null hypothesis $H_0 : \mu = 0$ against the alternative hypothesis, $H_1 : \mu \neq 0$.

The relevant *Pythagorean breakup* is

$$\|\mathbf{y}\|^2 = (\mathbf{y}\cdot\mathbf{U}_1)^2 + (\mathbf{y}\cdot\mathbf{U}_2)^2$$

or,

$$117 = 112.5 + 4.5$$

as illustrated in Figure 2.5.

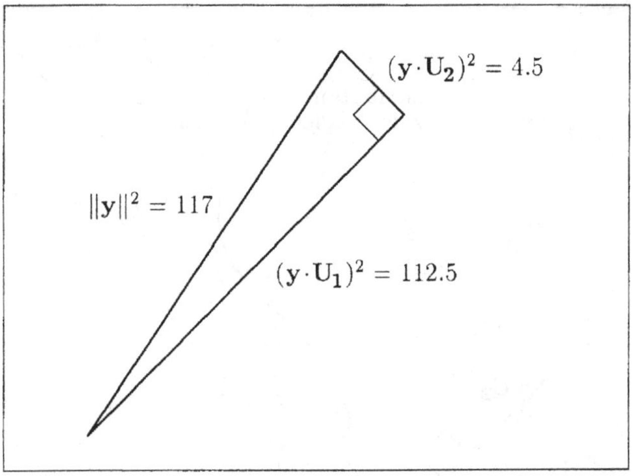

Figure 2.5: The Pythagorean breakup of the squared length of the observation vector into the sum of the squared lengths of its projections onto an appropriate set of orthogonal directions.

To test the hypothesis we simply check whether the squared projection length $(\mathbf{y}\cdot\mathbf{U_1})^2$ is similar to, or considerably greater than, the squared projection length $(\mathbf{y}\cdot\mathbf{U_2})^2$. The resulting test statistic is

$$F = \frac{(\mathbf{y}\cdot\mathbf{U_1})^2}{(\mathbf{y}\cdot\mathbf{U_2})^2} = \frac{112.5}{4.5} = 25.0$$

Is this observed F value large or small? Now if $\mu = 0$, the projection lengths $\mathbf{y}\cdot\mathbf{U_1}$ and $\mathbf{y}\cdot\mathbf{U_2}$ both come from an $N(0,\sigma^2)$ distribution, so the test statistic $F = (\mathbf{y}\cdot\mathbf{U_1})^2/(\mathbf{y}\cdot\mathbf{U_2})^2$ comes from the $F_{1,1}$ distribution defined in Appendix B. Hence to see whether our observed value of 25 is large or small, we compare it with the percentiles of the $F_{1,1}$ distribution as given in Table T.2. The 90, 95 and 99 percentiles are 40, 161 and 4052, respectively, so our observed value of 25 is not large, even when compared with the 90 percentile (or 10% *critical value*). Hence we conclude that there is insufficient evidence to reject the idea that the population mean is zero. That is, we have not shown beyond reasonable doubt that there is any difference in mean height between the male and female twins in our study population.

Paired Samples t Test

By taking the square root, our test statistic $F = (\mathbf{y}\cdot\mathbf{U_1})^2/(\mathbf{y}\cdot\mathbf{U_2})^2 = A^2/B^2$ can be transformed to the form given in Chapter 1:

$$t = \frac{\mathbf{y}\cdot\mathbf{U_1}}{|\,\mathbf{y}\cdot\mathbf{U_2}\,|} = \frac{A}{B} = \frac{15/\sqrt{2}}{3/\sqrt{2}} = 5.0$$

As already stated, if $\mu = 0$ the projection lengths $\mathbf{y}\cdot\mathbf{U}_1$ and $\mathbf{y}\cdot\mathbf{U}_2$ both come from an $N(0, \sigma^2)$ distribution. Hence the test statistic $t = \mathbf{y}\cdot\mathbf{U}_1 / |\mathbf{y}\cdot\mathbf{U}_2|$ comes from the t_1 distribution defined in Appendix B. To see whether our observed t value of 5.0 is large or small, we compare it with the percentiles of the t_1 distribution as given in Table T.2. The 95, 97.5 and 99.5 percentiles of this two-sided distribution are 6.31, 12.71 and 63.66, respectively, so our observed value of 5.0 is not large, even when compared with the 95 percentile (or 10% critical value). Hence we again conclude that we have found insufficient evidence to reject the idea that the population mean is zero.

Comments

In the above analysis the direction $\mathbf{U}_1$ is used to estimate the population mean μ, while the direction $\mathbf{U}_2$ is used to estimate the population variance σ^2. In the long run the projection length $\mathbf{y}\cdot\mathbf{U}_1 = \sqrt{2}\bar{y}$ averages to $\sqrt{2}\mu$, so $\bar{y} = 7.5$ serves as our estimate of μ. That is, the *best estimate* of the true mean difference in height between male and female twins is 7.5 in. However, our analysis also tells us that this estimate is not very precise, with a true mean difference in height of zero also being plausible. Also, in the long run the squared projection length $(\mathbf{y}\cdot\mathbf{U}_2)^2$ averages to σ^2, so $(\mathbf{y}\cdot\mathbf{U}_2)^2 = 4.5$ serves as our estimate of σ^2.

In addition, it can be shown that in the long run the squared projection length $(\mathbf{y}\cdot\mathbf{U}_1)^2$ averages to $2\mu^2 + \sigma^2$. Hence if $\mu \neq 0$ the quantity $(\mathbf{y}\cdot\mathbf{U}_1)^2$ is inflated, and our test statistic $(\mathbf{y}\cdot\mathbf{U}_1)^2/(\mathbf{y}\cdot\mathbf{U}_2)^2$ can be thought of as an inflated estimate of σ^2 divided by an unbiased estimate of σ^2.

2.3 Sample Size of Three

What do we do differently for a sample size of three? The main difference is that we now have two coordinate axis directions available for estimation of the population variance, σ^2. To show how this affects our method we rework §2.2 using as our new example a random sample consisting of the sixth, eleventh and thirteenth pairs of twins in Table 2.1. The corresponding differences in height between male and female are 7.5, 3 and 9 in., respectively. The resulting *observation vector* is

$$\mathbf{y} = \begin{bmatrix} 7.5 \\ 3 \\ 9 \end{bmatrix}$$

The Basic Idea

As in §2.2 the basic idea is that if the average difference in height between male and female twins μ is zero, then our data point $(7.5, 3, 9)$ will be part of the scatter shown in Figure 2.6(a), while if μ is nonzero our data point will be part of the scatter shown in Figure 2.6(b). To decide between the possibilities $\mu = 0$ and $\mu \neq 0$ our method is to compare the squared distance of our data point $(7.5, 3, 9)$ up the equiangular line with the average of the two squared lengths of its projections in the two directions perpendicular to the line.

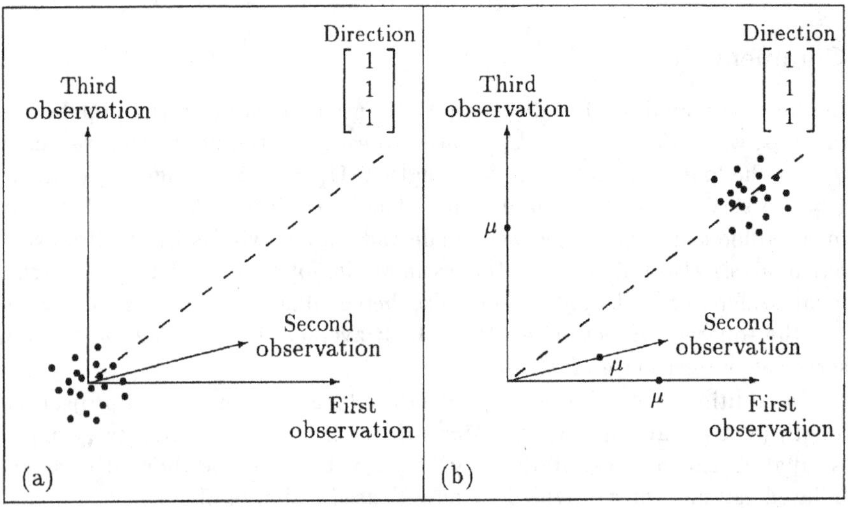

Figure 2.6: Scatters of data points representing many repetitions of our study. In the case (a) $\mu = 0$ the scatter is centered on the origin, while in the case (b) $\mu \neq 0$ the scatter is centered on the point (μ, μ, μ).

An Appropriate Coordinate System

A set of orthogonal coordinate axes for 3-space which is tailored to our purpose is

$$\mathbf{U}_1 = \frac{1}{\sqrt{3}}\begin{bmatrix}1\\1\\1\end{bmatrix}, \quad \mathbf{U}_2 = \frac{1}{\sqrt{2}}\begin{bmatrix}-1\\1\\0\end{bmatrix}, \quad \mathbf{U}_3 = \frac{1}{\sqrt{6}}\begin{bmatrix}-1\\-1\\2\end{bmatrix}$$

Here $\mathbf{U}_1 = [1,1,1]^T/\sqrt{3}$ is the *direction associated with the parameter of interest* μ.

The Orthogonal Decomposition

We now project our observation vector $\mathbf{y} = [7.5, 3, 9]^T$ onto each of our coordinate axes. The first projection vector is

$$(\mathbf{y}\cdot\mathbf{U}_1)\mathbf{U}_1 = \left(\begin{bmatrix}7.5\\3\\9\end{bmatrix}\cdot\frac{1}{\sqrt{3}}\begin{bmatrix}1\\1\\1\end{bmatrix}\right)\frac{1}{\sqrt{3}}\begin{bmatrix}1\\1\\1\end{bmatrix} = 6.5\begin{bmatrix}1\\1\\1\end{bmatrix}$$

The second projection vector is

$$(\mathbf{y}\cdot\mathbf{U}_2)\mathbf{U}_2 = \left(\begin{bmatrix}7.5\\3\\9\end{bmatrix}\cdot\frac{1}{\sqrt{2}}\begin{bmatrix}-1\\1\\0\end{bmatrix}\right)\frac{1}{\sqrt{2}}\begin{bmatrix}-1\\1\\0\end{bmatrix} = \begin{bmatrix}2.25\\-2.25\\0\end{bmatrix}$$

The third projection vector is

$$(\mathbf{y}\cdot\mathbf{U}_3)\mathbf{U}_3 = \left(\begin{bmatrix}7.5\\3\\9\end{bmatrix}\cdot\frac{1}{\sqrt{6}}\begin{bmatrix}-1\\-1\\2\end{bmatrix}\right)\frac{1}{\sqrt{6}}\begin{bmatrix}-1\\-1\\2\end{bmatrix} = \begin{bmatrix}-1.25\\-1.25\\2.50\end{bmatrix}$$

That is, our orthogonal decomposition is

$$\begin{aligned}\mathbf{y} &= (\mathbf{y}\cdot\mathbf{U}_1)\mathbf{U}_1 + (\mathbf{y}\cdot\mathbf{U}_2)\mathbf{U}_2 + (\mathbf{y}\cdot\mathbf{U}_3)\mathbf{U}_3\\ \begin{bmatrix}7.5\\3\\9\end{bmatrix} &= 6.5\begin{bmatrix}1\\1\\1\end{bmatrix} + \begin{bmatrix}2.25\\-2.25\\0\end{bmatrix} + \begin{bmatrix}-1.25\\-1.25\\2.50\end{bmatrix}\end{aligned}$$

as shown in Figure 2.7. This simplifies to the fitted model

$$\begin{array}{ccccc}\begin{bmatrix}7.5\\3\\9\end{bmatrix} & = & 6.5\begin{bmatrix}1\\1\\1\end{bmatrix} & + & \begin{bmatrix}1.0\\-3.5\\2.5\end{bmatrix}\\ \mathbf{y} & = & \bar{\mathbf{y}} & + & (\mathbf{y}-\bar{\mathbf{y}})\\ \text{observation vector} & & \text{model vector} & & \text{error vector}\end{array}$$

where $\bar{y} = 6.5$ in. is the best estimate of the difference in height between male and female twins. This decomposition is as also shown in Figure 2.7. (Note that we are using $(\mathbf{y} - \bar{\mathbf{y}})$ to refer to the vector of differences between the observations and the sample mean.)

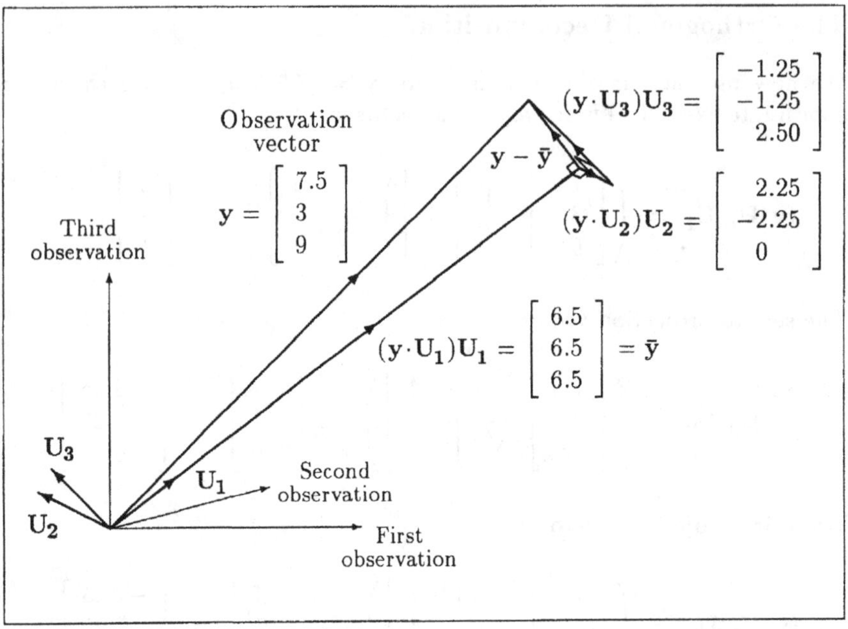

Figure 2.7: The orthogonal decomposition of the observation vector in terms of the three basic orthogonal projections, $(\mathbf{y}\cdot\mathbf{U}_i)\mathbf{U}_i$, and in terms of the model vector, $\bar{\mathbf{y}}$, plus error vector, $\mathbf{y}-\bar{\mathbf{y}}$.

Distributions of the Projection Lengths

To prepare for testing whether the true mean difference in height could plausibly be zero, we need to know the distributions of the projection lengths $\mathbf{y}\cdot\mathbf{U}_1$, $\mathbf{y}\cdot\mathbf{U}_2$ and $\mathbf{y}\cdot\mathbf{U}_3$.

$$\text{Now} \qquad \mathbf{y}\cdot\mathbf{U}_1 = \begin{bmatrix} y_1 \\ y_2 \\ y_3 \end{bmatrix} \cdot \frac{1}{\sqrt{3}} \begin{bmatrix} 1 \\ 1 \\ 1 \end{bmatrix} = \frac{y_1+y_2+y_3}{\sqrt{3}}$$

where y_1, y_2 and y_3 are independent observations from a normal distribution with mean μ and variance σ^2.

Proceeding as in §2.2, we find that the mean of $\mathbf{y}\cdot\mathbf{U}_1$ is $\sqrt{3}\mu$, and the variance of $\mathbf{y}\cdot\mathbf{U}_1$ is σ^2. That is, $\mathbf{y}\cdot\mathbf{U}_1$ comes from an $N(\sqrt{3}\mu,\sigma^2)$ distribution. Similarly, we find that $\mathbf{y}\cdot\mathbf{U}_2$ and $\mathbf{y}\cdot\mathbf{U}_3$ both come from an $N(0,\sigma^2)$ distribution.

As in §2.2, the point to note is that the distribution of $\mathbf{y}\cdot\mathbf{U}_1$ is centered on a potentially nonzero quantity whereas the distributions of $\mathbf{y}\cdot\mathbf{U}_2$ and $\mathbf{y}\cdot\mathbf{U}_3$ are always centered on zero.

Testing the Hypothesis

We now investigate the study objective: Is the true mean difference in height between male and female twins zero? For our hypothesis test we check whether the squared projection length $(\mathbf{y}\cdot\mathbf{U}_1)^2$ is similar to, or considerably greater than, the average of the squared projection lengths for the "error" directions, $[(\mathbf{y}\cdot\mathbf{U}_2)^2 + (\mathbf{y}\cdot\mathbf{U}_3)^2]/2$, where $\mathbf{U}_2$ and $\mathbf{U}_3$ are the coordinate axes spanning the error space.

The associated Pythagorean breakup is

$$\|\mathbf{y}\|^2 = (\mathbf{y}\cdot\mathbf{U}_1)^2 + (\mathbf{y}\cdot\mathbf{U}_2)^2 + (\mathbf{y}\cdot\mathbf{U}_3)^2$$

or,

$$146.25 = 126.75 + 10.125 + 9.375$$

as illustrated in Figure 2.8.

The resulting test statistic is

$$F = \frac{(\mathbf{y}\cdot\mathbf{U}_1)^2}{[(\mathbf{y}\cdot\mathbf{U}_2)^2 + (\mathbf{y}\cdot\mathbf{U}_3)^2]/2} = \frac{126.75}{19.5/2} = 13.0$$

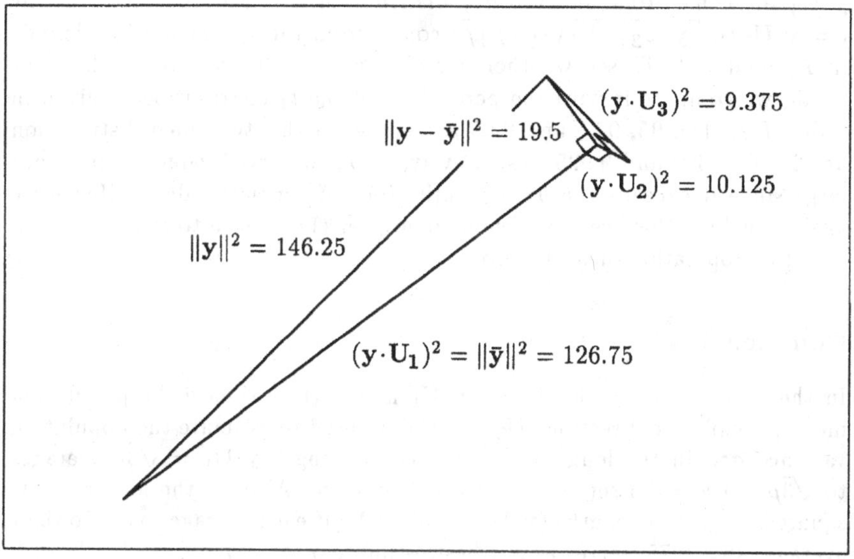

Figure 2.8: The Pythagorean breakup of the squared length of the observation vector as the sum of the squared lengths of the three orthogonal projections, $(\mathbf{y}\cdot\mathbf{U}_i)^2$, and as the sum of the squared lengths of the fitted model and error vectors, $\|\bar{\mathbf{y}}\|^2$ and $\|\mathbf{y} - \bar{\mathbf{y}}\|^2$.

Is this observed F value large or small? Now if $\mu = 0$, the projection lengths $\mathbf{y}\cdot\mathbf{U_1}$, $\mathbf{y}\cdot\mathbf{U_2}$ and $\mathbf{y}\cdot\mathbf{U_3}$ all come from an $N(0, \sigma^2)$ distribution, so the test statistic $(\mathbf{y}\cdot\mathbf{U_1})^2/\{[(\mathbf{y}\cdot\mathbf{U_2})^2 + (\mathbf{y}\cdot\mathbf{U_3})^2]/2\}$ comes from the $F_{1,2}$ distribution defined in Appendix B. Hence to see whether our observed F value of 13 is large or small, we compare it with the 90, 95 and 99 percentiles of the $F_{1,2}$ distribution, 8.5, 18.5 and 98.5, respectively. The result is that our observed value of 13 is somewhat unusual, since it is greater than the 90 percentile of 8.5. Hence we conclude that we have found some (weak) evidence to reject the idea that the population mean is zero. That is, the data suggests that there may be, on average, a difference in height between the male and female twins in our study population.

Paired Samples t Test

Note that our test statistic $F = (\mathbf{y}\cdot\mathbf{U_1})^2/\{[(\mathbf{y}\cdot\mathbf{U_2})^2 + (\mathbf{y}\cdot\mathbf{U_3})^2]/2\}$ $= A^2/[B^2/2] = 13.0$ can be transformed to

$$t = \frac{\mathbf{y}\cdot\mathbf{U_1}}{\sqrt{[(\mathbf{y}\cdot\mathbf{U_2})^2 + (\mathbf{y}\cdot\mathbf{U_3})^2]/2}} = \frac{A}{B/\sqrt{2}} = \frac{11.2583}{4.4159/\sqrt{2}} = 3.606$$

As already stated, if $\mu = 0$ the projection lengths $\mathbf{y}\cdot\mathbf{U_1}$, $\mathbf{y}\cdot\mathbf{U_2}$ and $\mathbf{y}\cdot\mathbf{U_3}$ all come from an $N(0, \sigma^2)$ distribution. Hence the test statistic $t = \mathbf{y}\cdot\mathbf{U_1}/\sqrt{[(\mathbf{y}\cdot\mathbf{U_2})^2 + (\mathbf{y}\cdot\mathbf{U_3})^2]/2}$ comes from the t_2 distribution defined in Appendix B. To see whether our observed t value of 3.606 is large or small, we compare it with the percentiles of the t_2 distribution as given in Table T.2. The 95, 97.5 and 99.5 percentiles of this two-sided distribution are 2.920, 4.303 and 9.925, respectively, so our observed value is somewhat large since it exceeds the 95 percentile (or 10% critical value). Hence we again conclude that we have found some (weak) evidence to reject the idea that the population mean is zero.

Comments

In the above analysis the direction $\mathbf{U_1}$ is used to estimate the population mean μ, while the directions $\mathbf{U_2}$ and $\mathbf{U_3}$ are used to estimate the population variance σ^2. In the long run the projection length $\mathbf{y}\cdot\mathbf{U_1} = \sqrt{3}\bar{y}$ averages to $\sqrt{3}\mu$, so $\bar{y} = 6.5$ serves as our estimate of μ. Also, in the long run the squared projection lengths $(\mathbf{y}\cdot\mathbf{U_2})^2$ and $(\mathbf{y}\cdot\mathbf{U_3})^2$ each average to σ^2, so their average, $s^2 = 9.75$, serves as our best estimate of σ^2.

In addition it can be shown that in the long run the squared projection length $(\mathbf{y}\cdot\mathbf{U_1})^2$ averages to $3\mu^2 + \sigma^2$. Hence if $\mu \neq 0$ the quantity $(\mathbf{y}\cdot\mathbf{U_1})^2$ is inflated, and our test statistic $(\mathbf{y}\cdot\mathbf{U_1})^2/s^2$ can be thought of as an inflated estimate of σ^2 divided by an unbiased estimate of σ^2.

2.4 General Case

To illustrate the method in its full generality we now analyze the full data set of nine height differences as given in Table 2.1. The corresponding *observation vector* is

$$\mathbf{y} = \begin{bmatrix} 11.8 \\ 7.5 \\ 6 \\ 7.9 \\ 4.9 \\ 4 \\ 3 \\ 10 \\ 9 \end{bmatrix}$$

The Basic Idea

The basic idea is the same as in previous sections. If the average difference in height between male and female twins μ is zero, then our data point will be part of the scatter shown in Figure 2.9(a), while if μ is nonzero our data point will be part of the scatter shown in Figure 2.9(b). To decide between the possibilities $\mu = 0$ and $\mu \neq 0$ our method is to compare the squared distance of our data point $(11.8, 7.5, 6, 7.9, 4.9, 4, 3, 10, 9)$ up the equiangular line with the average of the eight squared lengths of its projections in the eight directions perpendicular to the line.

An Appropriate Coordinate System

An appropriate set of orthogonal coordinate axes for 9-space is

$$\begin{array}{ccccccc} \mathbf{U_1} & \mathbf{U_2} & \mathbf{U_3} & \cdots & \mathbf{U_7} & \mathbf{U_8} & \mathbf{U_9} \\ \frac{\begin{bmatrix} 1 \\ 1 \\ 1 \\ 1 \\ 1 \\ 1 \\ 1 \\ 1 \\ 1 \end{bmatrix}}{\sqrt{9}} & \frac{\begin{bmatrix} -1 \\ 1 \\ 0 \\ 0 \\ 0 \\ 0 \\ 0 \\ 0 \\ 0 \end{bmatrix}}{\sqrt{2}} & \frac{\begin{bmatrix} -1 \\ -1 \\ 2 \\ 0 \\ 0 \\ 0 \\ 0 \\ 0 \\ 0 \end{bmatrix}}{\sqrt{6}} & \cdots & \frac{\begin{bmatrix} -1 \\ -1 \\ -1 \\ -1 \\ -1 \\ -1 \\ 6 \\ 0 \\ 0 \end{bmatrix}}{\sqrt{42}} & \frac{\begin{bmatrix} -1 \\ -1 \\ -1 \\ -1 \\ -1 \\ -1 \\ -1 \\ 7 \\ 0 \end{bmatrix}}{\sqrt{56}} & \frac{\begin{bmatrix} -1 \\ -1 \\ -1 \\ -1 \\ -1 \\ -1 \\ -1 \\ -1 \\ 8 \end{bmatrix}}{\sqrt{72}} \end{array}$$

Here $\mathbf{U_1} = [1, 1, 1, 1, 1, 1, 1, 1, 1]^T/\sqrt{9}$ is the *direction associated with the parameter of interest* μ.

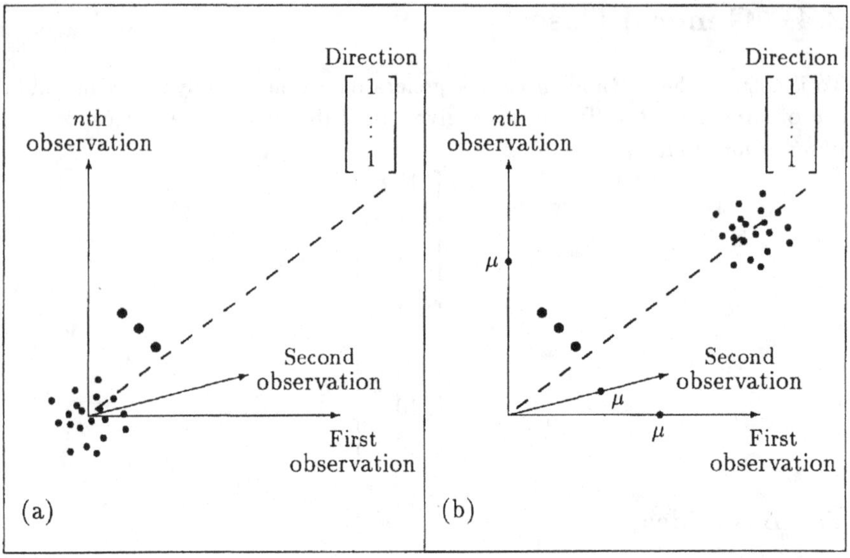

Figure 2.9: Scatters of data points representing many repetitions of our study. In the case (a) $\mu = 0$ the scatter is centered on the origin, while in the case (b) $\mu \neq 0$ the scatter is centered on the point $(\mu, \ldots, \mu)$. Note that the three large dots signify the $n - 3$ axes which we have omitted from the diagram.

The Orthogonal Decomposition

We now project our observation vector $\mathbf{y} = [11.8, 7.5, 6, 7.9, 4.9, 4, 3, 10, 9]^T$ onto each of our coordinate axes. The resulting orthogonal decomposition is

$$\mathbf{y} = (\mathbf{y}\cdot\mathbf{U}_1)\mathbf{U}_1 + (\mathbf{y}\cdot\mathbf{U}_2)\mathbf{U}_2 + \cdots + (\mathbf{y}\cdot\mathbf{U}_9)\mathbf{U}_9$$

$$\begin{bmatrix} 11.8 \\ 7.5 \\ 6 \\ 7.9 \\ 4.9 \\ 4 \\ 3 \\ 10 \\ 9 \end{bmatrix} = \begin{bmatrix} 7.1222 \\ 7.1222 \\ 7.1222 \\ 7.1222 \\ 7.1222 \\ 7.1222 \\ 7.1222 \\ 7.1222 \\ 7.1222 \end{bmatrix} + \begin{bmatrix} 2.15 \\ -2.15 \\ 0 \\ 0 \\ 0 \\ 0 \\ 0 \\ 0 \\ 0 \end{bmatrix} + \cdots + \begin{bmatrix} -0.2347 \\ -0.2347 \\ -0.2347 \\ -0.2347 \\ -0.2347 \\ -0.2347 \\ -0.2347 \\ -0.2347 \\ 1.8778 \end{bmatrix}$$

This simplifies to the fitted model

$$\begin{array}{ccccc} \mathbf{y} & = & \bar{\mathbf{y}} & + & (\mathbf{y}-\bar{\mathbf{y}}) \\ \begin{bmatrix} 11.8 \\ 7.5 \\ 6 \\ 7.9 \\ 4.9 \\ 4 \\ 3 \\ 10 \\ 9 \end{bmatrix} & = & \begin{bmatrix} 7.1222 \\ 7.1222 \\ 7.1222 \\ 7.1222 \\ 7.1222 \\ 7.1222 \\ 7.1222 \\ 7.1222 \\ 7.1222 \end{bmatrix} & + & \begin{bmatrix} 4.6778 \\ 0.3778 \\ -1.1222 \\ 0.7778 \\ -2.2222 \\ -3.1222 \\ -4.1222 \\ 2.8778 \\ 1.8778 \end{bmatrix} \\ \text{observation vector} & & \text{model vector} & & \text{error vector} \end{array}$$

as shown in Figure 2.10.

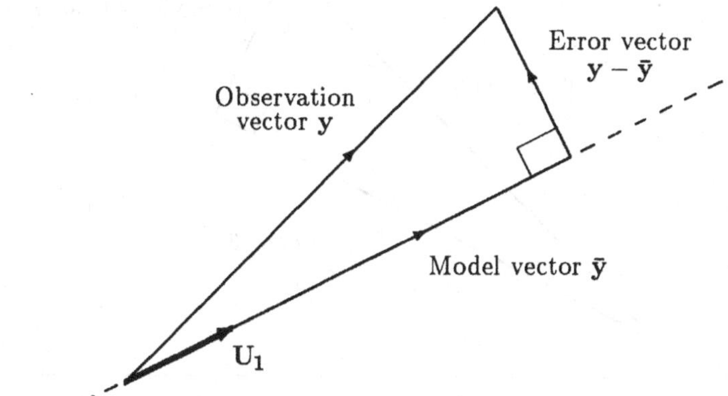

Figure 2.10: The orthogonal decomposition of the observation vector for the general paired samples case.

Distributions of the Projection Lengths

Proceeding as in §2.2 and §2.3, we find that the mean of $\mathbf{y}\cdot\mathbf{U}_1$ is $\sqrt{9}\mu$, and the variance of $\mathbf{y}\cdot\mathbf{U}_1$ is σ^2. That is, $\mathbf{y}\cdot\mathbf{U}_1$ comes from an $N(\sqrt{9}\mu,\sigma^2)$ distribution. Similarly, we find that $\mathbf{y}\cdot\mathbf{U}_2,\ldots,\mathbf{y}\cdot\mathbf{U}_9$ all come from an $N(0,\sigma^2)$ distribution.

As previously, the point to note is that the distribution of $\mathbf{y}\cdot\mathbf{U}_1$ is centered on a potentially nonzero quantity whereas the distributions of $\mathbf{y}\cdot\mathbf{U}_2,\ldots,\mathbf{y}\cdot\mathbf{U}_9$ are always centered on zero.

Testing the Hypothesis

We now investigate the study objective: Is the true mean difference in height between male and female twins zero? For our hypothesis test we check whether the squared projection length $(\mathbf{y}\cdot\mathbf{U}_1)^2$ is similar to, or considerably greater than, the average of the squared projection lengths for the "error" directions, $[(\mathbf{y}\cdot\mathbf{U}_2)^2 + \cdots + (\mathbf{y}\cdot\mathbf{U}_9)^2]/8$.

The associated Pythagorean breakup is

$$\|\mathbf{y}\|^2 = (\mathbf{y}\cdot\mathbf{U}_1)^2 + (\mathbf{y}\cdot\mathbf{U}_2)^2 + \cdots + (\mathbf{y}\cdot\mathbf{U}_9)^2$$

or,

$$\|\mathbf{y}\|^2 = \|\bar{\mathbf{y}}\|^2 + \|\mathbf{y} - \bar{\mathbf{y}}\|^2$$

or,

$$523.91 = 456.53 + 67.38$$

as illustrated in Figure 2.11.

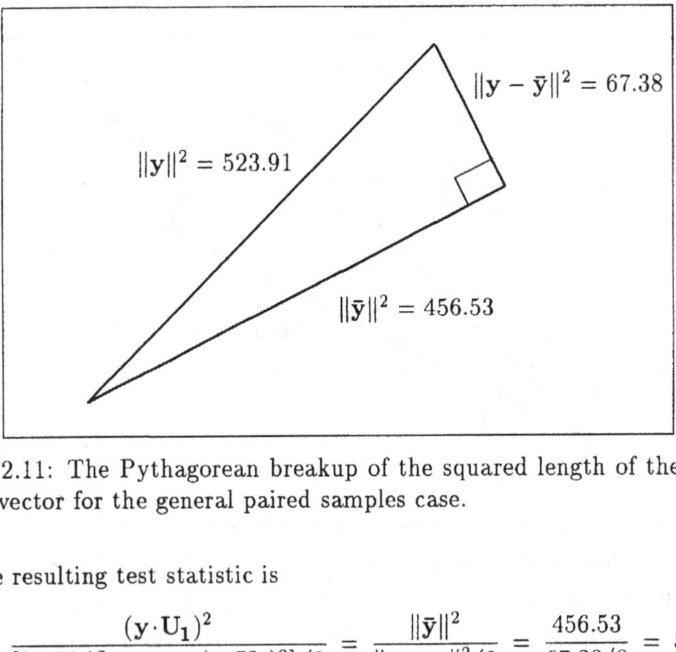

Figure 2.11: The Pythagorean breakup of the squared length of the observation vector for the general paired samples case.

The resulting test statistic is

$$F = \frac{(\mathbf{y}\cdot\mathbf{U}_1)^2}{[(\mathbf{y}\cdot\mathbf{U}_2)^2 + \cdots + (\mathbf{y}\cdot\mathbf{U}_9)^2]/8} = \frac{\|\bar{\mathbf{y}}\|^2}{\|\mathbf{y} - \bar{\mathbf{y}}\|^2/8} = \frac{456.53}{67.38/8} = 54.20$$

If $\mu = 0$ this comes from the $F_{1,8}$ distribution. Since the observed F value (54.20) is larger than the 99 percentile of the $F_{1,8}$ distribution (11.26) we reject the hypothesis that $\mu = 0$, and conclude that there is very strong evidence to suggest that male twins are taller than female twins. Note that this conclusion differs from those obtained in §2.2 and §2.3. This illustrates the increase in the *power* of the test associated with increasing the sample size from two or three up to nine. (The power is the probability of detecting a nonzero mean pairwise difference.)

Paired Samples t Test

Note that our test statistic $F = (\mathbf{y}\cdot\mathbf{U}_1)^2/\{[(\mathbf{y}\cdot\mathbf{U}_2)^2 + \cdots + (\mathbf{y}\cdot\mathbf{U}_9)^2]/8\} = \|\bar{\mathbf{y}}\|^2/\,[\|\mathbf{y}-\bar{\mathbf{y}}\|^2/8] = 54.20$ can be transformed to

$$t = \frac{\mathbf{y}\cdot\mathbf{U}_1}{\sqrt{[(\mathbf{y}\cdot\mathbf{U}_2)^2 + \cdots + (\mathbf{y}\cdot\mathbf{U}_9)^2]/8}} = \frac{\|\bar{\mathbf{y}}\|}{\|\mathbf{y}-\bar{\mathbf{y}}\|/\sqrt{8}} = \frac{\sqrt{9}\times 7.1222}{\sqrt{67.38/8}} = 7.363$$

If $\mu = 0$ this comes from the t_8 distribution. As previously, the size of our observed value constitutes very strong evidence against the notion that the population mean could be zero.

Note also that the usual algebraic expression for t is

$$t = \frac{\|\bar{\mathbf{y}}\|}{\|\mathbf{y}-\bar{\mathbf{y}}\|/\sqrt{n-1}} = \frac{\sqrt{n}\bar{y}}{\sqrt{\sum_{i=1}^{n}(y_i-\bar{y})^2/(n-1)}} = \frac{\bar{y}}{s/\sqrt{n}}$$

where s is the square root of the sample variance, $s^2 = \|\mathbf{y}-\bar{\mathbf{y}}\|^2/(n-1)$, and where $n = 9$ in this example.

Comments

As in §2.2 and §2.3, the "model space" direction $\mathbf{U}_1$ is used to estimate the population mean μ, while the "error space" directions $\mathbf{U}_2, \ldots, \mathbf{U}_9$ are used to estimate the population variance σ^2.

To summarize our analysis of the full data set, we now have a best estimate of 7.1 in. for the mean difference in height between male and female for mixed-sex twins over 16 years of age living in New Zealand in 1992, with very strong evidence that the underlying population mean difference is non zero. That is, "grown up" male twins have been shown to be taller than "grown up" female twins.

2.5 Summary

To summarize, in the paired samples case the direction in n-space of special interest is the equiangular direction. If the mean paired difference μ is in fact different from zero, then this is the direction for which the length of the projection is increased. More fully, if μ is nonzero the length of the projection of the vector of observations onto the equiangular direction will be inflated, while the corresponding projection lengths for all directions perpendicular to the equiangular direction will be unaffected. Thus to distinguish between the cases $\mu = 0$ and $\mu \neq 0$ we divide the squared projection length for the equiangular direction by the average of the squared projection lengths for a full set of $n-1$ perpendicular directions; if the resulting ratio is small we decide that μ could be zero, while if the ratio is large we decide that μ could not be zero.

In general, the projection of the vector of observations $\mathbf{y} = [y_1, \ldots, y_n]^T$ onto the equiangular direction $\mathbf{U}_1 = [1, \ldots, 1]^T/\sqrt{n}$ is the vector $\bar{\mathbf{y}} = [\bar{y}, \ldots, \bar{y}]^T$. This leads to the orthogonal decomposition $\mathbf{y} = \bar{\mathbf{y}} + (\mathbf{y} - \bar{\mathbf{y}})$ shown in Figure 2.12. Here the "error vector" $(\mathbf{y} - \bar{\mathbf{y}})$ is a sum of $(n-1)$ orthogonal projection vectors, one for each of the orthogonal coordinate axes spanning the error space.

The resulting Pythagorean breakup is $||\mathbf{y}||^2 = ||\bar{\mathbf{y}}||^2 + ||\mathbf{y} - \bar{\mathbf{y}}||^2$. This leads to the test statistic

$$F = \frac{||\bar{\mathbf{y}}||^2}{||\mathbf{y} - \bar{\mathbf{y}}||^2/(n-1)} = \frac{n\bar{y}^2}{s^2}$$

This test statistic follows an $F_{1,n-1}$ distribution if $\mu = 0$, so we reject the hypothesis that $\mu = 0$ if the observed F value exceeds the 95 percentile of this distribution.

Note that the equivalent paired samples t test is

$$t = \frac{||\bar{\mathbf{y}}||}{||\mathbf{y} - \bar{\mathbf{y}}||/\sqrt{n-1}} = \frac{\sqrt{n}\bar{y}}{s}$$

For an analysis of the examples in this chapter using the angle θ as the test statistic, the reader is referred to Appendix D. For computer methods refer to Appendix C. For estimation of the 95% confidence interval for the population mean μ refer to pp. 76, 77, 82, 87, 538, 539 and 540 of Saville and Wood (1991).

For additional reading on the geometry of the paired samples case, the reader can refer to Chapter 5 of Saville and Wood (1991), or Box, Hunter and Hunter (1978), pp. 197–203.

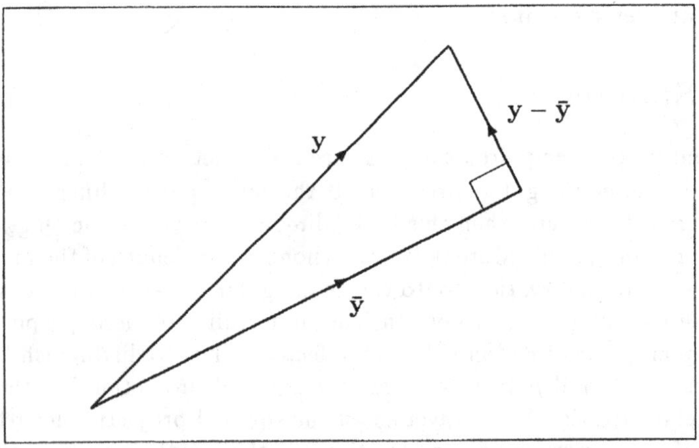

Figure 2.12: The orthogonal decomposition of the observation vector for the paired samples case.

Class Exercise

Twenty-six agricultural researchers were used as guinea pigs to determine whether heartbeat rate is increased by exercise. Each person took their heartbeat after ten minutes of sitting, and again after two minutes of running on the spot. On both occasions the number of beats per 30 seconds was recorded. The changes in heartbeat per 30 seconds are given in Table 2.2.

Person	Change	Person	Change	Person	Change
1	7	11	9	21	16
2	10	12	8	22	12
3	14	13	13	23	9
4	0	14	8	24	0
5	13	15	10	25	12
6	5	16	17	26	13
7	7	17	7		
8	7	18	14		
9	6	19	5		
10	20	20	4		

Table 2.2: Changes in the heartbeat rate per 30 seconds for 26 agricultural researchers.

The data were collected during a lunch-time talk at the Canterbury Agriculture and Science Centre, Lincoln, New Zealand. For social reasons not everyone ran on the spot as vigorously as requested!

(a) Each member of the class is asked to use the random numbers given in Table T.1 to select a sample of size four from the data in Table 2.2. The four changes in heartbeat make up the observation vector $\mathbf{y}$. (If the teacher would like more active class participation, the class can be broken into groups of size four, with each group generating its own data.)

The question to be answered is: Does running affect heartbeat? Assuming no prior knowledge of this subject, we shall carry out a 2-tailed test, H_0: $\mu = 0$ versus H_1: $\mu \neq 0$, where μ is the mean change in heartbeat for an infinitely large theoretical population of people of similar background.

An appropriate coordinate system is

$$\mathbf{U}_1 = \frac{1}{\sqrt{4}}\begin{bmatrix}1\\1\\1\\1\end{bmatrix}, \mathbf{U}_2 = \frac{1}{\sqrt{2}}\begin{bmatrix}1\\-1\\0\\0\end{bmatrix}, \mathbf{U}_3 = \frac{1}{\sqrt{6}}\begin{bmatrix}1\\1\\-2\\0\end{bmatrix}, \mathbf{U}_4 = \frac{1}{\sqrt{12}}\begin{bmatrix}1\\1\\1\\-3\end{bmatrix}$$

Here the first axis direction is the equiangular direction, which spans the "model space." The other three directions span the "error space," and can be chosen in an infinite variety of ways. For our choice, we first chose the error axis $\mathbf{U_2}$ by thinking of a comparison of the first and second observations. Then we chose the second error axis $\mathbf{U_3}$ to correspond to a comparison of the third observation with the average of the first two observations. Lastly, we chose the remaining error axis $\mathbf{U_4}$ to correspond to a comparison of the fourth observation with the average of the first three observations.

(b) Each class member is asked to calculate the scalars $\mathbf{y}\cdot\mathbf{U_1}$, $\mathbf{y}\cdot\mathbf{U_2}$, $\mathbf{y}\cdot\mathbf{U_3}$ and $\mathbf{y}\cdot\mathbf{U_4}$ using his or her own observation vector. The teacher can then tabulate the results for $\mathbf{y}\cdot\mathbf{U_1}$ in a histogram, and for $\mathbf{y}\cdot\mathbf{U_2}$, $\mathbf{y}\cdot\mathbf{U_3}$ and $\mathbf{y}\cdot\mathbf{U_4}$ in three further histograms. These histograms will approximate the distributions of the projection lengths $\mathbf{y}\cdot\mathbf{U}_i$, for $i = 1, 2, 3, 4$.

Questions:

1. Which of the projection lengths appear to have mean zero?
2. Is the variability similar between the histograms?

(c) Each class member is then asked to calculate

$$F = \frac{(\mathbf{y}\cdot\mathbf{U_1})^2}{((\mathbf{y}\cdot\mathbf{U_2})^2 + (\mathbf{y}\cdot\mathbf{U_3})^2 + (\mathbf{y}\cdot\mathbf{U_4})^2)/3}$$

and compare his or her answer with the 95 percentile of the $F_{1,3}$ distribution.

The teacher can then tabulate the class results in the form of a histogram. Note: This histogram should be saved for comparison with the class results to be obtained in Chapter 3.

Question:

Does the test have adequate "power" with a sample size of four? That is, was the null hypothesis rejected for most class members' data sets?

(d) As a further exercise, each class member can now calculate the projection vectors $(\mathbf{y}\cdot\mathbf{U_1})\mathbf{U_1}$, $(\mathbf{y}\cdot\mathbf{U_2})\mathbf{U_2}$, $(\mathbf{y}\cdot\mathbf{U_3})\mathbf{U_3}$ and $(\mathbf{y}\cdot\mathbf{U_4})\mathbf{U_4}$. He or she can then confirm that $(\mathbf{y}\cdot\mathbf{U_1})\mathbf{U_1} = \bar{\mathbf{y}}$, the mean vector, and that $(\mathbf{y}\cdot\mathbf{U_2})\mathbf{U_2} + (\mathbf{y}\cdot\mathbf{U_3})\mathbf{U_3} + (\mathbf{y}\cdot\mathbf{U_4})\mathbf{U_4} = (\mathbf{y} - \bar{\mathbf{y}})$, the error vector. He or she can also write out the orthogonal decomposition $\mathbf{y} = \bar{\mathbf{y}} + (\mathbf{y} - \bar{\mathbf{y}})$, and sketch a right-angled triangle of vectors to represent this decomposition.

(e) Using this new decomposition, each class member can now recalculate

$$F = \frac{\|\bar{\mathbf{y}}\|^2}{\|\mathbf{y} - \bar{\mathbf{y}}\|^2/3}$$

to confirm that this formula yields the same answer as in (c).

Exercises

Note that solutions to exercises marked with an * are given in Appendix E.

(2.1)* In §2.3 there were an infinite number of alternative pairs of orthogonal coordinate axes ($\mathbf{U_2}$, $\mathbf{U_3}$) which we could have used to span the "error space." One such choice is

$$\mathbf{U_2} = \frac{1}{\sqrt{42}}\begin{bmatrix} -4 \\ -1 \\ 5 \end{bmatrix}, \quad \mathbf{U_3} = \frac{1}{\sqrt{14}}\begin{bmatrix} 2 \\ -3 \\ 1 \end{bmatrix}$$

(a) Calculate the projections (meaning vectors) of our observation vector $\mathbf{y} = [7.5, 3, 9]^T$ onto these alternative axis directions, and show that

$$(\mathbf{y}\cdot\mathbf{U_2})\mathbf{U_2} + (\mathbf{y}\cdot\mathbf{U_3})\mathbf{U_3} = \mathbf{y} - \bar{\mathbf{y}}$$

as in §2.3.

(b) Also show that

$$F = \frac{(\mathbf{y}\cdot\mathbf{U_1})^2}{[(\mathbf{y}\cdot\mathbf{U_2})^2 + (\mathbf{y}\cdot\mathbf{U_3})^2]/2} = 13.0$$

as with our previous choice of $\mathbf{U_2}$ and $\mathbf{U_3}$.

(2.2) An experiment was set up to see whether nitrogen fertilizer was beneficial, detrimental, or had no effect on the growth of an alfalfa crop. Five pairs of plots were laid out, with nitrogen applied to just one plot of each pair. The resulting percentage decreases in production of the fertilized, compared to the unfertilized for the five pairs, were

8, 7, 5, 6 and 9%

(a) Take these values for the observation vector $\mathbf{y}$ and project $\mathbf{y}$ onto the coordinate system

$$\frac{1}{\sqrt{5}}\begin{bmatrix} 1 \\ 1 \\ 1 \\ 1 \\ 1 \end{bmatrix}, \quad \frac{1}{\sqrt{30}}\begin{bmatrix} 3 \\ 3 \\ -2 \\ -2 \\ -2 \end{bmatrix}, \quad \frac{1}{\sqrt{2}}\begin{bmatrix} 1 \\ -1 \\ 0 \\ 0 \\ 0 \end{bmatrix}, \quad \frac{1}{\sqrt{2}}\begin{bmatrix} 0 \\ 0 \\ 1 \\ -1 \\ 0 \end{bmatrix}, \quad \frac{1}{\sqrt{6}}\begin{bmatrix} 0 \\ 0 \\ 1 \\ 1 \\ -2 \end{bmatrix}$$

writing out each projection vector. Check your working by substituting into the orthogonal decomposition: $\mathbf{y}$ = sum of projection vectors.

(b) Calculate the squared lengths of these projections, and check that Pythagoras' Theorem is obeyed.

(c) Use your answers in (b) to calculate the F test of the hypothesis H_0: $\mu = 0$ against the alternative H_1: $\mu \neq 0$. What do you conclude?

(d) Recalculate this F test using the breakdown $||\mathbf{y}||^2 = ||\bar{\mathbf{y}}||^2 + ||\mathbf{y} - \bar{\mathbf{y}}||^2$.

(2.3)* (a) Write down an orthogonal coordinate system for 8-space which includes the unit vector $[1,1,1,1,1,1,1,1]^T/\sqrt{8}$. Write down a second such system.

(b) Write down two orthogonal coordinate systems for 5-space which include the unit vector $[1,1,1,1,1]/\sqrt{5}$. Do not use the system given in Exercise 2.2.

(2.4) A study was carried out by the Christchurch School of Medicine in 1993 to compare the changes in blood plasma selenium levels of babies fed either on unsupplemented milk formula or on a formula supplemented with selenium (Darlow et al., 1995). Blood samples were taken from the umbilical cord of each baby at birth, and at one and three months of age. The resulting data for the unsupplemented group of babies are shown in the table, by courtesy of Professor Christine Winterbourn. Note that all three data values are present for less than half of the babies.

Unsupplemented	Plasma selenium levels (μg/L)		
Baby no.	Birth	1 month	3 months
1		20	21
2	46	20	
3		19	
4		28	21
5		21	
6	43	24	21
7		21	
8	37	19	42
9		23	29
10		21	46
11	27	13	
12		23	25
13	29	18	40
14	49	21	26
15	40	26	47

(a) We shall firstly examine the changes in selenium level of this group of babies from birth to one month of age. For this time period write down the change in selenium level (birth minus one month) for the babies for whom samples were taken at both sampling times. Rewrite these values as an observation vector $\mathbf{y}$.

(b) Calculate the orthogonal decomposition $\mathbf{y} = \bar{\mathbf{y}} + (\mathbf{y} - \bar{\mathbf{y}})$.

(c) Test the hypothesis of no change in mean selenium level ($\mu = 0$) over the time period by writing out the corresponding Pythagorean breakup and calculating the appropriate F value. What is your conclusion?

(d) Repeat (a) to (c) for the time period one to three months of age, subtracting the one month value from the three month value to minimize the number of negative values.

(2.5) In order to monitor the weight changes in a herd of calves, six randomly chosen calves were individually identified with ear tags and weighed monthly. The weights, in kilograms, on two successive dates were:

Calf tag no.	Weight (22/2/85)	Weight (22/3/85)
635	125	128
123	115	126
715	115	124
817	112	119
125	142	152
347	110	118

We assume that the *changes* in weight come from a normal distribution $N(\mu, \sigma^2)$, where μ and σ^2 are the mean and variance of the weight changes for the herd of calves.

(a) Write down the observation vector $\mathbf{y}$ comprising the six weight changes, a unit vector $\mathbf{U}_1$ spanning the one-dimensional model space, and unit vectors $\mathbf{U}_2$ to $\mathbf{U}_6$ spanning the error space.

(b) Fit the model by expressing $\mathbf{y}$ in the form

$$\mathbf{y} = (\mathbf{y}\cdot\mathbf{U}_1)\mathbf{U}_1 + (\mathbf{y}\cdot\mathbf{U}_2)\mathbf{U}_2 + \cdots + (\mathbf{y}\cdot\mathbf{U}_6)\mathbf{U}_6$$

(c) Rewrite the model in the form

$$\mathbf{y} = \bar{\mathbf{y}} + (\mathbf{y} - \bar{\mathbf{y}})$$

That is, express the observation vector $\mathbf{y}$ as a sum of the model vector $\bar{\mathbf{y}}$, together with the error vector $\mathbf{y} - \bar{\mathbf{y}}$.

(d) Draw a vector diagram showing the relationship of $\mathbf{y}$, $\bar{\mathbf{y}}$ and $\mathbf{y} - \bar{\mathbf{y}}$. Label the vectors with their numerical values.

(e) Calculate the squared lengths of the three vectors $\|\mathbf{y}\|^2$, $\|\bar{\mathbf{y}}\|^2$ and $\|\mathbf{y} - \bar{\mathbf{y}}\|^2$, and write out the resulting Pythagorean breakup.

(f) Test the hypothesis $H_0\colon \mu = 0$ against the alternative $H_1\colon \mu \neq 0$. Is there evidence of any weight change in the herd during the month?

(2.6)* The residents of West Melton, 22 km inland from Christchurch, New Zealand, were curious about how their annual rainfall compared with that at the meteorological station at Darfield, a further 20 km inland. Dr Ivor Lewis, a West Melton resident, had recorded rainfall for several years, so his records were obtained and compared with those from Darfield. In the table we present the rainfall for five calendar years (January to December) for each area. Dr Lewis is thanked for the use of his data.

	Rainfall (mm)	
Year	Darfield	West Melton
1985	692	604
1986	1053	927
1987	681	638
1988	508	400
1989	797	739

(a) For each year, calculate the difference in rainfall between Darfield and West Melton (as D–WM) and express this as a percentage of the Darfield rainfall. Write out the resulting percentage differences as an observation vector $\mathbf{y}$.

(b) Test the hypothesis "mean % difference = 0" by fitting the model $\mathbf{y} = \bar{\mathbf{y}} + (\mathbf{y} - \bar{\mathbf{y}})$ and calculating the appropriate F value. Is there evidence of a difference in mean rainfall between the two areas, and how strong is the evidence?

(c) Also calculate the t test statistic using the formula

$$t = \frac{\pm\|\bar{\mathbf{y}}\|}{\|\mathbf{y} - \bar{\mathbf{y}}\|/\sqrt{n-1}}$$

Compare your calculated value with the percentiles of the appropriate t reference distribution in Table T.2. Do your results agree with those in (b)? Are your test statistics linked via the formula $F = t^2$?

(d) What is your best estimate of the long-term mean percentage difference in rainfall between the two areas? Over the period 1919–1980 Darfield averaged 801 mm of rainfall per annum; what is your best estimate of the corresponding average for West Melton? Suppose that in a subsequent year Darfield recorded 927 mm of rainfall; what is your best estimate of the West Melton rainfall for this year?

(2.7)* Symptoms of lead poisoning were detected in two users of an indoor small bore rifle range (George et al., 1993). Red cell lead levels, in units of μmol/L, were therefore collected for 32 small bore shooters at the end (END) of the six month winter indoor shooting season and prior to (PRE) the start of the next shooting season. Southern Hemisphere dates were September, 1990 for END samples and March, 1991 for PRE samples. The resulting data are presented in the table below by courtesy of Trevor Walmsley. Note that some data sorting has been carried out (on the basis of PRE data values). Note also that this exercise could prove tedious for readers who do not have access to computers (see Appendix C for computing methods).

Shooter	Lead levels		Shooter	Lead levels	
number	END	PRE	number	END	PRE
1	2.54	0.85	17	2.72	1.41
2	1.85	0.90	18	3.02	1.49
3	1.02	0.90	19	2.93	1.57
4	1.42	0.95	20	2.18	1.61
5	1.39	0.96	21	2.05	1.69
6	2.43	0.97	22	3.64	1.76
7	1.22	1.04	23	2.54	1.77
8	1.84	1.07	24	3.36	1.77
9	2.88	1.11	25	3.80	1.78
10	1.02	1.17	26	3.02	1.78
11	2.60	1.21	27	5.32	1.83
12	2.19	1.27	28	3.61	2.09
13	1.31	1.29	29	4.11	2.21
14	2.06	1.30	30	3.64	2.41
15	2.90	1.38	31	4.33	2.41
16	3.51	1.40	32	4.48	2.45

(a) Calculate the change in red cell lead level (END − PRE) during the Southern Hemisphere summer recess for each of the 32 shooters. These values make up the observation vector $\mathbf{y}$.

(b) Calculate the mean change in red cell lead level $\bar{y}$ and the Pythagorean breakup $||\mathbf{y}||^2 = ||\bar{\mathbf{y}}||^2 + ||\mathbf{y} - \bar{\mathbf{y}}||^2$.

(c) Use this breakup to calculate the F test statistic. What is your conclusion?

(d) Calculate the corresponding t value using the formula

$$t = \frac{\pm||\bar{\mathbf{y}}||}{||\mathbf{y} - \bar{\mathbf{y}}||/\sqrt{n-1}}$$

What are the percentiles of the appropriate t reference distribution? Do you reach the same conclusion as in (c)?

(2.8) A plant breeder has carried out nine trials to evaluate a promising new cultivar of barley when his director queries him as to whether the percentage screenings of the new cultivar is different from the industry average of 10%. The percentage screenings is the percent by weight of grains which are small enough to fall through a standard screen.

The percentage screenings in the nine trials were as follows:

$$S = \text{percentage screenings} = 12, 6, 10, 7, 7, 12, 5, 8, 9$$

Assume these values come from a normal distribution with mean $\mu_S = 10$ and variance σ^2. We wish to test the hypothesis $H_0\colon \mu_S = 10$ against the alternative $H_1\colon \mu_S \neq 10$.

(a) Transform the problem into one we can solve by calculating $Y = 10 - S$ for each trial. Take the resulting values as your observation vector $\mathbf{y}$.

(b) Test the hypothesis $H_0\colon \mu_Y = 0$ against the alternative $H_1\colon \mu_Y \neq 0$. This is equivalent to testing $H_0\colon \mu_S = 10$ against $H_1\colon \mu_S \neq 10$. What can you conclude?

Chapter 3

Independent Samples

Independent samples arise when there is no natural pairing of the observations. For example, we may randomly select five females and five males and measure their heights, or measure the heartbeats of six joggers and of six resting people. As in Chapter 2 our interest is normally in the average *difference* in height between the females and males in our study population, or in the average *difference* in heartbeat between resting and jogging humans.

More formally, we are interested in comparing the means, μ_1 and μ_2, of two study populations. For the analysis we assume that both populations are normally distributed and that the two population variances are the same ($\sigma_1^2 = \sigma_2^2 = \sigma^2$).

In §3.1 of this chapter we shall describe a data set consisting of the heights of female and màle students in a first-year university class in California. In §3.2 we analyze a small subset of this data set, consisting of the heights of two female and two male students. In §3.3 we analyze a slightly larger subset, consisting of the heights of four female and four male students. The chapter closes with a summary in §3.4, a class exercise and general exercises.

3.1 Heights of Students

While one of the authors was a visiting lecturer in the Department of Agronomy and Range Science at the University of California at Davis, he taught a first-year statistics course code-named ASM150 (Agricultural Science and Management). To generate data for a comparison of two independent samples he sent around the class a sheet of paper with a request that each student record on it their height, in inches, along with their sex. The resulting data, shown in Figure 3.1, are the heights of 49 females and 77 males.

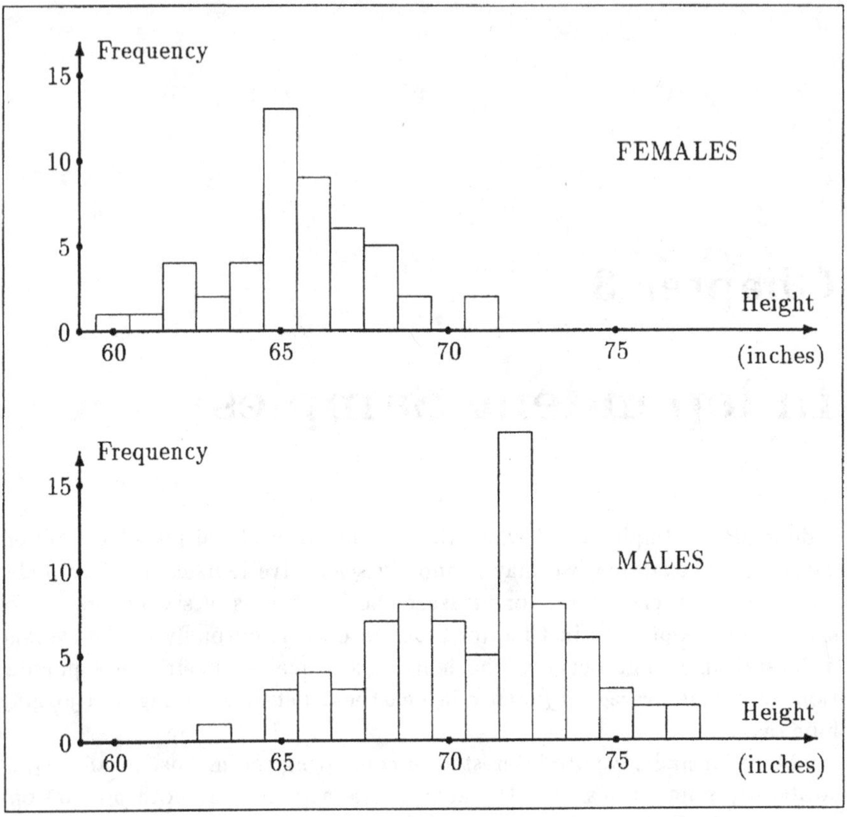

Figure 3.1: Histograms of heights of female and male students in the ASM150 class.

An interesting feature of Figure 3.1(a) is the large number of males who wrote down their height as 72 in. (6 ft), and the relatively small number who recorded their height as 71 in. Presumably 6 ft is a more popular height among males than 5 ft 11 in.!

In considering the usefulness of these data, we must try to define the populations to which the results may be extrapolated, and also consider whether the method of data collection has any inherent problems other than the one just mentioned.

One choice of definition of the underlying study populations is the populations of female and male students at the University of California at Davis in the fall of 1984. Is it reasonable to regard the respondents as random samples of these populations? One point which comes to mind is that the ASM150 course contained many "conscripts," meaning students who would not have enrolled of their own free will, since it was a required course for quite a range of degrees. This means it is more reasonable to treat the students in the class as a sample of the overall student body than if this had not been the case.

The method of data collection did not ensure anonymity of the responses since the person next receiving the sheet of paper could easily glance back over the responses of his or her neighbors. This may have meant that people who were embarrassed about their tallness or shortness may either have not responded or modified their response. In terms of nonresponse, the number of responses represents 70–80% of the students in the class on that particular day; this is reasonably good.

In spite of any misgivings we may have, we now treat our two samples of female and male heights as random samples from the populations of female and male heights for students attending the University of California at Davis in the fall of 1984.

3.2 Samples of Size Two

In this section we apply our geometric method to the case where we wish to compare the means of two populations, and have at our disposal two independent random samples of size two. For our example we shall use the data shown in Table 3.1; these data are the heights of two females and two males from Figure 3.1.

	Heights (inches)		Means
Females	63	65	64
Males	69	74	71.5

Table 3.1: Heights, in inches, of two females and two males from the ASM150 class, together with the two sample means.

The resulting *observation vector* is

$$\mathbf{y} = \begin{bmatrix} 63 \\ 65 \\ 69 \\ 74 \end{bmatrix}$$

Objective

The study objective is as follows:

> To test whether there is, on average, a difference in height between the female and male students attending the University of California at Davis (UCD) in the fall of 1984.

The Basic Idea

Figure 3.2 explains the basic idea for our new situation. If the female and male students are on average the same in height ($\mu_1 = \mu_2$), then our observation vector will be part of a scatter centered on the point (μ, μ, μ, μ), where $\mu_1 = \mu_2 = \mu$, as shown in Figure 3.2(a). However, if they differ in height ($\mu_1 \neq \mu_2$), then our observation vector will be part of a scatter centered on the point $(\mu_1, \mu_1, \mu_2, \mu_2)$, as shown in Figure 3.2(b). In other words, if $\mu_1 = \mu_2$ the scatter is centered on the equiangular line, whereas if $\mu_1 \neq \mu_2$ the scatter is moved away from the equiangular line in the direction

$$\begin{bmatrix} \mu_1 \\ \mu_1 \\ \mu_2 \\ \mu_2 \end{bmatrix} - \begin{bmatrix} \mu \\ \mu \\ \mu \\ \mu \end{bmatrix} = \begin{bmatrix} \mu_1 - \mu \\ \mu_1 - \mu \\ \mu_2 - \mu \\ \mu_2 - \mu \end{bmatrix} = \begin{bmatrix} (\mu_1 - \mu_2)/2 \\ (\mu_1 - \mu_2)/2 \\ (\mu_2 - \mu_1)/2 \\ (\mu_2 - \mu_1)/2 \end{bmatrix} = \frac{\mu_2 - \mu_1}{2} \begin{bmatrix} -1 \\ -1 \\ 1 \\ 1 \end{bmatrix}$$

since $\mu = (\mu_1 + \mu_2)/2$.

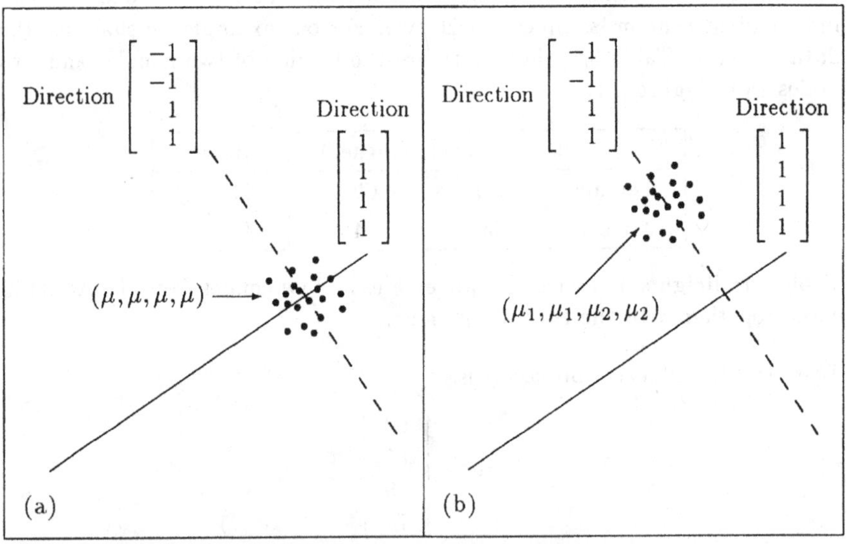

Figure 3.2: Clouds of data points representing many repetitions of our height study. In the case (a) $\mu_1 = \mu_2$ the cloud is centered on the point (μ, μ, μ, μ), while in the case (b) $\mu_1 \neq \mu_2$ the cloud is centered on the point $(\mu_1, \mu_1, \mu_2, \mu_2)$. That is, if $\mu_1 = \mu_2$ the cloud is centered on the equiangular line, while if $\mu_1 \neq \mu_2$ the cloud is moved away from the line in the direction $[-1, -1, 1, 1]^T$.

An Appropriate Coordinate System

For our purposes, then, an appropriate set of coordinate axes for 4-space is

$$\mathbf{U_1} = \frac{1}{\sqrt{4}}\begin{bmatrix} 1 \\ 1 \\ 1 \\ 1 \end{bmatrix}, \quad \mathbf{U_2} = \frac{1}{\sqrt{4}}\begin{bmatrix} -1 \\ -1 \\ 1 \\ 1 \end{bmatrix}, \quad \mathbf{U_3} = \frac{1}{\sqrt{2}}\begin{bmatrix} -1 \\ 1 \\ 0 \\ 0 \end{bmatrix}, \quad \mathbf{U_4} = \frac{1}{\sqrt{2}}\begin{bmatrix} 0 \\ 0 \\ -1 \\ 1 \end{bmatrix}$$

Here $\mathbf{U_1} = [1,1,1,1]^T/\sqrt{4}$ is the direction associated with the overall mean, μ, while $\mathbf{U_2} = [-1,-1,1,1]^T/\sqrt{4}$ is the *direction associated with the difference of interest* $(\mu_2 - \mu_1)$.

The Orthogonal Decomposition

We now project our observation vector $\mathbf{y} = [63, 65, 69, 74]^{\mathrm{T}}$ onto each of our coordinate axes. The first projection vector is

$$(\mathbf{y}\cdot\mathbf{U_1})\mathbf{U_1} = \left(\begin{bmatrix} 63 \\ 65 \\ 69 \\ 74 \end{bmatrix} \cdot \frac{1}{\sqrt{4}}\begin{bmatrix} 1 \\ 1 \\ 1 \\ 1 \end{bmatrix}\right)\frac{1}{\sqrt{4}}\begin{bmatrix} 1 \\ 1 \\ 1 \\ 1 \end{bmatrix} = 67.75\begin{bmatrix} 1 \\ 1 \\ 1 \\ 1 \end{bmatrix}$$

The second projection vector is

$$(\mathbf{y}\cdot\mathbf{U_2})\mathbf{U_2} = \left(\begin{bmatrix} 63 \\ 65 \\ 69 \\ 74 \end{bmatrix} \cdot \frac{1}{\sqrt{4}}\begin{bmatrix} -1 \\ -1 \\ 1 \\ 1 \end{bmatrix}\right)\frac{1}{\sqrt{4}}\begin{bmatrix} -1 \\ -1 \\ 1 \\ 1 \end{bmatrix} = 3.75\begin{bmatrix} -1 \\ -1 \\ 1 \\ 1 \end{bmatrix}$$

The third projection vector is

$$(\mathbf{y}\cdot\mathbf{U_3})\mathbf{U_3} = \left(\begin{bmatrix} 63 \\ 65 \\ 69 \\ 74 \end{bmatrix} \cdot \frac{1}{\sqrt{2}}\begin{bmatrix} -1 \\ 1 \\ 0 \\ 0 \end{bmatrix}\right)\frac{1}{\sqrt{2}}\begin{bmatrix} -1 \\ 1 \\ 0 \\ 0 \end{bmatrix} = \begin{bmatrix} -1 \\ 1 \\ 0 \\ 0 \end{bmatrix}$$

The fourth projection vector is

$$(\mathbf{y}\cdot\mathbf{U_4})\mathbf{U_4} = \left(\begin{bmatrix} 63 \\ 65 \\ 69 \\ 74 \end{bmatrix} \cdot \frac{1}{\sqrt{2}}\begin{bmatrix} 0 \\ 0 \\ -1 \\ 1 \end{bmatrix}\right)\frac{1}{\sqrt{2}}\begin{bmatrix} 0 \\ 0 \\ -1 \\ 1 \end{bmatrix} = 2.5\begin{bmatrix} 0 \\ 0 \\ -1 \\ 1 \end{bmatrix}$$

The resulting orthogonal decomposition of the observation vector is

$$\begin{aligned}
\mathbf{y} &= (\mathbf{y}\cdot\mathbf{U}_1)\mathbf{U}_1 + (\mathbf{y}\cdot\mathbf{U}_2)\mathbf{U}_2 + (\mathbf{y}\cdot\mathbf{U}_3)\mathbf{U}_3 + (\mathbf{y}\cdot\mathbf{U}_4)\mathbf{U}_4 \\
\begin{bmatrix} 63 \\ 65 \\ 69 \\ 74 \end{bmatrix} &= (271/\sqrt{4})\mathbf{U}_1 + (15/\sqrt{4})\mathbf{U}_2 + (2/\sqrt{2})\mathbf{U}_3 + (5/\sqrt{2})\mathbf{U}_4 \\
\begin{bmatrix} 63 \\ 65 \\ 69 \\ 74 \end{bmatrix} &= \begin{bmatrix} 67.75 \\ 67.75 \\ 67.75 \\ 67.75 \end{bmatrix} + \begin{bmatrix} -3.75 \\ -3.75 \\ 3.75 \\ 3.75 \end{bmatrix} + \begin{bmatrix} -1 \\ 1 \\ 0 \\ 0 \end{bmatrix} + \begin{bmatrix} 0 \\ 0 \\ -2.5 \\ 2.5 \end{bmatrix}
\end{aligned}$$

This simplifies to the fitted model

$$\underset{\text{observation vector}}{\begin{bmatrix} 63 \\ 65 \\ 69 \\ 74 \end{bmatrix}} = \underset{\text{overall mean vector}}{\begin{bmatrix} 67.75 \\ 67.75 \\ 67.75 \\ 67.75 \end{bmatrix}} + \underset{\text{treatment vector}}{\begin{bmatrix} -3.75 \\ -3.75 \\ 3.75 \\ 3.75 \end{bmatrix}} + \underset{\text{error vector}}{\begin{bmatrix} -1 \\ 1 \\ -2.5 \\ 2.5 \end{bmatrix}}$$

That is, we have broken our observation vector $\mathbf{y}$ into three orthogonal components, an *overall mean vector*, a *treatment vector*, and an *error vector*, as shown in Figure 3.3. The term "treatment" is borrowed from the experimental sciences, where populations of study are artificially generated by way of experimental treatments; we use it here as a convenient abbreviation for "differences between populations of study," a phrase too long for our purposes (if this is unclear, see also page 52).

Distributions of the Projection Lengths

To prepare for testing whether the true mean difference in height could plausibly be zero, we now work out the distributions of the projection lengths $\mathbf{y}\cdot\mathbf{U}_1$, $\mathbf{y}\cdot\mathbf{U}_2$, $\mathbf{y}\cdot\mathbf{U}_3$ and $\mathbf{y}\cdot\mathbf{U}_4$.

Firstly,
$$\mathbf{y}\cdot\mathbf{U}_1 = \begin{bmatrix} y_{11} \\ y_{12} \\ y_{21} \\ y_{22} \end{bmatrix} \cdot \frac{1}{\sqrt{4}} \begin{bmatrix} 1 \\ 1 \\ 1 \\ 1 \end{bmatrix} = \frac{y_{11} + y_{12} + y_{21} + y_{22}}{\sqrt{4}}$$

where we use the symbols y_{11} and y_{12} to refer to the first and second observations from the first study population (with mean μ_1 and variance σ^2), and the symbols y_{21} and y_{22} to refer to the first and second observations from

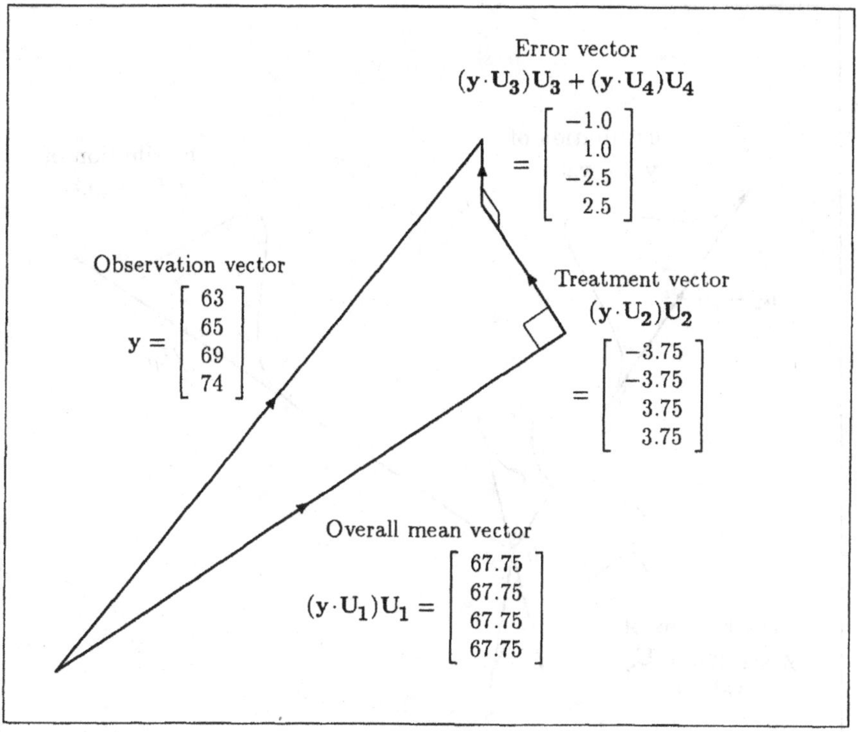

Figure 3.3: Orthogonal decomposition of the observation vector for the case of independent samples of size two from two study populations. Note that for reasons of visibility the figure is not exactly to scale!

the second study population (with mean μ_2 and variance σ^2).

$$\text{Hence the mean of} \quad \mathbf{y}\cdot\mathbf{U_1} = \frac{\mu_1+\mu_1+\mu_2+\mu_2}{\sqrt{4}} = \frac{4\mu}{\sqrt{4}} = \sqrt{4}\mu$$

as illustrated in Figure 3.4.

$$\text{Secondly,} \quad \mathbf{y}\cdot\mathbf{U_2} = \begin{bmatrix} y_{11} \\ y_{12} \\ y_{21} \\ y_{22} \end{bmatrix} \cdot \frac{1}{\sqrt{4}} \begin{bmatrix} -1 \\ -1 \\ 1 \\ 1 \end{bmatrix} = \frac{-y_{11}-y_{12}+y_{21}+y_{22}}{\sqrt{4}}$$

$$\text{Hence the mean of} \quad \mathbf{y}\cdot\mathbf{U_2} = \frac{-\mu_1-\mu_1+\mu_2+\mu_2}{\sqrt{4}} = \mu_2-\mu_1$$

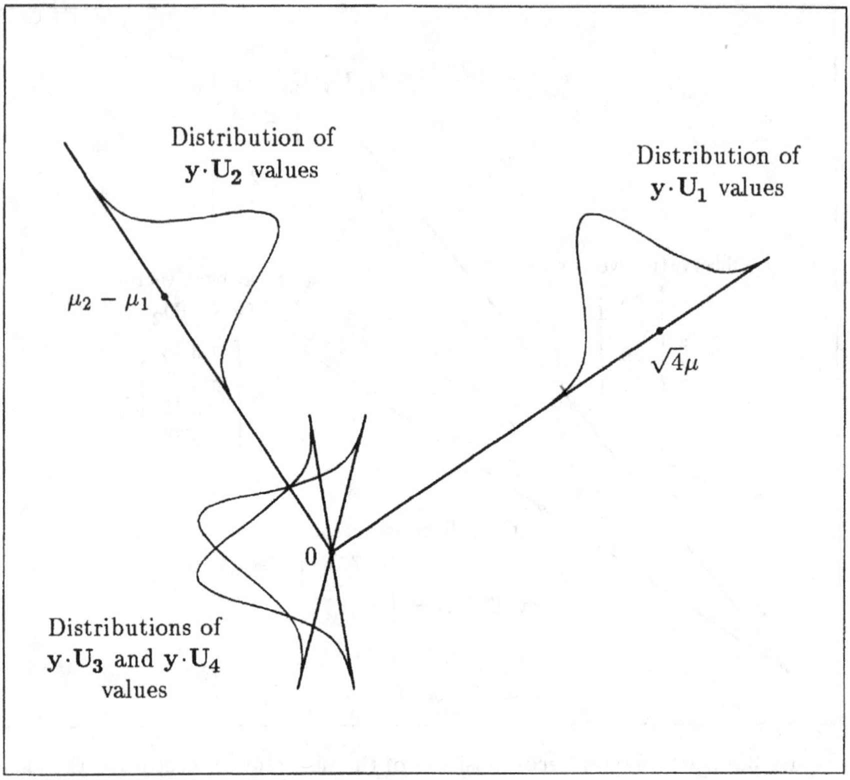

Figure 3.4: Pictorial representation of the distributions of the projection lengths $\mathbf{y}\cdot\mathbf{U_1}$, $\mathbf{y}\cdot\mathbf{U_2}$, $\mathbf{y}\cdot\mathbf{U_3}$ and $\mathbf{y}\cdot\mathbf{U_4}$. We have clearly used artistic license in attempting to depict four mutually orthogonal directions on a two-dimensional piece of paper!

$$\text{Thirdly,}\quad \mathbf{y}\cdot\mathbf{U_3} = \begin{bmatrix} y_{11} \\ y_{12} \\ y_{21} \\ y_{22} \end{bmatrix} \cdot \frac{1}{\sqrt{2}} \begin{bmatrix} -1 \\ 1 \\ 0 \\ 0 \end{bmatrix} = \frac{-y_{11}+y_{12}}{\sqrt{2}}$$

$$\text{Hence the mean of}\quad \mathbf{y}\cdot\mathbf{U_3} = \frac{-\mu_1+\mu_1}{\sqrt{2}} = 0$$

$$\text{Lastly,}\quad \mathbf{y}\cdot\mathbf{U_4} = \begin{bmatrix} y_{11} \\ y_{12} \\ y_{21} \\ y_{22} \end{bmatrix} \cdot \frac{1}{\sqrt{2}} \begin{bmatrix} 0 \\ 0 \\ -1 \\ 1 \end{bmatrix} = \frac{-y_{21}+y_{22}}{\sqrt{2}}$$

$$\text{Hence the mean of}\quad \mathbf{y}\cdot\mathbf{U_4} = \frac{-\mu_2+\mu_2}{\sqrt{2}} = 0$$

In addition, it can be readily shown that all four projection lengths have variance σ^2. Thus $\mathbf{y}\cdot\mathbf{U_1}$ comes from an $N(\sqrt{4}\mu, \sigma^2)$ distribution, $\mathbf{y}\cdot\mathbf{U_2}$ comes from an $N(\mu_2 - \mu_1, \sigma^2)$ distribution, and $\mathbf{y}\cdot\mathbf{U_3}$ and $\mathbf{y}\cdot\mathbf{U_4}$ both come from an $N(0, \sigma^2)$ distribution, as illustrated in Figure 3.4.

The point to note is that the distributions of $\mathbf{y}\cdot\mathbf{U_1}$ and $\mathbf{y}\cdot\mathbf{U_2}$ are centered on potentially nonzero quantities whereas the distributions of $\mathbf{y}\cdot\mathbf{U_3}$ and $\mathbf{y}\cdot\mathbf{U_4}$ are always centered on zero.

Testing the Hypothesis

We now proceed to investigate the study objective: Is the true mean difference in height between females and males zero? More formally, we wish to test the null hypothesis $H_0\colon \mu_1 = \mu_2$ against the alternative hypothesis, $H_1\colon \mu_1 \neq \mu_2$.

The relevant *Pythagorean breakup* is

$$
\begin{aligned}
\|\mathbf{y}\|^2 &= (\mathbf{y}\cdot\mathbf{U_1})^2 + (\mathbf{y}\cdot\mathbf{U_2})^2 + (\mathbf{y}\cdot\mathbf{U_3})^2 + (\mathbf{y}\cdot\mathbf{U_4})^2 \\
\text{or,}\qquad 18431 &= 18360.25 + 56.25 + 2 + 12.5
\end{aligned}
$$

as illustrated in Figure 3.5.

To test the hypothesis we simply check whether the squared projection length $(\mathbf{y}\cdot\mathbf{U_2})^2$ is similar to, or considerably greater than, the average of the squared projection lengths $(\mathbf{y}\cdot\mathbf{U_3})^2$ and $(\mathbf{y}\cdot\mathbf{U_4})^2$. The resulting test statistic is

$$
F = \frac{(\mathbf{y}\cdot\mathbf{U_2})^2}{[(\mathbf{y}\cdot\mathbf{U_3})^2 + (\mathbf{y}\cdot\mathbf{U_4})^2]/2} = \frac{56.25}{[2 + 12.5]/2} = \frac{56.25}{7.25} = 7.76
$$

Is this observed F value large or small? Now if $\mu_1 = \mu_2$, the projection lengths $\mathbf{y}\cdot\mathbf{U_2}$, $\mathbf{y}\cdot\mathbf{U_3}$ and $\mathbf{y}\cdot\mathbf{U_4}$ all come from an $N(0, \sigma^2)$ distribution, so the test statistic $F = (\mathbf{y}\cdot\mathbf{U_2})^2/\{[(\mathbf{y}\cdot\mathbf{U_3})^2 + (\mathbf{y}\cdot\mathbf{U_4})^2]/2\}$ comes from the $F_{1,2}$ distribution defined in Appendix B. Hence to see whether our observed value of 7.76 is large or small, we compare it with the percentiles of the $F_{1,2}$ distribution as given in Table T.2. The 90, 95 and 99 percentiles are 8.5, 18.5 and 98.5, respectively, so our observed value of 7.76 is not large, even when compared with the 90 percentile (or 10% critical value). Hence we conclude that we have found insufficient evidence to reject the idea that the two population means are identical. That is, we have not shown beyond reasonable doubt that there is, on average, any difference in height between the female and male students in our study populations.

Note that in the independent samples case the direction $\mathbf{U_1}$ is no longer of interest in terms of our hypothesis test, with its place being taken by $\mathbf{U_2}$.

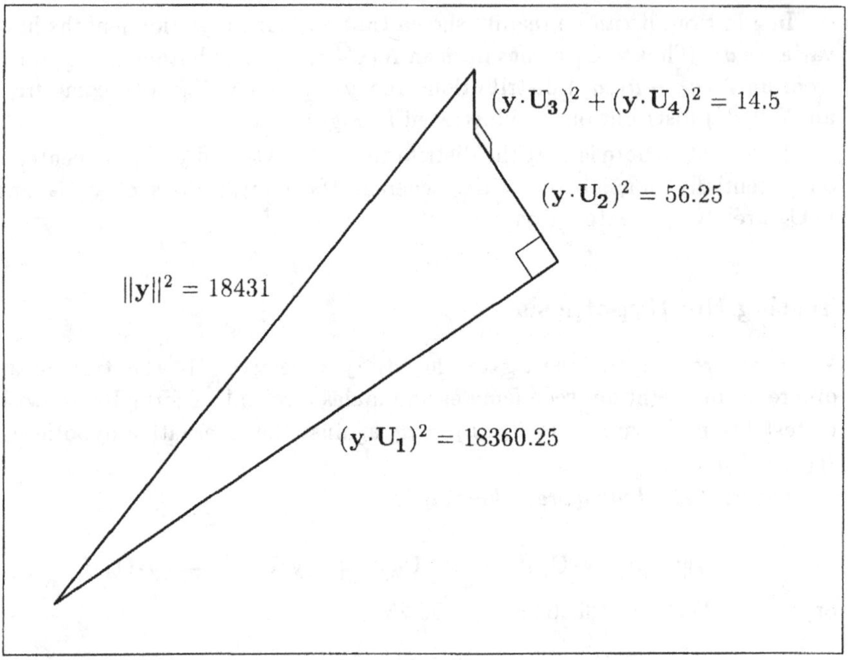

Figure 3.5: The Pythagorean breakup of the squared length of the observation vector as the sum of the squared lengths of the four orthogonal projections, $(\mathbf{y}\cdot\mathbf{U_i})^2$. This diagram is not to scale.

Independent Samples t Test

Note that our test statistic $F = (\mathbf{y}\cdot\mathbf{U_2})^2/\{[(\mathbf{y}\cdot\mathbf{U_3})^2 + (\mathbf{y}\cdot\mathbf{U_4})^2]/2\}$ can be rewritten using the notation given in Chapter 1 as $A^2/(B^2/2)$, where B is the length of the error vector. By taking square roots we can also arrive at the equivalent test statistic

$$t = \frac{\mathbf{y}\cdot\mathbf{U_2}}{\sqrt{[(\mathbf{y}\cdot\mathbf{U_3})^2 + (\mathbf{y}\cdot\mathbf{U_4})^2]/2}} = \frac{A}{B/\sqrt{2}} = \frac{15\sqrt{4}}{\sqrt{(2+12.5)/2}} = 2.785$$

As already stated, if $\mu_1 = \mu_2$ the projection lengths $\mathbf{y}\cdot\mathbf{U_2}$, $\mathbf{y}\cdot\mathbf{U_3}$ and $\mathbf{y}\cdot\mathbf{U_4}$ all come from an $N(0,\sigma^2)$ distribution. Hence the test statistic $t = \mathbf{y}\cdot\mathbf{U_2}/\sqrt{[(\mathbf{y}\cdot\mathbf{U_3})^2 + (\mathbf{y}\cdot\mathbf{U_4})^2]/2}$ comes from the t_2 distribution defined in Appendix B. To see whether our observed t value of 2.785 is large or small, we compare it with the percentiles of the t_2 distribution as given in Table

T.2. The 95, 97.5 and 99.5 percentiles of this two-sided distribution are 2.920, 4.303 and 9.925, respectively, with the 5, 2.5 and 0.5 percentiles being the negative of these values. Our observed value of 2.785 is not large since it lies within the 90% "normal range," bounded by the 5 and 95 percentiles of −2.920 and 2.920, respectively. Hence we again conclude that we have found insufficient evidence to reject the idea that the population means are identical.

Comments

In the above analysis the direction $\mathbf{U_1}$ is used to estimate the mean of the two population means μ, the direction $\mathbf{U_2}$ is used to estimate the difference between the two population means $\mu_2 - \mu_1$, and the directions $\mathbf{U_3}$ and $\mathbf{U_4}$ are used to estimate the population variance σ^2. In the long run the projection length $\mathbf{y}\cdot\mathbf{U_1} = \sqrt{4}\bar{y}$ averages to $\sqrt{4}\mu$, so $\bar{y} = 67.75$ serves as our estimate of μ. Similarly, the projection length $\mathbf{y}\cdot\mathbf{U_2} = \bar{y}_2 - \bar{y}_1$ averages to $\mu_2 - \mu_1$, so $\bar{y}_2 - \bar{y}_1 = 71.5 - 64 = 7.5$ serves as our estimate of $\mu_2 - \mu_1$. Also the squared projection lengths $(\mathbf{y}\cdot\mathbf{U_3})^2$ and $(\mathbf{y}\cdot\mathbf{U_4})^2$ both average to σ^2, so $s^2 = [(\mathbf{y}\cdot\mathbf{U_3})^2 + (\mathbf{y}\cdot\mathbf{U_4})^2]/2 = (2 + 12.5)/2 = 7.25$ serves as our best estimate of σ^2. (As an aside, we have introduced new notation; $\bar{y}_1$ and $\bar{y}_2$ for the two sample means, and $\bar{y}$ for the overall mean.)

In addition, it can be shown that in the long run the squared projection length $(\mathbf{y}\cdot\mathbf{U_1})^2$ averages to $4\mu^2+\sigma^2$, while $(\mathbf{y}\cdot\mathbf{U_2})^2$ averages to $(\mu_2-\mu_1)^2+\sigma^2$. Hence if $\mu_1 \neq \mu_2$ the quantity $(\mathbf{y}\cdot\mathbf{U_2})^2$ is inflated, and our test statistic $(\mathbf{y}\cdot\mathbf{U_2})^2/s^2$ can be thought of as an inflated estimate of σ^2 divided by an unbiased estimate of σ^2.

3.3 General Case

Our full data set in Figure 3.1 has unequal sample sizes, with 49 female and 77 male heights. This is unfortunate, since we want to introduce just the simplest case in this primer. We therefore content ourselves with illustrating the general case using slightly increased sample sizes of four. (For completeness the curious reader will be led through the full, unequal sample size analysis in the last exercise in this chapter.)

In our new example we shall use the data shown in Table 3.2; these data are the heights of four males and four females, randomly selected from Figure 3.1.

	Heights (inches)				Means
Females	60	64	65	68	64.25
Males	70	69	77	71	71.75

Table 3.2: Heights, in inches, of four females and four males from the ASM150 class, together with the two sample means. By coincidence the sample means again differ by 7.5 inches!

The resulting *observation vector* is

$$\mathbf{y} = \begin{bmatrix} 60 \\ 64 \\ 65 \\ 68 \\ 70 \\ 69 \\ 77 \\ 71 \end{bmatrix}$$

The study objective is as previously stated in §3.2.

The Basic Idea

The basic idea is the same as in §3.2. If the male and female students are on average the same in height ($\mu_1 = \mu_2$), then our observation vector will be part of a scatter centered on the point $(\mu, \mu, \mu, \mu, \mu, \mu, \mu, \mu)$, as shown in Figure 3.6(a). However, if they differ in height ($\mu_1 \neq \mu_2$), then our observation vector will be part of a scatter centered on the point $(\mu_1, \mu_1, \mu_1, \mu_1, \mu_2, \mu_2, \mu_2, \mu_2)$, as shown in Figure 3.6(b). In other words, if $\mu_1 = \mu_2$ the scatter is centered on the equiangular line, whereas if $\mu_1 \neq \mu_2$ the scatter is moved away from the equiangular line in the direction

$$\begin{bmatrix} \mu_1 \\ \mu_1 \\ \mu_1 \\ \mu_1 \\ \mu_2 \\ \mu_2 \\ \mu_2 \\ \mu_2 \end{bmatrix} - \begin{bmatrix} \mu \\ \mu \\ \mu \\ \mu \\ \mu \\ \mu \\ \mu \\ \mu \end{bmatrix} = \begin{bmatrix} \mu_1 - \mu \\ \mu_1 - \mu \\ \mu_1 - \mu \\ \mu_1 - \mu \\ \mu_2 - \mu \\ \mu_2 - \mu \\ \mu_2 - \mu \\ \mu_2 - \mu \end{bmatrix} = \begin{bmatrix} (\mu_1 - \mu_2)/2 \\ (\mu_1 - \mu_2)/2 \\ (\mu_1 - \mu_2)/2 \\ (\mu_1 - \mu_2)/2 \\ (\mu_2 - \mu_1)/2 \\ (\mu_2 - \mu_1)/2 \\ (\mu_2 - \mu_1)/2 \\ (\mu_2 - \mu_1)/2 \end{bmatrix} = \frac{\mu_2 - \mu_1}{2} \begin{bmatrix} -1 \\ -1 \\ -1 \\ -1 \\ 1 \\ 1 \\ 1 \\ 1 \end{bmatrix}$$

since $\mu = (\mu_1 + \mu_2)/2$.

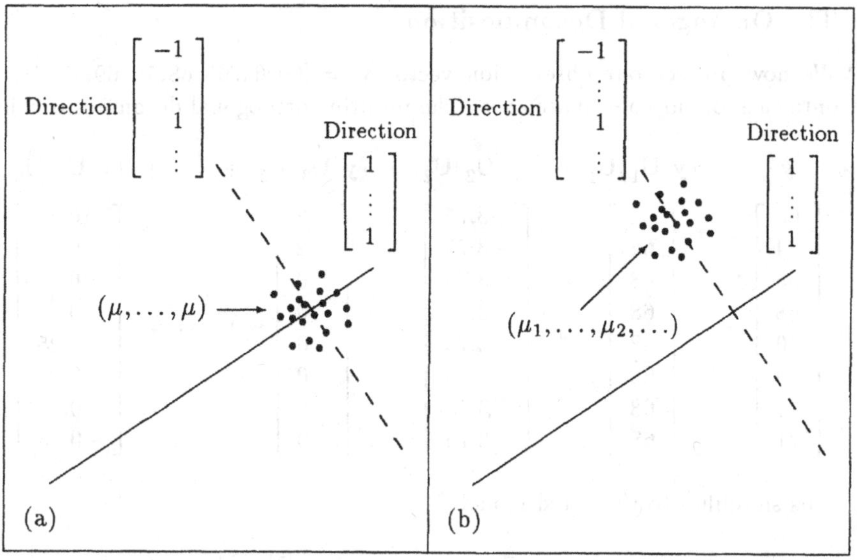

Figure 3.6: Clouds of data points representing many repetitions of our height study. In the case (a) $\mu_1 = \mu_2$ the cloud is centered on the point $(\mu, \ldots, \mu)$, while in the case (b) $\mu_1 \neq \mu_2$ the cloud is centered on the point $(\mu_1, \ldots, \mu_2, \ldots)$. That is, if $\mu_1 = \mu_2$ the cloud is centered on the equiangular line, while if $\mu_1 \neq \mu_2$ the cloud is moved away from the line in the direction $[-1, \ldots, 1, \ldots]^{\mathrm{T}}$.

An Appropriate Coordinate System

An appropriate set of coordinate axes for 8-space is therefore

$$
\mathbf{U_1} = \frac{\begin{bmatrix} 1\\1\\1\\1\\1\\1\\1\\1 \end{bmatrix}}{\sqrt{8}} \quad
\mathbf{U_2} = \frac{\begin{bmatrix} -1\\-1\\-1\\-1\\1\\1\\1\\1 \end{bmatrix}}{\sqrt{8}} \quad
\mathbf{U_3} = \frac{\begin{bmatrix} -1\\1\\0\\0\\0\\0\\0\\0 \end{bmatrix}}{\sqrt{2}} \quad
\mathbf{U_4} = \frac{\begin{bmatrix} -1\\-1\\2\\0\\0\\0\\0\\0 \end{bmatrix}}{\sqrt{6}} \quad
\mathbf{U_5} = \frac{\begin{bmatrix} -1\\-1\\-1\\3\\0\\0\\0\\0 \end{bmatrix}}{\sqrt{12}} \quad
\mathbf{U_6} = \frac{\begin{bmatrix} 0\\0\\0\\0\\-1\\1\\0\\0 \end{bmatrix}}{\sqrt{2}} \quad
\mathbf{U_7} = \frac{\begin{bmatrix} 0\\0\\0\\0\\-1\\-1\\2\\0 \end{bmatrix}}{\sqrt{6}} \quad
\mathbf{U_8} = \frac{\begin{bmatrix} 0\\0\\0\\0\\-1\\-1\\-1\\3 \end{bmatrix}}{\sqrt{12}}
$$

Here $\mathbf{U_2} = [-1, -1, -1, -1, 1, 1, 1, 1]^T/\sqrt{8}$ is the *direction associated with the difference of interest*, $\mu_2 - \mu_1$.

The Orthogonal Decomposition

We now project our observation vector $\mathbf{y} = [60, 64, 65, 68, 70, 69, 77, 71]^T$ onto each of our coordinate axes. The resulting orthogonal decomposition is

$$\begin{array}{ccccccccccc} \mathbf{y} & = & (\mathbf{y}\cdot\mathbf{U}_1)\mathbf{U}_1 & + & (\mathbf{y}\cdot\mathbf{U}_2)\mathbf{U}_2 & + & (\mathbf{y}\cdot\mathbf{U}_3)\mathbf{U}_3 & + & \cdots & + & (\mathbf{y}\cdot\mathbf{U}_8)\mathbf{U}_8 \\ \begin{bmatrix} 60 \\ 64 \\ 65 \\ 68 \\ 70 \\ 69 \\ 77 \\ 71 \end{bmatrix} & = & \begin{bmatrix} 68 \\ 68 \\ 68 \\ 68 \\ 68 \\ 68 \\ 68 \\ 68 \end{bmatrix} & + & \begin{bmatrix} -3.75 \\ -3.75 \\ -3.75 \\ -3.75 \\ 3.75 \\ 3.75 \\ 3.75 \\ 3.75 \end{bmatrix} & + & \begin{bmatrix} -2 \\ 2 \\ 0 \\ 0 \\ 0 \\ 0 \\ 0 \\ 0 \end{bmatrix} & + & \cdots & + & \begin{bmatrix} 0 \\ 0 \\ 0 \\ 0 \\ 0.25 \\ 0.25 \\ 0.25 \\ -0.75 \end{bmatrix} \end{array}$$

This simplifies to the fitted model

$$\begin{array}{ccccccc} \mathbf{y} & = & \bar{\mathbf{y}} & + & (\bar{\mathbf{y}}_i - \bar{\mathbf{y}}) & + & (\mathbf{y} - \bar{\mathbf{y}}_i) \\ \begin{bmatrix} 60 \\ 64 \\ 65 \\ 68 \\ 70 \\ 69 \\ 77 \\ 71 \end{bmatrix} & = & \begin{bmatrix} 68 \\ 68 \\ 68 \\ 68 \\ 68 \\ 68 \\ 68 \\ 68 \end{bmatrix} & + & \begin{bmatrix} -3.75 \\ -3.75 \\ -3.75 \\ -3.75 \\ 3.75 \\ 3.75 \\ 3.75 \\ 3.75 \end{bmatrix} & + & \begin{bmatrix} -4.25 \\ -0.25 \\ 0.75 \\ 3.75 \\ -1.75 \\ -2.75 \\ 5.25 \\ -0.75 \end{bmatrix} \\ \text{observation vector} & & \text{overall mean vector} & & \text{treatment vector} & & \text{error vector} \end{array}$$

as shown in Figure 3.7. Notice here that we have slipped in the abbreviations $\bar{\mathbf{y}}$, which denotes the vector with the overall mean in all positions, $(\bar{\mathbf{y}}_i - \bar{\mathbf{y}})$, which denotes the vector of differences between sample means and the overall mean, and $(\mathbf{y} - \bar{\mathbf{y}}_i)$, which denotes the vector of differences between the observations and their respective sample means.

Distributions of the Projection Lengths

Proceeding as in §3.2, we find that $\mathbf{y}\cdot\mathbf{U}_1$ comes from an $N(\sqrt{8}\mu, \sigma^2)$ distribution, $\mathbf{y}\cdot\mathbf{U}_2$ comes from an $N(\sqrt{2}(\mu_2 - \mu_1), \sigma^2)$ distribution, and $\mathbf{y}\cdot\mathbf{U}_3$, $\mathbf{y}\cdot\mathbf{U}_4$, $\mathbf{y}\cdot\mathbf{U}_5$, $\mathbf{y}\cdot\mathbf{U}_6$, $\mathbf{y}\cdot\mathbf{U}_7$ and $\mathbf{y}\cdot\mathbf{U}_8$ all come from an $N(0, \sigma^2)$ distribution.

Again, the point to note is that the distributions of $\mathbf{y}\cdot\mathbf{U}_1$ and $\mathbf{y}\cdot\mathbf{U}_2$ are centered on potentially nonzero quantities whereas the distributions of $\mathbf{y}\cdot\mathbf{U}_3$, ..., $\mathbf{y}\cdot\mathbf{U}_8$ are always centered on zero.

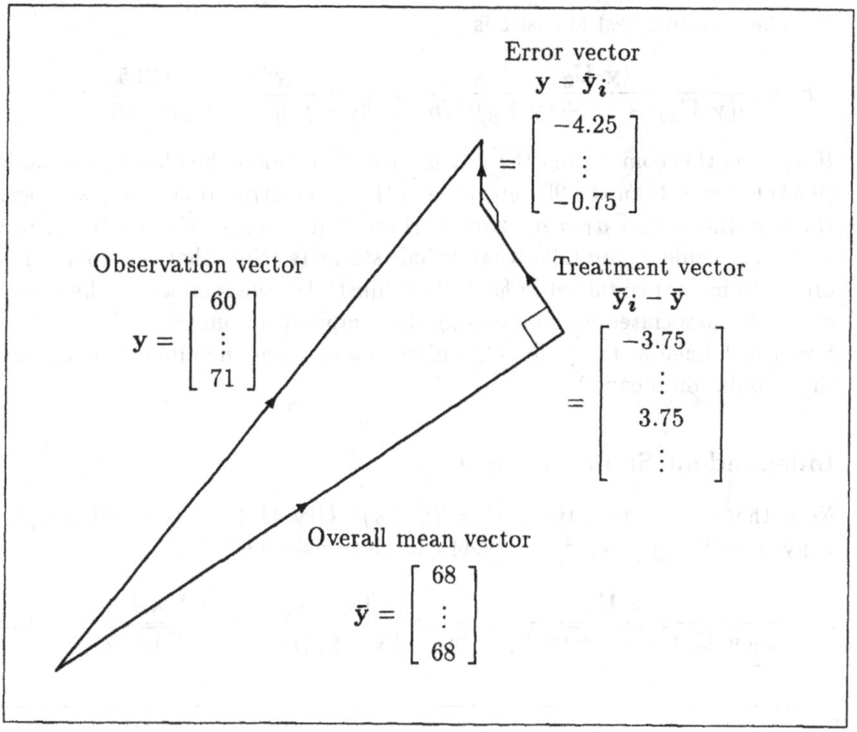

Figure 3.7: Orthogonal decomposition of the observation vector for the case of independent samples of size four from two study populations. This figure is not to scale.

Testing the Hypothesis

We now investigate the study objective: Is the true mean difference in height between male and female students zero? For our hypothesis test we check whether the squared projection length $(\mathbf{y}\cdot\mathbf{U}_2)^2$ is similar to, or considerably greater than, the average of the squared projection lengths for the "error" directions, $[(\mathbf{y}\cdot\mathbf{U}_3)^2 + \cdots + (\mathbf{y}\cdot\mathbf{U}_8)^2]/6$.

The associated Pythagorean breakup is

$$\|\mathbf{y}\|^2 = (\mathbf{y}\cdot\mathbf{U}_1)^2 + (\mathbf{y}\cdot\mathbf{U}_2)^2 + (\mathbf{y}\cdot\mathbf{U}_3)^2 + \cdots + (\mathbf{y}\cdot\mathbf{U}_8)^2$$

or, $$\|\mathbf{y}\|^2 = \|\bar{\mathbf{y}}\|^2 + \|\bar{\mathbf{y}}_i - \bar{\mathbf{y}}\|^2 + \|\mathbf{y} - \bar{\mathbf{y}}_i\|^2$$

or, $$37176 = 36992 + 112.5 + 71.5$$

as illustrated in Figure 3.8.

The resulting test statistic is

$$F = \frac{(\mathbf{y}\cdot\mathbf{U}_2)^2}{[(\mathbf{y}\cdot\mathbf{U}_3)^2 + \cdots + (\mathbf{y}\cdot\mathbf{U}_8)^2]/6} = \frac{\|\bar{\mathbf{y}}_i - \bar{\mathbf{y}}\|^2}{\|\mathbf{y} - \bar{\mathbf{y}}_i\|^2/6} = \frac{112.5}{71.5/6} = 9.44$$

If $\mu_1 = \mu_2$ this comes from the $F_{1,6}$ distribution. Since the observed F value (9.44) is larger than the 95 percentile of the $F_{1,6}$ distribution (5.99) we reject the hypothesis that $\mu_1 = \mu_2$, and conclude that there is evidence to suggest that male students are taller than female students. Note that this conclusion differs from that obtained in §3.2. This illustrates the increase in the power of the test associated with increasing the sample size from two to four. (Here power is defined as the probability of detecting a genuine difference between the population means.)

Independent Samples t Test

Note that our test statistic $F = (\mathbf{y}\cdot\mathbf{U}_2)^2/\{[(\mathbf{y}\cdot\mathbf{U}_3)^2 + \cdots + (\mathbf{y}\cdot\mathbf{U}_8)^2]/6\}$ $= \|\bar{\mathbf{y}}_i - \bar{\mathbf{y}}\|^2/\left[\|\mathbf{y} - \bar{\mathbf{y}}_i\|^2/6\right] = 9.44$ can be transformed to

$$t = \frac{\mathbf{y}\cdot\mathbf{U}_2}{\sqrt{[(\mathbf{y}\cdot\mathbf{U}_3)^2 + \cdots + (\mathbf{y}\cdot\mathbf{U}_8)^2]/6}} = \frac{\|\bar{\mathbf{y}}_i - \bar{\mathbf{y}}\|}{\|\mathbf{y} - \bar{\mathbf{y}}_i\|/\sqrt{6}} = \frac{\sqrt{8}\times 3.75}{\sqrt{71.5/6}} = 3.073$$

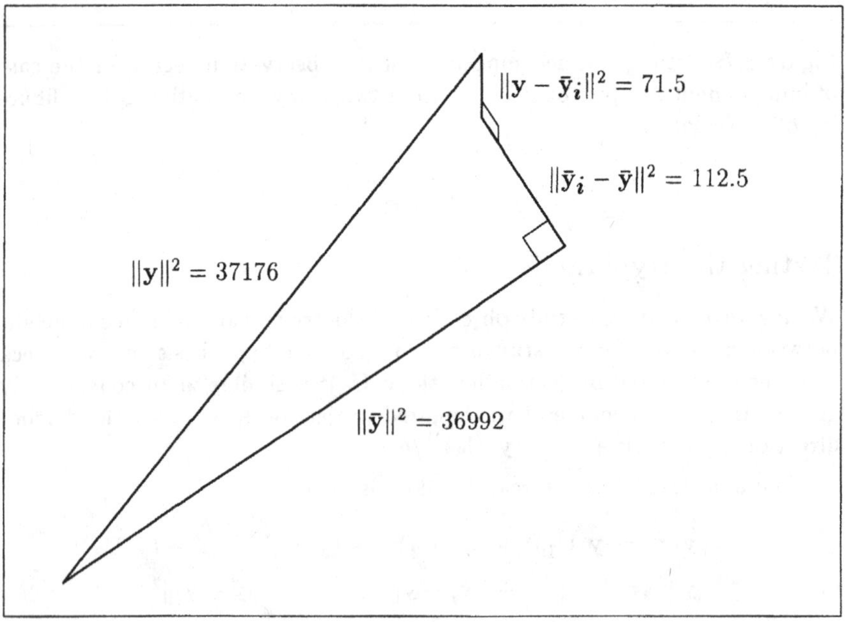

Figure 3.8: The Pythagorean breakup of the squared length of the observation vector for the general independent samples case.

If $\mu_1 = \mu_2$ this comes from the t_6 distribution. As previously, the size of our test statistic constitutes evidence against the notion that the two population means could be equal.

Note also that the usual algebraic expression for t is

$$t = \frac{\|\bar{\mathbf{y}}_i - \bar{\mathbf{y}}\|}{\|\mathbf{y} - \bar{\mathbf{y}}_i\| / \sqrt{2(n-1)}} = \frac{\sqrt{2n}(\bar{y}_2 - \bar{y}_1)/2}{\sqrt{\sum_{i=1}^{2}\sum_{j=1}^{n}(y_{ij} - \bar{y}_i)^2 / [2(n-1)]}} = \frac{\bar{y}_2 - \bar{y}_1}{\sqrt{2}s/\sqrt{n}}$$

where s is the square root of the pooled variance estimate, $s^2 = \|\mathbf{y} - \bar{\mathbf{y}}_i\|^2 / [2(n-1)]$, and where $n = 4$ in this example.

Comments

As in §3.2, the "model space" directions $\mathbf{U_1}$ and $\mathbf{U_2}$ are used to estimate the population means μ_1 and μ_2 (or, equivalently, their mean and the difference between them), while the "error space" directions $\mathbf{U_3}, \ldots, \mathbf{U_8}$ are used to estimate the common population variance σ^2.

To summarize our reanalysis with a sample size of four, we now have best estimates of 64.3 in. and 71.8 in. for mean female and male heights, respectively, with strong evidence of a difference between the underlying population means. For an analysis of the full data set, the reader is referred to Exercise 3.10 and its solution in Appendix E.

3.4 Summary

To summarize, in the independent samples case the direction of special interest is the direction $\mathbf{U_2} = [-1, \ldots, 1, \ldots]^T / \sqrt{2n}$. If there is a difference between the population means μ_1 and μ_2, then this is the direction for which the length of the projection is increased. More fully, if $\mu_1 \neq \mu_2$ the length of the projection of the vector of observations onto the direction $\mathbf{U_2}$ will be inflated relative to the projection lengths for the "error" directions $\mathbf{U_3}, \ldots, \mathbf{U_{2n}}$. Thus to distinguish between the cases $\mu_1 = \mu_2$ and $\mu_1 \neq \mu_2$ we divide the squared projection length for the $\mathbf{U_2}$ direction by the average of the squared projection lengths for the $2(n-1)$ perpendicular directions in the "error space"; if the resulting ratio is small we decide that μ_1 could be equal to μ_2, while if the ratio is large we decide that μ_1 could not be equal to μ_2.

In general, the projection of the vector of observations, $\mathbf{y} = [y_{11}, \ldots, y_{21}, \ldots]^T$, onto the direction $\mathbf{U_2} = [-1, \ldots, 1, \ldots]^T / \sqrt{2n}$ is the vector $\bar{\mathbf{y}}_i - \bar{\mathbf{y}} = [(\bar{y}_1 - \bar{y}), \ldots, (\bar{y}_2 - \bar{y}), \ldots]^T$. This leads to the orthogonal decomposition $\mathbf{y} = \bar{\mathbf{y}} + (\bar{\mathbf{y}}_i - \bar{\mathbf{y}}) + (\mathbf{y} - \bar{\mathbf{y}}_i)$ shown in Figure 3.9. Here the "error vector" $(\mathbf{y} - \bar{\mathbf{y}}_i)$ is a sum of $2(n-1)$ orthogonal projection vectors, one for each of the orthogonal coordinate axes spanning the error space.

The resulting Pythagorean breakup is

$$||\mathbf{y}||^2 = ||\bar{\mathbf{y}}||^2 + ||\bar{\mathbf{y}}_i - \bar{\mathbf{y}}||^2 + ||\mathbf{y} - \bar{\mathbf{y}}_i||^2$$

This leads to the test statistic

$$F = \frac{||\bar{\mathbf{y}}_i - \bar{\mathbf{y}}||^2}{||\mathbf{y} - \bar{\mathbf{y}}_i||^2 / [2(n-1)]} = \frac{2n\left[(\bar{y}_2 - \bar{y}_1)/2\right]^2}{s^2} = \frac{n(\bar{y}_2 - \bar{y}_1)^2}{2s^2}$$

This test statistic follows an $F_{1,2(n-1)}$ distribution if $\mu_1 = \mu_2$, so we reject the hypothesis that $\mu_1 = \mu_2$ if the observed F value exceeds the 95 percentile of this distribution.

The equivalent independent samples t test is

$$t = \frac{||\bar{\mathbf{y}}_i - \bar{\mathbf{y}}||}{||\mathbf{y} - \bar{\mathbf{y}}_i|| / \sqrt{2(n-1)}} = \frac{\sqrt{n}(\bar{y}_2 - \bar{y}_1)}{\sqrt{2}s}$$

For an analysis of the examples in this chapter using the angle θ as the test statistic, the reader is referred to Appendix D. For computer methods refer to Appendix C. For estimation of the 95% confidence interval for the difference $(\mu_2 - \mu_1)$, refer to pp. 109, 110, 118, 124, 538, 539 and 540 of Saville and Wood (1991).

For more reading on the geometry of the independent samples case, the reader can refer to Chapter 6 and Appendix A of Saville and Wood (1991).

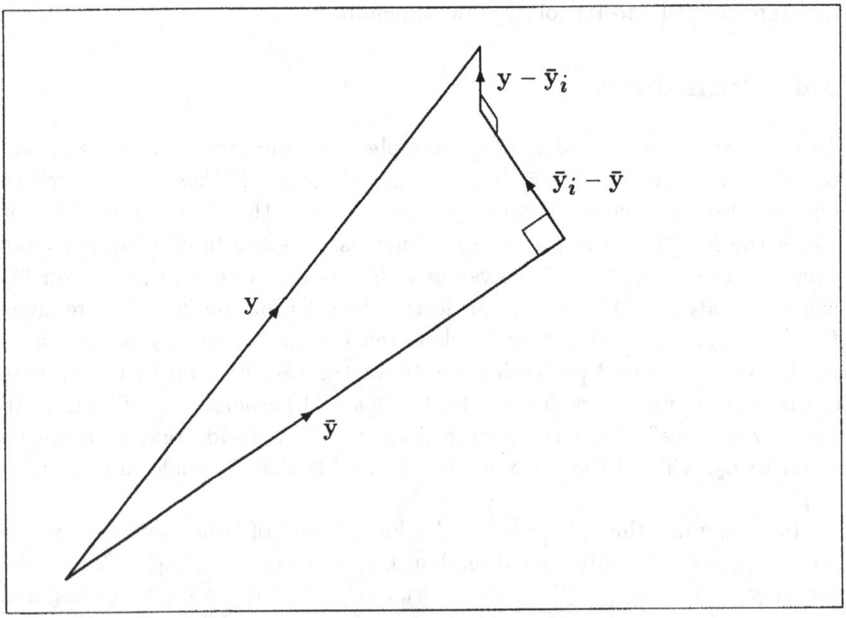

Figure 3.9: The orthogonal decomposition of the observation vector for the independent samples case.

Class Exercise

In the class exercise of Chapter 2 the *change* in heartbeat, before and after running, was recorded for each person. For the exercise in this chapter we shall consider a different design for the study. Under the new design we pretend that we have 52 people who were randomly divided into two groups of size 26. In the first group the people recorded their heartbeats, over 30 seconds, after 10 minutes of sitting still. In the second group the people recorded their heartbeat immediately after 2 minutes of running on the spot. The resulting data are given in the following two tables:

(a) Group 1: People at rest

Person	No. of beats	Person	No. of beats	Person	No. of beats
1	24	11	26	21	40
2	40	12	35	22	29
3	34	13	26	23	35
4	35	14	30	24	32
5	31	15	32	25	41
6	37	16	38	26	41
7	40	17	31		
8	33	18	32		
9	35	19	37		
10	27	20	27		

(b) Group 2: People after running

Person	No. of beats	Person	No. of beats	Person	No. of beats
1	34	11	38	21	47
2	40	12	39	22	38
3	57	13	43	23	43
4	46	14	44	24	40
5	44	15	44	25	37
6	38	16	42	26	48
7	44	17	27		
8	60	18	43		
9	45	19	54		
10	47	20	44		

The above data are actually the heartbeats, before and after running, of the 26 agricultural researchers from the class exercise in Chapter 2. However, we have jumbled each set of 26 data values, so the pairing has been lost.

(a) Each member of the class is asked to use the random numbers in Table T.1 to select a sample of size three from each of the above two groups. A more active class atmosphere can be created by breaking the class into groups of size six, and having each group generate its own data, after randomly allocating members to "rest" and "running." Either way, the six heartbeat numbers which are produced make up the observation vector $\mathbf{y}$. The question of interest is: "Does running affect heartbeat?" To answer this question we formally test the hypothesis H_0: $\mu_1 = \mu_2$ versus H_1: $\mu_1 \neq \mu_2$.

For our purposes an appropriate coordinate system is

$$\mathbf{U}_1 = \frac{\begin{bmatrix}1\\1\\1\\1\\1\\1\end{bmatrix}}{\sqrt{6}} \quad \mathbf{U}_2 = \frac{\begin{bmatrix}-1\\-1\\-1\\1\\1\\1\end{bmatrix}}{\sqrt{6}} \quad \mathbf{U}_3 = \frac{\begin{bmatrix}-1\\1\\0\\0\\0\\0\end{bmatrix}}{\sqrt{2}} \quad \mathbf{U}_4 = \frac{\begin{bmatrix}-1\\-1\\2\\0\\0\\0\end{bmatrix}}{\sqrt{6}} \quad \mathbf{U}_5 = \frac{\begin{bmatrix}0\\0\\0\\-1\\1\\0\end{bmatrix}}{\sqrt{2}} \quad \mathbf{U}_6 = \frac{\begin{bmatrix}0\\0\\0\\-1\\-1\\2\end{bmatrix}}{\sqrt{6}}$$

Here $\mathbf{U}_1$ is associated with the overall size of the observations, $\mathbf{U}_2$ is associated with the difference in heartbeat rate between people resting and after running, and $\mathbf{U}_3, \ldots, \mathbf{U}_6$ are associated with the random variation within the two study populations.

(b) Each class member is asked to calculate the scalars $\mathbf{y}\cdot\mathbf{U}_1$, $\mathbf{y}\cdot\mathbf{U}_2$, $\mathbf{y}\cdot\mathbf{U}_3$, $\mathbf{y}\cdot\mathbf{U}_4$, $\mathbf{y}\cdot\mathbf{U}_5$ and $\mathbf{y}\cdot\mathbf{U}_6$, using his or her own observation vector. The teacher can then plot the class results in six histograms (with a common scale). These histograms will approximate the distributions of the projection lengths $\mathbf{y}\cdot\mathbf{U}_1, \ldots, \mathbf{y}\cdot\mathbf{U}_6$.

Questions:

1. Which of the projection lengths appear to have mean zero?
2. Is the variability similar between the histograms?

(c) Each class member is now asked to calculate the test statistic

$$F = \frac{(\mathbf{y}\cdot\mathbf{U}_2)^2}{\left[(\mathbf{y}\cdot\mathbf{U}_3)^2 + (\mathbf{y}\cdot\mathbf{U}_4)^2 + (\mathbf{y}\cdot\mathbf{U}_5)^2 + (\mathbf{y}\cdot\mathbf{U}_6)^2\right]/4}$$

and compare his or her answer with the 95 percentile of the $F_{1,4}$ distribution.

The teacher can then plot the class results in the form of a histogram.

Questions:

1. Does the test have adequate "power" with two samples of size three? That is, what was the percentage of F values which exceeded the 95 percentile of the $F_{1,4}$ distribution?

2. Compare your latest histogram with the histogram of F values which you obtained in the class exercise in Chapter 2. Which of the two designs (four paired samples or two independent samples of size three) gave the greatest proportion of rejections of the null hypothesis?

(d) Each class member can now calculate the projection vectors $(\mathbf{y}\cdot\mathbf{U}_1)\mathbf{U}_1, \ldots, (\mathbf{y}\cdot\mathbf{U}_6)\mathbf{U}_6$, and obtain the orthogonal decomposition

$$\mathbf{y} = (\mathbf{y}\cdot\mathbf{U}_1)\mathbf{U}_1 + \cdots + (\mathbf{y}\cdot\mathbf{U}_6)\mathbf{U}_6$$

He or she can then confirm that $(\mathbf{y}\cdot\mathbf{U}_1)\mathbf{U}_1 = \bar{\mathbf{y}}$, the overall mean vector, that $(\mathbf{y}\cdot\mathbf{U}_2)\mathbf{U}_2 = (\bar{\mathbf{y}}_i - \bar{\mathbf{y}})$, the treatment vector, and that $(\mathbf{y}\cdot\mathbf{U}_3)\mathbf{U}_3 + \cdots + (\mathbf{y}\cdot\mathbf{U}_6)\mathbf{U}_6 = (\mathbf{y} - \bar{\mathbf{y}}_i)$, the error vector. He or she can also write out the orthogonal decomposition $\mathbf{y} = \bar{\mathbf{y}} + (\bar{\mathbf{y}}_i - \bar{\mathbf{y}}) + (\mathbf{y} - \bar{\mathbf{y}}_i)$, and draw a sketch to represent this decomposition.

(e) Using this new decomposition, each class member can now recalculate

$$F = \frac{\|\bar{\mathbf{y}}_i - \bar{\mathbf{y}}\|^2}{\|\mathbf{y} - \bar{\mathbf{y}}_i\|^2/4}$$

to confirm that this formula yields the same answer as in (c).

Exercises

Note that solutions to exercises marked with an * are given in Appendix E.

(3.1)* (a) In §3.2, rewrite the orthogonal decomposition

$$\mathbf{y} = \bar{\mathbf{y}} + (\bar{\mathbf{y}}_i - \bar{\mathbf{y}}) + (\mathbf{y} - \bar{\mathbf{y}}_i)$$

in the alternative form

$$(\mathbf{y} - \bar{\mathbf{y}}) = (\bar{\mathbf{y}}_i - \bar{\mathbf{y}}) + (\mathbf{y} - \bar{\mathbf{y}}_i)$$

by subtracting the overall mean vector from both sides of the equation. Redraw Figure 3.3, adding the vector $(\mathbf{y} - \bar{\mathbf{y}})$ and highlighting the triangle which represents our new decomposition.

(b) Calculate the associated Pythagorean breakup

$$\|\mathbf{y} - \bar{\mathbf{y}}\|^2 = \|\bar{\mathbf{y}}_i - \bar{\mathbf{y}}\|^2 + \|\mathbf{y} - \bar{\mathbf{y}}_i\|^2$$

and recalculate the test statistic using the formula

$$F = \frac{\|\bar{\mathbf{y}}_i - \bar{\mathbf{y}}\|^2}{\|\mathbf{y} - \bar{\mathbf{y}}_i\|^2/2}$$

You should get the same answer as in §3.2.

Note that this is the Pythagorean breakup which forms the basis of the "analysis of variance" table given in standard statistical textbooks.

(3.2) An experiment was established to determine whether boron fertilizer affects the seed yield of alfalfa. The experiment consisted of six plots marked out in a field of alfalfa. Three plots received an application of boron fertilizer and three plots were unfertilized control plots. The two treatments, "control" and "boron," were assigned to the plots in a completely random manner.

The seed yields, in kg/ha, obtained from each plot were as follows:

Treatment	Replicate 1	Replicate 2	Replicate 3
Control	61	65	63
Boron	76	73	70

(a) Write down the observation vector $\mathbf{y}$.

(b) Using the orthogonal coordinate system

$$\mathbf{U}_1 = \frac{\begin{bmatrix} 1 \\ 1 \\ 1 \\ 1 \\ 1 \\ 1 \end{bmatrix}}{\sqrt{6}} \quad \mathbf{U}_2 = \frac{\begin{bmatrix} -1 \\ -1 \\ -1 \\ 1 \\ 1 \\ 1 \end{bmatrix}}{\sqrt{6}} \quad \mathbf{U}_3 = \frac{\begin{bmatrix} -1 \\ 1 \\ 0 \\ 0 \\ 0 \\ 0 \end{bmatrix}}{\sqrt{2}} \quad \mathbf{U}_4 = \frac{\begin{bmatrix} -1 \\ -1 \\ 2 \\ 0 \\ 0 \\ 0 \end{bmatrix}}{\sqrt{6}} \quad \mathbf{U}_5 = \frac{\begin{bmatrix} 0 \\ 0 \\ 0 \\ -1 \\ 1 \\ 0 \end{bmatrix}}{\sqrt{2}} \quad \mathbf{U}_6 = \frac{\begin{bmatrix} 0 \\ 0 \\ 0 \\ -1 \\ -1 \\ 2 \end{bmatrix}}{\sqrt{6}}$$

calculate the squared length of the projection of $\mathbf{y}$ onto each of the directions $\mathbf{U}_1, \ldots, \mathbf{U}_6$.

(c) Assume the observations are independent samples from normal populations with true means μ_1 and μ_2, and a common variance σ^2. Test the hypothesis $\mu_1 = \mu_2$ using an F test obtained by dividing $(\mathbf{y} \cdot \mathbf{U}_2)^2$ by the average of $(\mathbf{y} \cdot \mathbf{U}_3)^2, \ldots, (\mathbf{y} \cdot \mathbf{U}_6)^2$. What is your conclusion?

(d) Write down the vector decomposition $\mathbf{y} = \bar{\mathbf{y}} + (\bar{\mathbf{y}}_i - \bar{\mathbf{y}}) + (\mathbf{y} - \bar{\mathbf{y}}_i)$. Recalculate the F value using the formula

$$F = \frac{||\bar{\mathbf{y}}_i - \bar{\mathbf{y}}||^2}{||\mathbf{y} - \bar{\mathbf{y}}_i||^2/4}$$

(3.3) Plasma selenium levels, in μg/L, were obtained at Christchurch Hospital in 1990 using excess blood from infants less than 12 months of age who were having blood samples taken for other diagnostic purposes. In the table we present the data for infants between 1 and 3 months of age who were either solely breast fed or solely milk formula fed. Data follows Dolamore et al. (1992), and is supplied by courtesy of Professor Christine Winterbourn, Christchurch School of Medicine.

Breast fed	36	32	58	51	67	43	
Formula fed	17	30	17	19	19	22	10

(a) Use the six values in the first group and the first six values in the second group to form an observation vector $\mathbf{y}$ in 12-space. Calculate the orthogonal decomposition $\mathbf{y} = \bar{\mathbf{y}} + (\bar{\mathbf{y}}_i - \bar{\mathbf{y}}) + (\mathbf{y} - \bar{\mathbf{y}}_i)$.

(b) Calculate the Pythagorean breakup

$$||\mathbf{y}||^2 = ||\bar{\mathbf{y}}||^2 + ||\bar{\mathbf{y}}_i - \bar{\mathbf{y}}||^2 + ||\mathbf{y} - \bar{\mathbf{y}}_i||^2$$

and the usual test statistic, F.

(c) Assuming the data are independent values from normal populations with means μ_1 and μ_2 and a common variance σ^2, test the hypothesis $\mu_1 = \mu_2$ by comparing the calculated F value with the appropriate percentiles from Table T.2. What is your conclusion?

Note that subsequent to this study two New Zealand milk formula manufacturers have started supplementing their formula with selenium. Data on the effects of this supplementation are given in Exercises 2.4 and 3.7.

(3.4)* An experiment was carried out to determine the effect of phosphate fertilizer application on the yield of a mixed ryegrass/white clover pasture. A uniform area of pasture was chosen, and six 8 m × 3 m experimental plots were marked out. These were assigned in a completely random manner to the two treatments:

1. No fertilizer application (control).

2. Phosphate (P) at the rate of 40 kg P/ha.

During the winter the phosphate was spread uniformly over the appropriate plots. Throughout the following growing season the central 6 m × 2 m of each plot was cut at monthly intervals, with clippings being weighed and returned to each plot, except for a small sample which was oven dried to determine the percentage of dry matter. The total dry matter yields for each plot, converted to tonnes/ha, were as follows:

1. Control	9.6	7.4	10.0
2. Phosphate	11.3	10.1	12.2

(a) Put these six values into an observation vector $\mathbf{y}$, and calculate the orthogonal decomposition $\mathbf{y} = \bar{\mathbf{y}} + (\bar{\mathbf{y}}_i - \bar{\mathbf{y}}) + (\mathbf{y} - \bar{\mathbf{y}}_i)$.

(b) Calculate the Pythagorean breakup

$$||\mathbf{y}||^2 = ||\bar{\mathbf{y}}||^2 + ||\bar{\mathbf{y}}_i - \bar{\mathbf{y}}||^2 + ||\mathbf{y} - \bar{\mathbf{y}}_i||^2$$

and the usual test statistic, F.

(c) Assuming the data are independent values from normal populations with means μ_1 and μ_2 and a common variance σ^2, test the hypothesis $\mu_1 = \mu_2$ by comparing the calculated F value with the appropriate percentiles from Table T.2. What is your conclusion?

(3.5) Vehicles offloading rubbish were weighed and classified according to method of transport (car, trailer, truck) and source of rubbish (residential, business) as they arrived at the landfill at Picton, at the northern end of the South Island of New Zealand. Upon departure they were reweighed, and the "load weight" calculated as the difference between the "in" and "out" weights. The following data are load weights for five residential and five business trailers in August, 1994, by courtesy of Mr John Larcombe of the Marlborough District Council.

Source	Load weight (kg)				
Residential	120	180	140	100	150
Business	70	410	40	90	540

(a) Make up an observation vector $\mathbf{y}$ and calculate the appropriate orthogonal decomposition $\mathbf{y} = \bar{\mathbf{y}} + (\bar{\mathbf{y}}_i - \bar{\mathbf{y}}) + (\mathbf{y} - \bar{\mathbf{y}}_i)$.

(b) Check for a difference in mean load weight between residential and business trailers by writing down the corresponding Pythagorean breakup and calculating the appropriate F test statistic. What is your conclusion?

(c) Calculate the corresponding t test statistic using the formula

$$t = \pm \frac{\|\bar{\mathbf{y}}_i - \bar{\mathbf{y}}\|}{\|\mathbf{y} - \bar{\mathbf{y}}_i\| / \sqrt{2(n-1)}}$$

Does this lead to the same conclusion as in (c)?

(3.6) An experiment was carried out to determine the tolerance of Golden Queen peach seedlings to the herbicide oxadiazon. Eight plots of seedlings were randomly allocated to the treatments:

1. No herbicide (control).
2. Oxadiazon at normal rate.

All plots were hand weeded throughout the experiment. At the end of the first season the average height of the seedlings in the eight plots were:

1. Control	80, 68, 62, 82 cm
2. Oxadiazon	62, 56, 68, 66 cm

(a) Take the observation vector $\mathbf{y} = [80, 68, 62, 82, 62, 56, 68, 66]^{\mathrm{T}}$ and calculate the orthogonal decomposition $\mathbf{y} = (\mathbf{y}\cdot\mathbf{U}_1)\mathbf{U}_1 + \cdots + (\mathbf{y}\cdot\mathbf{U}_8)\mathbf{U}_8$, where $\mathbf{U}_1, \ldots, \mathbf{U}_8$ are the unit vectors

$$
\begin{array}{cccccccc}
\mathbf{U}_1 & \mathbf{U}_2 & \mathbf{U}_3 & \mathbf{U}_4 & \mathbf{U}_5 & \mathbf{U}_6 & \mathbf{U}_7 & \mathbf{U}_8 \\
\begin{bmatrix} 1 \\ 1 \\ 1 \\ 1 \\ 1 \\ 1 \\ 1 \\ 1 \end{bmatrix} &
\begin{bmatrix} 1 \\ 1 \\ 1 \\ 1 \\ -1 \\ -1 \\ -1 \\ -1 \end{bmatrix} &
\begin{bmatrix} 1 \\ 1 \\ -1 \\ -1 \\ 0 \\ 0 \\ 0 \\ 0 \end{bmatrix} &
\begin{bmatrix} 1 \\ -1 \\ 0 \\ 0 \\ 0 \\ 0 \\ 0 \\ 0 \end{bmatrix} &
\begin{bmatrix} 0 \\ 0 \\ 1 \\ -1 \\ 0 \\ 0 \\ 0 \\ 0 \end{bmatrix} &
\begin{bmatrix} 0 \\ 0 \\ 0 \\ 0 \\ 1 \\ 1 \\ -1 \\ -1 \end{bmatrix} &
\begin{bmatrix} 0 \\ 0 \\ 0 \\ 0 \\ 1 \\ -1 \\ 0 \\ 0 \end{bmatrix} &
\begin{bmatrix} 0 \\ 0 \\ 0 \\ 0 \\ 0 \\ 0 \\ 1 \\ -1 \end{bmatrix} \\
\hline
\sqrt{8} & \sqrt{8} & \sqrt{4} & \sqrt{2} & \sqrt{2} & \sqrt{4} & \sqrt{2} & \sqrt{2}
\end{array}
$$

(b) Calculate the corresponding Pythagorean breakup

$$\|\mathbf{y}\|^2 = (\mathbf{y}\cdot\mathbf{U}_1)^2 + \cdots + (\mathbf{y}\cdot\mathbf{U}_8)^2$$

(c) Assume the observations are independent samples from normal populations with true means μ_1 and μ_2, and a common variance σ^2. Test H_0: $\mu_1 = \mu_2$ against H_1: $\mu_1 \neq \mu_2$ using an F test obtained by dividing $(\mathbf{y}\cdot\mathbf{U}_2)^2$ by the average of $(\mathbf{y}\cdot\mathbf{U}_3)^2, \ldots, (\mathbf{y}\cdot\mathbf{U}_8)^2$. What is your conclusion?

(d) Recalculate the F value using the vector decomposition $\mathbf{y} = \bar{\mathbf{y}} + (\bar{\mathbf{y}}_i - \bar{\mathbf{y}}) + (\mathbf{y} - \bar{\mathbf{y}}_i)$ and the formula

$$F = \frac{\|\bar{\mathbf{y}}_i - \bar{\mathbf{y}}\|^2}{\|\mathbf{y} - \bar{\mathbf{y}}_i\|^2/6}$$

(3.7)* In Exercise 2.4 we presented data on plasma selenium levels for a group of babies who were being fed with an unsupplemented milk formula. In the following table we present the corresponding data for the second group of babies, fed with a milk formula supplemented with selenium. Data is by courtesy of Professor Christine Winterbourn.

(a) We shall now compare the average change in selenium level from birth to 1 month of age between the two groups of babies. For both groups calculate this change (birth minus 1 month) for the babies for whom samples were taken at both sampling times. Take just the *first five* of these changes from each group and make up a 10-long observation vector $\mathbf{y}$.

(b) Calculate the orthogonal decomposition $\mathbf{y} = \bar{\mathbf{y}} + (\bar{\mathbf{y}}_i - \bar{\mathbf{y}}) + (\mathbf{y} - \bar{\mathbf{y}}_i)$.

Supplemented	Plasma selenium levels (μg/L)		
Baby no.	Birth	1 month	3 months
1	55	40	73
2		48	27
3		40	52
4	41	54	
5		52	60
6		51	74
7		33	
8		34	
9		28	31
10	40	39	66
11		53	73
12		47	72
13	46	45	77
14	31	36	62
15		47	47
16	32	57	44
17		51	56
18	33	50	83
19	48	43	84
20		35	48

(c) Test the hypothesis that the mean changes in selenium level were the same for the two groups ($\mu_1 = \mu_2$) by writing out the corresponding Pythagorean breakup and calculating the appropriate F value. What is your conclusion?

(d) Repeat (a) to (c) for the time period 1 to 3 months of age, subtracting the 1 month value from the 3 month value and using the *first ten* values from each group.

(3.8) *The virtues of tagging*
In Exercise 2.3, suppose that on each date six calves were randomly selected and weighed, with no individual identification. Reanalyze the data treating the 12 calf weights as two independent samples of size six, as follows:

(a) Write down the orthogonal decomposition $\mathbf{y} = \bar{\mathbf{y}} + (\bar{\mathbf{y}}_i - \bar{\mathbf{y}}) + (\mathbf{y} - \bar{\mathbf{y}}_i)$.

(b) Calculate the corresponding Pythagorean breakup

$$||\mathbf{y}||^2 = ||\bar{\mathbf{y}}||^2 + ||\bar{\mathbf{y}}_i - \bar{\mathbf{y}}||^2 + ||\mathbf{y} - \bar{\mathbf{y}}_i||^2$$

(c) Test the hypothesis $\mu_1 = \mu_2$ by calculating the test statistic

$$F = \frac{||\bar{\mathbf{y}}_i - \bar{\mathbf{y}}||^2}{||\mathbf{y} - \bar{\mathbf{y}}_i||^2/10}$$

and comparing it with the 90, 95 and 99 percentiles of the appropriate distribution.

(d) How do your conclusions compare with those obtained in Exercise 2.3? What are the virtues of tagging? (In other words, why is a paired samples design more powerful than an independent samples design?)

(3.9) In Exercise 3.6 the coordinate system $\mathbf{U}_3, \ldots, \mathbf{U}_8$ for the error space, whilst a perfectly good system, did not follow the pattern given elsewhere in this chapter. An alternative system following this pattern is

$$\mathbf{U}_3 = \frac{\begin{bmatrix} -1 \\ 1 \\ 0 \\ 0 \\ 0 \\ 0 \\ 0 \\ 0 \end{bmatrix}}{\sqrt{2}} \quad \mathbf{U}_4 = \frac{\begin{bmatrix} -1 \\ -1 \\ 2 \\ 0 \\ 0 \\ 0 \\ 0 \\ 0 \end{bmatrix}}{\sqrt{6}} \quad \mathbf{U}_5 = \frac{\begin{bmatrix} -1 \\ -1 \\ -1 \\ 3 \\ 0 \\ 0 \\ 0 \\ 0 \end{bmatrix}}{\sqrt{12}} \quad \mathbf{U}_6 = \frac{\begin{bmatrix} 0 \\ 0 \\ 0 \\ 0 \\ -1 \\ 1 \\ 0 \\ 0 \end{bmatrix}}{\sqrt{2}} \quad \mathbf{U}_7 = \frac{\begin{bmatrix} 0 \\ 0 \\ 0 \\ 0 \\ -1 \\ -1 \\ 2 \\ 0 \end{bmatrix}}{\sqrt{6}} \quad \mathbf{U}_8 = \frac{\begin{bmatrix} 0 \\ 0 \\ 0 \\ 0 \\ -1 \\ -1 \\ -1 \\ 3 \end{bmatrix}}{\sqrt{12}}$$

Calculate $s^2 = [(\mathbf{y}\cdot\mathbf{U}_3)^2 + \cdots + (\mathbf{y}\cdot\mathbf{U}_8)^2]/6$ using the new coordinate system to show that the result is the same as that obtained using the old coordinate system.

(3.10)* We now lead the reader through the analysis of the full data set of 49 female and 77 male heights displayed in Figure 3.1. Note that this exercise could prove tedious for readers who do not have access to computers (see Appendix C for computing methods). First, we redisplay the two histograms with the frequency of each height added (Figure 3.10) for the convenience of the reader.

In this unequally replicated example the *direction associated with the difference of interest*, $\mu_2 - \mu_1$, is

$$\mathbf{U}_2 = \frac{1}{\sqrt{\frac{1}{49} + \frac{1}{77}}} \begin{bmatrix} -1/49 \\ \vdots \\ 1/77 \\ \vdots \end{bmatrix}$$

This is because the projection length $\mathbf{y}\cdot\mathbf{U}_2$ is equal to $\sqrt{\frac{1}{49} + \frac{1}{77}}(\bar{y}_2 - \bar{y}_1)$, which averages, in the long run, to a multiple of $(\mu_2 - \mu_1)$.

Therefore an appropriate coordinate system for our unequal sample size analysis is

$$\mathbf{U_1} = \frac{1}{\sqrt{126}}\begin{bmatrix} 1 \\ \vdots \\ 1 \\ \vdots \end{bmatrix}, \quad \mathbf{U_2} = \frac{1}{\sqrt{\frac{1}{49}+\frac{1}{77}}}\begin{bmatrix} -1/49 \\ \vdots \\ 1/77 \\ \vdots \end{bmatrix}, \quad \mathbf{U_3}, \ldots, \mathbf{U_{126}}$$

where we shall not bother to write down the 124 coordinate axes which span the error space.

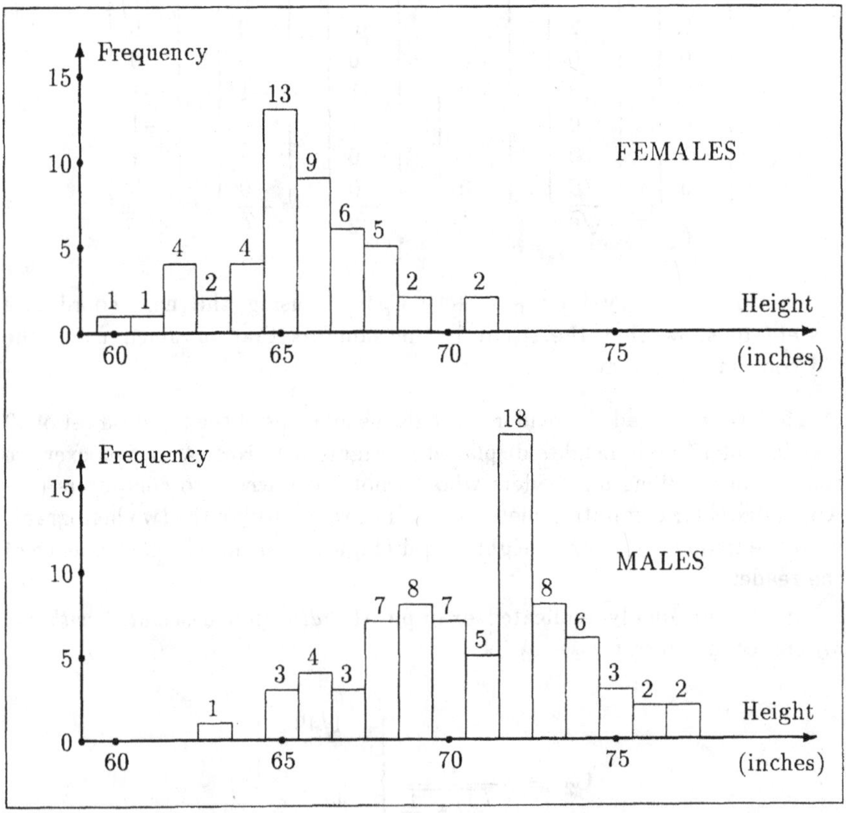

Figure 3.10: Histograms of heights of female and male students in the ASM150 class, with the addition of the frequency of occurrence of each height.

(a) The reader can now project their 126-long observation vector onto the first two coordinate axes, $\mathbf{U}_1$ and $\mathbf{U}_2$. Do the resulting projection vectors look familiar in terms of $\bar{\mathbf{y}}$ and $(\bar{\mathbf{y}}_i - \bar{\mathbf{y}})$?

(b) The reader can now write out the full orthogonal decomposition

$$\mathbf{y} = (\mathbf{y}\cdot\mathbf{U}_1)\mathbf{U}_1 + (\mathbf{y}\cdot\mathbf{U}_2)\mathbf{U}_2 + \text{error vector}$$

where the error vector is obtained by subtraction. Does the error vector also look familiar? What is it in terms of $\mathbf{y}$, $\bar{\mathbf{y}}$ and $\bar{\mathbf{y}}_i$?

(c) Write down the associated Pythagorean breakup.

(d) To test the hypothesis $\mu_1 = \mu_2$ the reader can now calculate the test statistic

$$F = \frac{(\mathbf{y}\cdot\mathbf{U}_2)^2}{[(\mathbf{y}\cdot\mathbf{U}_3)^2 + \cdots + (\mathbf{y}\cdot\mathbf{U}_{126})^2]/124} = \frac{(\mathbf{y}\cdot\mathbf{U}_2)^2}{\|\text{error vector}\|^2/124}$$

using the latter of the two forms! How conclusive is the evidence for a difference in height between female and male UCD students?

Note that the above analysis follows the method explained further in Appendix A of Saville and Wood (1991).

Chapter 4

Several Independent Samples (Analysis of Variance)

In this chapter we generalize the results of the last chapter to the case of independent samples from more than two populations (this is the simplest case of what is usually called an "analysis of variance"). For example, we may have cholesterol levels of one hundred randomly selected 50-year-olds for each of the states of the United States of America, or we may have grain yields from six experimental plots for each of ten wheat cultivars. In this more general case of k independent samples, we can specify more independent comparisons between the population means than in the two sample case, where there is only one possible comparison. For example, in the case of the cholesterol levels we may decide: (1) to contrast coastal with landlocked states; and (2) to contrast east coast states with west coast states. In the case of grain yields of wheat cultivars we may decide: (1) to contrast locally bred cultivars with imported cultivars; (2) to contrast European imports with Mexican imports; and (3) to contrast short-stemmed Mexican cultivars with long-stemmed Mexican cultivars.

More formally, we are interested in comparisons, or *contrasts*, between the means, $\mu_1, \ldots, \mu_k$, of the k study populations. For the analysis we assume that all k populations are normally distributed and that the population variances are all the same ($\sigma_1^2 = \cdots = \sigma_k^2 = \sigma^2$).

In §4.1 of this chapter we shall describe a data set consisting of the blood plasma selenium levels of adults, full-term babies and premature babies in Christchurch, New Zealand (a low selenium community). In §4.2 we analyze a small subset of this data set, consisting of the selenium levels of two adults, two full-term babies and two premature babies. In §4.3 we go on to analyze the full data set, and in §4.4 we discuss the general case, involving varying numbers of populations and different types of contrasts. The chapter closes with a summary in §4.5, a class exercise and general exercises.

4.1 Selenium Levels

The soils in the South Island of New Zealand are deficient in the trace element selenium. Until this was realized, young livestock grazing on this land suffered from poor growth rates and, in some cases, sudden death. The animal health problem disappeared when dosing with selenium was universally adopted. However, the human population was still only receiving about half of the recommended daily allowance of selenium. In addition, there was an exceptionally high rate of sudden infant death (cot death) in the South Island of New Zealand. This prompted a group of locally based

Sample number	Blood plasma selenium level (μg/L) Adults	Full-term babies	Premature babies
1	60	24	31
2	30	40	28
3	60	28	35
4	81	36	12
5	52	32	27
6	57	36	42
7	54	40	37
8	67	28	40
9	66	32	16
10	72	40	16
11	64	28	17
12	92	36	18
13	84	40	20
14	43	32	30
15	76	40	27
16	48	47	38
17	62	40	40
18	80	24	39
19	100	43	28
20	110	32	27
21	66	40	21
22	87	51	20
23	115	28	8
24	88	28	32

Table 4.1: Plasma selenium levels, in micrograms per liter, for 24 adults, full-term babies and premature neonatal babies in blood samples collected during 1989 and 1990 in Christchurch, New Zealand. Data by courtesy of Professor Christine Winterbourn, Christchurch School of Medicine.

medical researchers, led by Professor Christine Winterbourn, to commence investigations into the selenium status of the human population, with a view to exploring the possibility of a link with cot death. As part of this project the blood plasma selenium levels were determined for three samples of humans consisting of 108 adults, 30 full-term babies and 24 premature neonatal babies, as described by Sluis et al. (1992). In Table 4.1 we present a subsample of these data, 24 values for each of the three categories (obtained by random selection from the original data in the case of categories one and two). The data in the table are also displayed in histograms in Figure 4.1.

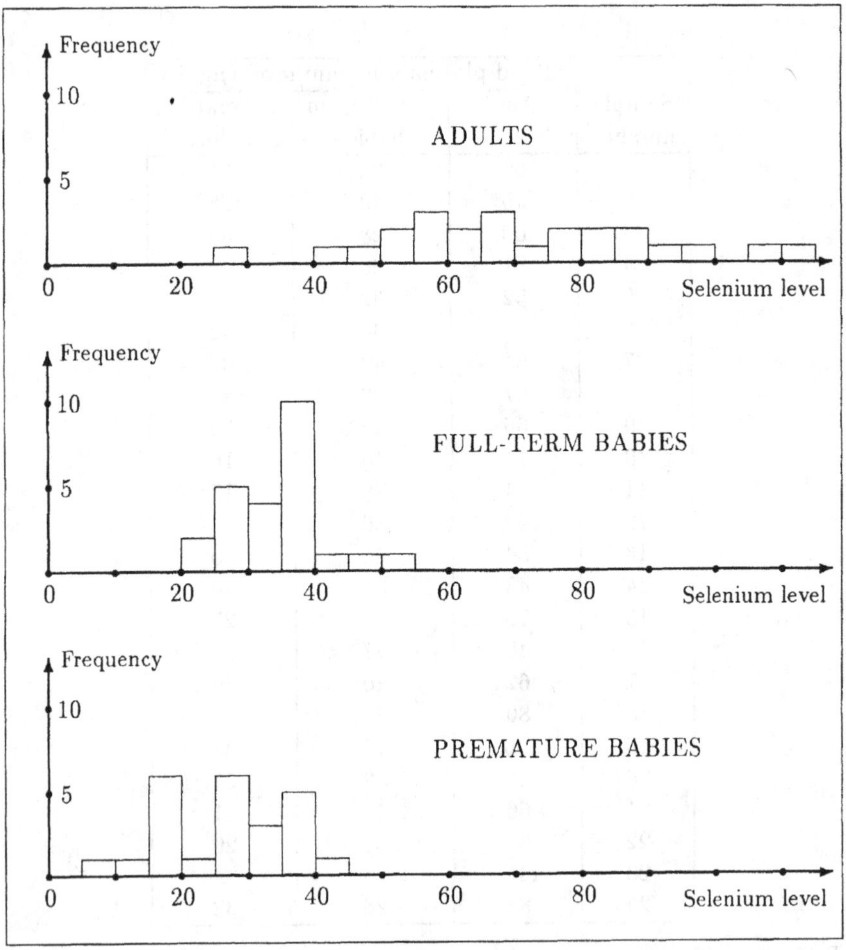

Figure 4.1: Histograms of blood plasma selenium levels (μg/L) for subsamples of 24 adults, full-term babies and premature neonatal babies. The class intervals are 0–5, 6–10, 11–15 and so on.

The observations in Table 4.1 are assumed to come from three normally distributed populations: the plasma selenium levels of healthy Christchurch adults, of full-term babies born at Christchurch Women's Hospital, and of premature babies in the neonatal unit of the same hospital, during 1989 and 1990. The population means are assumed to be μ_1, μ_2 and μ_3, and the variance, σ^2, is assumed to be the same in the three populations. Note that the latter assumption is not very reasonable since the first population appears to be more variable than the other two populations (Figure 4.1); however, to keep things simple we shall ignore this problem for the moment, coming back to it after we have established our basic method.

4.2 Samples of Size Two

Again in the interests of simplicity, we now take three random samples of size two from Table 4.1 and work through the analysis of this miniature data set. The data we shall use are given in Table 4.2:

Sample number	Blood plasma selenium level (μg/L) Adults	Full-term babies	Premature babies
1	84	28	40
2	52	28	30

Table 4.2: Blood plasma selenium levels, in micrograms per liter, for three samples of size two randomly selected from Table 4.1.

The resulting *observation vector* is

$$\mathbf{y} = \begin{bmatrix} 84 \\ 52 \\ 28 \\ 28 \\ 40 \\ 30 \end{bmatrix}$$

Objectives

The study objectives are as follows:

- To estimate the mean selenium levels for the three populations of study.
- To test whether selenium levels differ between adults and babies.
- To test whether selenium levels differ between full-term and premature babies.

That is, we are interested in estimating the population means, μ_1, μ_2 and μ_3, and in testing whether the *contrasts* $[\mu_1 - (\mu_2 + \mu_3)/2]$ (adults versus babies) and $(\mu_3 - \mu_2)$ (full-term versus premature babies) differ from zero.

The Basic Idea

Figure 4.2 explains the basic idea. Four cases can occur. Firstly, if there are no differences in mean selenium level between adults, full-term babies and premature babies ($\mu_1 = \mu_2 = \mu_3 = \mu$), then our observation vector will be part of a scatter centered on the point $(\mu, \mu, \mu, \mu, \mu, \mu)$ as shown in Figure 4.2(a).

Secondly, if adults and babies differ in mean selenium level ($\mu_1 \neq (\mu_2 + \mu_3)/2$), but full-term and premature babies have the same mean selenium level ($\mu_2 = \mu_3$), then our observation vector will be part of a scatter centered on the point $(\mu_1, \mu_1, \mu_2, \mu_2, \mu_2, \mu_2)$ as shown in Figure 4.2(b). In this case, the scatter is moved away from the equiangular line in the direction $[-2, -2, 1, 1, 1, 1]^T$ since

$$\begin{bmatrix} \mu_1 \\ \mu_1 \\ \mu_2 \\ \mu_2 \\ \mu_2 \\ \mu_2 \end{bmatrix} - \begin{bmatrix} \mu \\ \mu \\ \mu \\ \mu \\ \mu \\ \mu \end{bmatrix} = \begin{bmatrix} \mu_1 - \mu \\ \mu_1 - \mu \\ \mu_2 - \mu \\ \mu_2 - \mu \\ \mu_2 - \mu \\ \mu_2 - \mu \end{bmatrix} = \begin{bmatrix} 2(\mu_1 - \mu_2)/3 \\ 2(\mu_1 - \mu_2)/3 \\ (\mu_2 - \mu_1)/3 \\ (\mu_2 - \mu_1)/3 \\ (\mu_2 - \mu_1)/3 \\ (\mu_2 - \mu_1)/3 \end{bmatrix} = \frac{\mu_1 - \mu_2}{3} \begin{bmatrix} 2 \\ 2 \\ -1 \\ -1 \\ -1 \\ -1 \end{bmatrix}$$

where we have substituted $\mu = (\mu_1 + \mu_2 + \mu_2)/3 = (\mu_1 + 2\mu_2)/3$.

Thirdly, if adults and babies have the same mean selenium level ($\mu_1 = (\mu_2 + \mu_3)/2$), but full-term and premature babies differ in mean selenium level ($\mu_2 \neq \mu_3$), then our observation vector will be part of a scatter centered on the point $((\mu_2 + \mu_3)/2, (\mu_2 + \mu_3)/2, \mu_2, \mu_2, \mu_3, \mu_3)$ as shown in Figure 4.2(c). In this case, the scatter is moved away from the equiangular line in the direction $[0, 0, -1, -1, 1, 1]^T$ since

$$\begin{bmatrix} (\mu_2 + \mu_3)/2 \\ (\mu_2 + \mu_3)/2 \\ \mu_2 \\ \mu_2 \\ \mu_3 \\ \mu_3 \end{bmatrix} - \begin{bmatrix} \mu \\ \mu \\ \mu \\ \mu \\ \mu \\ \mu \end{bmatrix} = \begin{bmatrix} 0 \\ 0 \\ (\mu_2 - \mu_3)/2 \\ (\mu_2 - \mu_3)/2 \\ (\mu_3 - \mu_2)/2 \\ (\mu_3 - \mu_2)/2 \end{bmatrix} = \frac{\mu_3 - \mu_2}{2} \begin{bmatrix} 0 \\ 0 \\ -1 \\ -1 \\ 1 \\ 1 \end{bmatrix}$$

where we have substituted $\mu = [(\mu_2 + \mu_3)/2 + \mu_2 + \mu_3]/3 = (\mu_2 + \mu_3)/2$.

Lastly, if adults and babies differ in mean selenium level ($\mu_1 \neq (\mu_2 + \mu_3)/2$), and full-term and premature babies also differ in mean selenium level ($\mu_2 \neq \mu_3$), then our observation vector will be part of a scatter centered on the point $(\mu_1, \mu_1, \mu_2, \mu_2, \mu_3, \mu_3)$ as shown in Figure 4.2(d). In this case, the scatter is moved away from the equiangular line in a direction which has components in both of the directions $[2, 2, -1, -1, -1, -1]^T$ and $[0, 0, -1, -1, 1, 1]^T$.

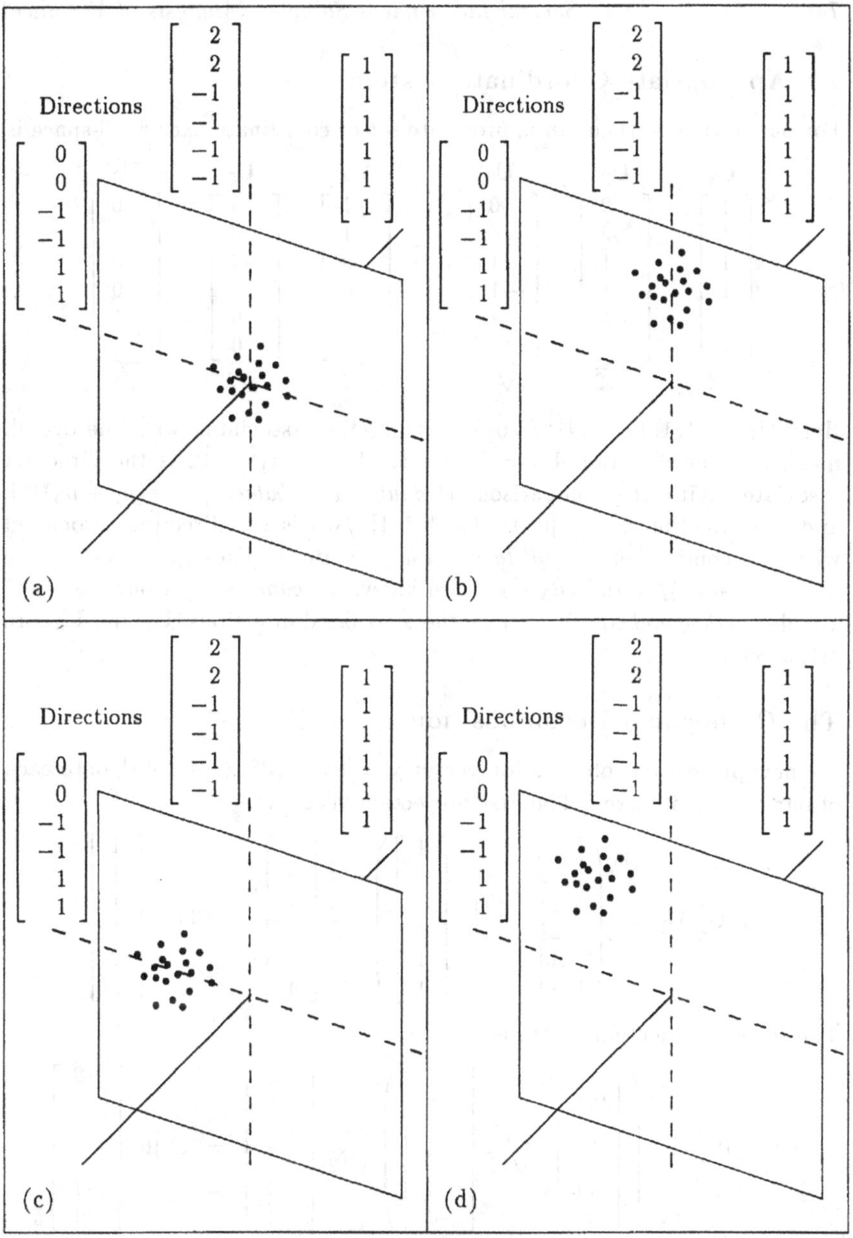

Figure 4.2: Clouds of data points representing many repetitions of our selenium study. If (a) $\mu_1 = \mu_2 = \mu_3$ the cloud is centered on the point $(\mu, \mu, \mu, \mu, \mu, \mu)$. If (b) $\mu_2 = \mu_3 (\neq \mu_1)$ the cloud is centered on the point $(\mu_1, \mu_1, \mu_2, \mu_2, \mu_2, \mu_2)$. If (c) $\mu_1 = (\mu_2 + \mu_3)/2$ but $\mu_2 \neq \mu_3$ the cloud is centered on the point $((\mu_2 + \mu_3)/2, (\mu_2 + \mu_3)/2, \mu_2, \mu_2, \mu_3, \mu_3)$. Lastly, and most generally, if (d) μ_1, μ_2 and μ_3 are all different the cloud is centered on the point $(\mu_1, \mu_1, \mu_2, \mu_2, \mu_3, \mu_3)$.

An Appropriate Coordinate System

For our purposes, then, an appropriate set of coordinate axes for 6-space is

$$\mathbf{U_1} = \frac{\begin{bmatrix}1\\1\\1\\1\\1\\1\end{bmatrix}}{\sqrt{6}} \quad \mathbf{U_2} = \frac{\begin{bmatrix}2\\2\\-1\\-1\\-1\\-1\end{bmatrix}}{\sqrt{12}} \quad \mathbf{U_3} = \frac{\begin{bmatrix}0\\0\\-1\\-1\\1\\1\end{bmatrix}}{\sqrt{4}} \quad \mathbf{U_4} = \frac{\begin{bmatrix}-1\\1\\0\\0\\0\\0\end{bmatrix}}{\sqrt{2}} \quad \mathbf{U_5} = \frac{\begin{bmatrix}0\\0\\-1\\1\\0\\0\end{bmatrix}}{\sqrt{2}} \quad \mathbf{U_6} = \frac{\begin{bmatrix}0\\0\\0\\0\\-1\\1\end{bmatrix}}{\sqrt{2}}$$

Here $\mathbf{U_1} = [1,1,1,1,1,1]^T/\sqrt{6}$ is the direction associated with the overall mean μ. The direction $\mathbf{U_2} = [2,2,-1,-1,-1,-1]^T/\sqrt{12}$ is the direction associated with the comparison of *adults with babies* $[\mu_1 - (\mu_2 + \mu_3)/2]$, and the direction $\mathbf{U_3} = [0,0,-1,-1,1,1]^T/\sqrt{4}$ is the direction associated with the comparison of *full-term with premature babies* $(\mu_3 - \mu_2)$. Here $[\mu_1 - (\mu_2 + \mu_3)/2]$ and $(\mu_3 - \mu_2)$ are known as *contrasts*. In our case they are also *orthogonal* contrasts since the associated directions, $\mathbf{U_2}$ and $\mathbf{U_3}$, are orthogonal.

The Orthogonal Decomposition

We now project our observation vector $\mathbf{y} = [84, 52, 28, 28, 40, 30]^T$ onto each of our coordinate axes. The first projection vector is

$$(\mathbf{y} \cdot \mathbf{U_1})\mathbf{U_1} = \left(\begin{bmatrix}84\\52\\28\\28\\40\\30\end{bmatrix} \cdot \frac{1}{\sqrt{6}}\begin{bmatrix}1\\1\\1\\1\\1\\1\end{bmatrix}\right)\frac{1}{\sqrt{6}}\begin{bmatrix}1\\1\\1\\1\\1\\1\end{bmatrix} = 43.667\begin{bmatrix}1\\1\\1\\1\\1\\1\end{bmatrix}$$

The second projection vector is

$$(\mathbf{y} \cdot \mathbf{U_2})\mathbf{U_2} = \left(\begin{bmatrix}84\\52\\28\\28\\40\\30\end{bmatrix} \cdot \frac{1}{\sqrt{12}}\begin{bmatrix}2\\2\\-1\\-1\\-1\\-1\end{bmatrix}\right)\frac{1}{\sqrt{12}}\begin{bmatrix}2\\2\\-1\\-1\\-1\\-1\end{bmatrix} = 12.167\begin{bmatrix}2\\2\\-1\\-1\\-1\\-1\end{bmatrix}$$

The third projection vector is

$$(\mathbf{y} \cdot \mathbf{U_3})\mathbf{U_3} = \left(\begin{bmatrix}84\\52\\28\\28\\40\\30\end{bmatrix} \cdot \frac{1}{\sqrt{4}}\begin{bmatrix}0\\0\\-1\\-1\\1\\1\end{bmatrix}\right)\frac{1}{\sqrt{4}}\begin{bmatrix}0\\0\\-1\\-1\\1\\1\end{bmatrix} = 3.5\begin{bmatrix}0\\0\\-1\\-1\\1\\1\end{bmatrix}$$

Here the projection lengths are $\mathbf{y}\cdot\mathbf{U}_1 = 262/\sqrt{6} = 106.96$, $\mathbf{y}\cdot\mathbf{U}_2 = 146/\sqrt{12} = 42.15$ and $\mathbf{y}\cdot\mathbf{U}_3 = 14/\sqrt{4} = 7$.

Similarly, the three remaining projection vectors are

$$(\mathbf{y}\cdot\mathbf{U}_4)\mathbf{U}_4 = -16\begin{bmatrix}-1\\1\\0\\0\\0\\0\end{bmatrix},\quad (\mathbf{y}\cdot\mathbf{U}_5)\mathbf{U}_5 = 0\begin{bmatrix}0\\0\\-1\\1\\0\\0\end{bmatrix}\quad \text{and } (\mathbf{y}\cdot\mathbf{U}_6)\mathbf{U}_6 = -5\begin{bmatrix}0\\0\\0\\0\\-1\\1\end{bmatrix}$$

Here the signed projection lengths are $\mathbf{y}\cdot\mathbf{U}_4 = -32/\sqrt{2} = -22.63$, $\mathbf{y}\cdot\mathbf{U}_5 = 0$ and $\mathbf{y}\cdot\mathbf{U}_6 = -10/\sqrt{2} = -7.07$.

The resulting orthogonal decomposition of the observation vector is

$$\mathbf{y} = (\mathbf{y}\cdot\mathbf{U}_1)\mathbf{U}_1 + (\mathbf{y}\cdot\mathbf{U}_2)\mathbf{U}_2 + (\mathbf{y}\cdot\mathbf{U}_3)\mathbf{U}_3 + (\mathbf{y}\cdot\mathbf{U}_4)\mathbf{U}_4 + (\mathbf{y}\cdot\mathbf{U}_5)\mathbf{U}_5 + (\mathbf{y}\cdot\mathbf{U}_6)\mathbf{U}_6$$

$$\mathbf{y} = 106.96\mathbf{U}_1 + 42.15\mathbf{U}_2 + 7\mathbf{U}_3 - 22.63\mathbf{U}_4 + 0\mathbf{U}_5 - 7.07\mathbf{U}_6$$

In vectors this is

$$\underset{\text{observation vector}}{\begin{bmatrix}84\\52\\28\\28\\40\\30\end{bmatrix}} = \underset{\text{overall mean vector}}{\begin{bmatrix}43.667\\43.667\\43.667\\43.667\\43.667\\43.667\end{bmatrix}} + \underset{\text{contrast vectors}}{\begin{bmatrix}24.333\\24.333\\-12.167\\-12.167\\-12.167\\-12.167\end{bmatrix} + \begin{bmatrix}0\\0\\-3.5\\-3.5\\3.5\\3.5\end{bmatrix}} + \underset{\text{error vector}}{\begin{bmatrix}16\\-16\\0\\0\\5\\-5\end{bmatrix}}$$

where we have collapsed the last three projection vectors into a single "error" vector. If we so desire, this can be simplified further to the fitted model

$$\begin{bmatrix}84\\52\\28\\28\\40\\30\end{bmatrix} = \begin{bmatrix}43.667\\43.667\\43.667\\43.667\\43.667\\43.667\end{bmatrix} + \begin{bmatrix}24.333\\24.333\\-15.667\\-15.667\\-8.667\\-8.667\end{bmatrix} + \begin{bmatrix}16\\-16\\0\\0\\5\\-5\end{bmatrix}$$

$$\underset{\text{observation vector}}{\mathbf{y}} = \underset{\text{overall mean vector}}{\bar{\mathbf{y}}} + \underset{\text{treatment vector}}{(\bar{\mathbf{y}}_i - \bar{\mathbf{y}})} + \underset{\text{error vector}}{(\mathbf{y} - \bar{\mathbf{y}}_i)}$$

although we often prefer the previous form.

To summarize, we have broken our observation vector $\mathbf{y}$ into three orthogonal components, an *overall mean vector*, two *contrast vectors*, and an *error vector*, as shown in Figure 4.3. Note that if we sum the contrast vectors we obtain the *treatment vector*, $(\bar{\mathbf{y}}_i - \bar{\mathbf{y}})$, shown in Figure 4.3.

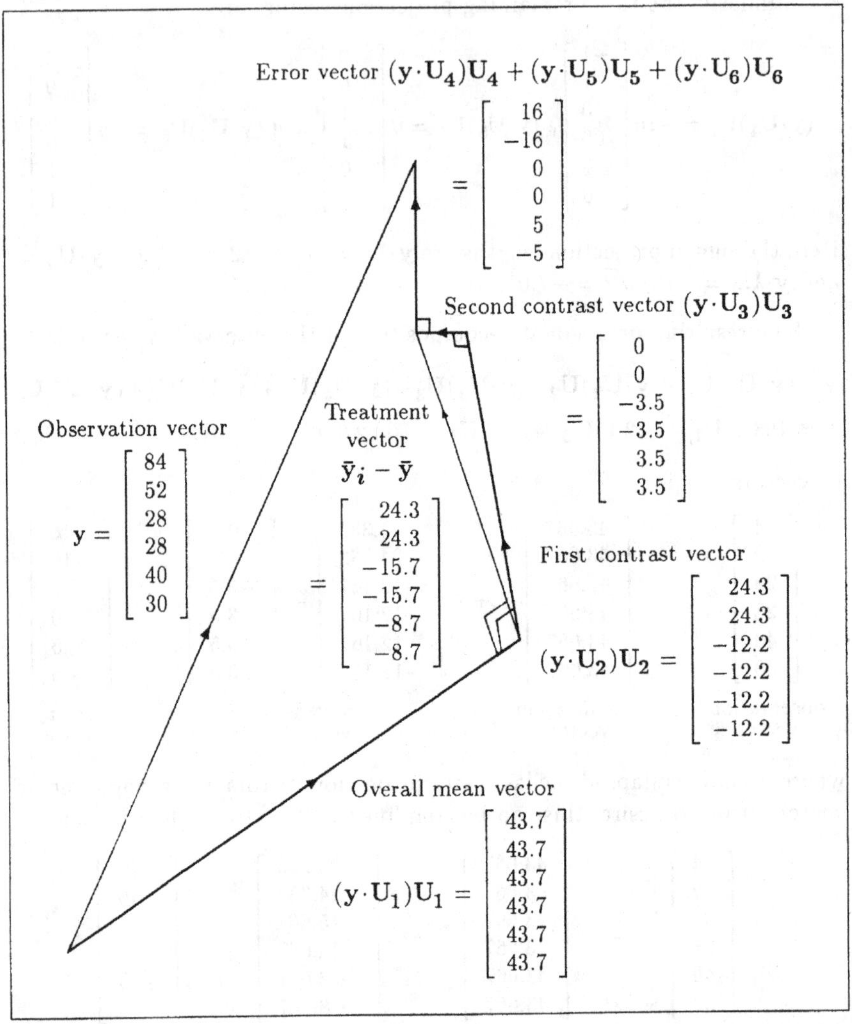

Figure 4.3: Orthogonal decomposition of the observation vector for the case of independent samples of size two from three study populations. Vectors have been rounded to one decimal place. Please do not be too critical of the writers' artistry; drawing a four-dimensional diagram on a two-dimensional surface is an impossible task!

To enable us to meet our first study objective, that of estimating the mean selenium levels for the three study populations, we further simplify our fitted model as follows:

$$\begin{bmatrix} 84 \\ 52 \\ 28 \\ 28 \\ 40 \\ 30 \end{bmatrix} = \begin{bmatrix} 68 \\ 68 \\ 28 \\ 28 \\ 35 \\ 35 \end{bmatrix} + \begin{bmatrix} 16 \\ -16 \\ 0 \\ 0 \\ 5 \\ -5 \end{bmatrix}$$

$$\underset{\substack{\text{observation}\\\text{vector}}}{\mathbf{y}} = \underset{\substack{\text{treatment mean}\\\text{vector}}}{\bar{\mathbf{y}}_i} + \underset{\substack{\text{error}\\\text{vector}}}{(\mathbf{y}-\bar{\mathbf{y}}_i)}$$

The resulting *least squares* estimates of μ_1, μ_2 and μ_3 are $\bar{y}_1 = 68$ (adults), $\bar{y}_2 = 28$ (full-term babies) and $\bar{y}_3 = 35$ (premature babies). The word "least squares," short for least squared distance, is used since the above estimates arise from the perpendicular projection of the observation vector onto the model space spanned by the unit vectors $\mathbf{U}_1$, $\mathbf{U}_2$ and $\mathbf{U}_3$; this projection minimizes the squared distance between the tip of the observation vector and the tip of a vector lying in the model space.

Distributions of the Projection Lengths

To prepare for testing whether our two contrasts could plausibly be zero, we now work out the distributions of the projection lengths $\mathbf{y}\cdot\mathbf{U}_1, \ldots, \mathbf{y}\cdot\mathbf{U}_6$. As in §3.2 the trick is to expand these lengths in terms of the observations $y_{11}, y_{12}, y_{21}, y_{22}, y_{31}$ and y_{32}, where y_{ij} denotes the jth observation from the ith study population. These expansions are

$$\begin{aligned}
\mathbf{y}\cdot\mathbf{U}_1 &= (y_{11} + y_{12} + y_{21} + y_{22} + y_{31} + y_{32})/\sqrt{6} \\
\mathbf{y}\cdot\mathbf{U}_2 &= (2y_{11} + 2y_{12} - y_{21} - y_{22} - y_{31} - y_{32})/\sqrt{12} \\
\mathbf{y}\cdot\mathbf{U}_3 &= (-y_{21} - y_{22} + y_{31} + y_{32})/\sqrt{4} \\
\mathbf{y}\cdot\mathbf{U}_4 &= (-y_{11} + y_{12})/\sqrt{2} \\
\mathbf{y}\cdot\mathbf{U}_5 &= (-y_{21} + y_{22})/\sqrt{2} \\
\mathbf{y}\cdot\mathbf{U}_6 &= (-y_{31} + y_{32})/\sqrt{2}
\end{aligned}$$

As in §3.2, it is then easy to show that $\mathbf{y}\cdot\mathbf{U}_1$ comes from an $N(\sqrt{6}\mu, \sigma^2)$ distribution, $\mathbf{y}\cdot\mathbf{U}_2$ comes from an $N\left[(2\mu_1 - \mu_2 - \mu_3)/\sqrt{3}, \sigma^2\right]$ distribution, $\mathbf{y}\cdot\mathbf{U}_3$ comes from an $N\left[(\mu_3 - \mu_2), \sigma^2\right]$ distribution, and $\mathbf{y}\cdot\mathbf{U}_4$, $\mathbf{y}\cdot\mathbf{U}_5$ and $\mathbf{y}\cdot\mathbf{U}_6$ all come from an $N(0, \sigma^2)$ distribution.

That is, the distributions of $\mathbf{y}\cdot\mathbf{U}_1$, $\mathbf{y}\cdot\mathbf{U}_2$ and $\mathbf{y}\cdot\mathbf{U}_3$ are centered on potentially nonzero quantities whereas the distributions of $\mathbf{y}\cdot\mathbf{U}_4$, $\mathbf{y}\cdot\mathbf{U}_5$ and $\mathbf{y}\cdot\mathbf{U}_6$ are always centered on zero.

Testing the Hypotheses

We now proceed to investigate our second and third study objectives, which involve testing:

- whether selenium levels differ between adults and babies;
- whether selenium levels differ between full-term and premature babies.

This involves testing whether our two contrasts differ from zero. More formally, we wish to test the null hypothesis $H_0: \mu_1 = (\mu_2 + \mu_3)/2$ against the alternative hypothesis, $H_1: \mu_1 \neq (\mu_2 + \mu_3)/2$, and we also wish to test the null hypothesis $H_0: \mu_2 = \mu_3$ against the alternative hypothesis, $H_1: \mu_2 \neq \mu_3$.

The relevant *Pythagorean breakup* is

$$\begin{aligned} \|\mathbf{y}\|^2 &= (\mathbf{y}\cdot\mathbf{U}_1)^2 + (\mathbf{y}\cdot\mathbf{U}_2)^2 + (\mathbf{y}\cdot\mathbf{U}_3)^2 + (\mathbf{y}\cdot\mathbf{U}_4)^2 + (\mathbf{y}\cdot\mathbf{U}_5)^2 + (\mathbf{y}\cdot\mathbf{U}_6)^2 \\ 13828 &= 11440.7 + 1776.3 + 49 + 512 + 0 + 50 \end{aligned}$$

as illustrated in Figure 4.4.

To test the first hypothesis we simply check whether the squared projection length $(\mathbf{y}\cdot\mathbf{U}_2)^2$ is similar to, or considerably greater than, the average of the squared projection lengths $(\mathbf{y}\cdot\mathbf{U}_4)^2$, $(\mathbf{y}\cdot\mathbf{U}_5)^2$ and $(\mathbf{y}\cdot\mathbf{U}_6)^2$. The resulting test statistic is

$$F' = \frac{(\mathbf{y}\cdot\mathbf{U}_2)^2}{[(\mathbf{y}\cdot\mathbf{U}_4)^2 + (\mathbf{y}\cdot\mathbf{U}_5)^2 + (\mathbf{y}\cdot\mathbf{U}_6)^2]/3} = \frac{1776.3}{[512 + 0 + 50]/3} = \frac{1776.3}{187.3} = 9.48$$

Is this value large or small? Now if $\mu_1 = (\mu_2 + \mu_3)/2$ the projection lengths $\mathbf{y}\cdot\mathbf{U}_2$, $\mathbf{y}\cdot\mathbf{U}_4$, $\mathbf{y}\cdot\mathbf{U}_5$ and $\mathbf{y}\cdot\mathbf{U}_6$ all come from an $N(0, \sigma^2)$ distribution, so the test statistic $(\mathbf{y}\cdot\mathbf{U}_2)^2/\{[(\mathbf{y}\cdot\mathbf{U}_4)^2 + (\mathbf{y}\cdot\mathbf{U}_5)^2 + (\mathbf{y}\cdot\mathbf{U}_6)^2]/3\}$ comes from the $F_{1,3}$ distribution defined in Appendix B. Hence to see whether our observed value of 9.48 is large or small, we compare it with the percentiles of the $F_{1,3}$ distribution as given in Table T.2. The 90, 95 and 99 percentiles are 5.54, 10.13 and 34.12 respectively, so our observed value of 9.48 is large when compared with the 90 percentile (or 10% critical value). That is, in our miniature study we have found some (weak) evidence to suggest that the populations of adults and babies differ in terms of their mean selenium levels.

Similarly, to test the second hypothesis we compare the squared projection length $(\mathbf{y}\cdot\mathbf{U}_3)^2$ with the average of the squared projection lengths $(\mathbf{y}\cdot\mathbf{U}_4)^2$, $(\mathbf{y}\cdot\mathbf{U}_5)^2$ and $(\mathbf{y}\cdot\mathbf{U}_6)^2$. The resulting test statistic is

$$F'' = \frac{(\mathbf{y}\cdot\mathbf{U}_3)^2}{[(\mathbf{y}\cdot\mathbf{U}_4)^2 + (\mathbf{y}\cdot\mathbf{U}_5)^2 + (\mathbf{y}\cdot\mathbf{U}_6)^2]/3} = \frac{49}{[512 + 0 + 50]/3} = \frac{49}{187.3} = 0.26$$

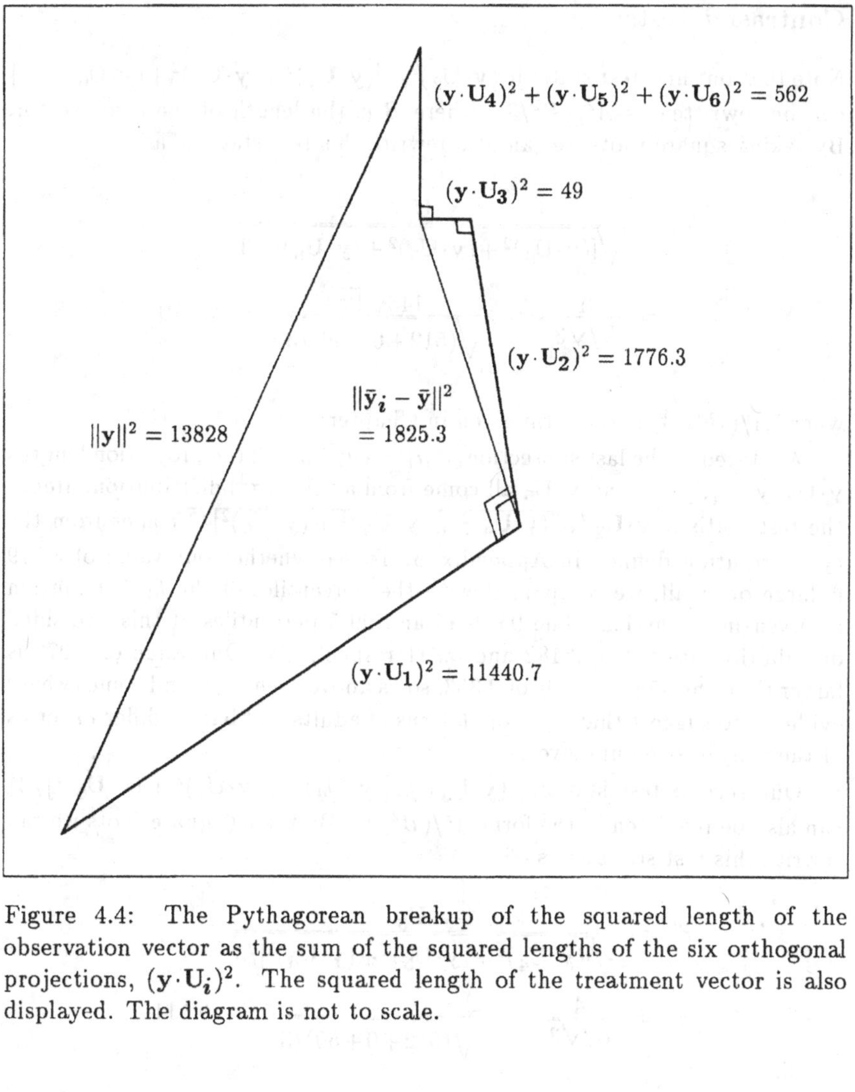

Figure 4.4: The Pythagorean breakup of the squared length of the observation vector as the sum of the squared lengths of the six orthogonal projections, $(\mathbf{y}\cdot\mathbf{U}_i)^2$. The squared length of the treatment vector is also displayed. The diagram is not to scale.

Is this value large or small? Now if $\mu_2 = \mu_3$ the projection lengths $\mathbf{y}\cdot\mathbf{U}_3$, $\mathbf{y}\cdot\mathbf{U}_4$, $\mathbf{y}\cdot\mathbf{U}_5$ and $\mathbf{y}\cdot\mathbf{U}_6$ all come from an $N(0,\sigma^2)$ distribution, so the test statistic $(\mathbf{y}\cdot\mathbf{U}_3)^2/\{[(\mathbf{y}\cdot\mathbf{U}_4)^2 + (\mathbf{y}\cdot\mathbf{U}_5)^2 + (\mathbf{y}\cdot\mathbf{U}_6)^2]/3\}$ comes from the $F_{1,3}$ distribution defined in Appendix B. In this case our observed value of 0.26 is small, even in comparison with the 90 percentile. That is, in our miniature study we have found no evidence to suggest that the populations of full-term and premature babies differ in terms of their mean selenium levels.

Contrast t Tests

Note that our first test statistic $(\mathbf{y}\cdot\mathbf{U_2})^2/\{[(\mathbf{y}\cdot\mathbf{U_4})^2+(\mathbf{y}\cdot\mathbf{U_5})^2+(\mathbf{y}\cdot\mathbf{U_6})^2]/3\}$ can be rewritten as $A^2/(B^2/3)$, where B is the length of the error vector. By taking square roots we can also rewrite this test statistic as

$$t' = \frac{\mathbf{y}\cdot\mathbf{U_2}}{\sqrt{[(\mathbf{y}\cdot\mathbf{U_4})^2+(\mathbf{y}\cdot\mathbf{U_5})^2+(\mathbf{y}\cdot\mathbf{U_6})^2]/3}}$$

$$= \frac{A}{B/\sqrt{3}} = \frac{146\sqrt{12}}{\sqrt{(512+0+50)/3}} = 3.079$$

where $A/(B/\sqrt{3})$ is the form given in Chapter 1.

As stated in the last subsection, if $\mu_1 = (\mu_2+\mu_3)/2$ the projection lengths $\mathbf{y}\cdot\mathbf{U_2}$, $\mathbf{y}\cdot\mathbf{U_4}$, $\mathbf{y}\cdot\mathbf{U_5}$ and $\mathbf{y}\cdot\mathbf{U_6}$ all come from an $N(0,\sigma^2)$ distribution. Hence the test statistic $\mathbf{y}\cdot\mathbf{U_2}/\sqrt{[(\mathbf{y}\cdot\mathbf{U_4})^2+(\mathbf{y}\cdot\mathbf{U_5})^2+(\mathbf{y}\cdot\mathbf{U_6})^2]/3}$ comes from the t_3 distribution defined in Appendix B. To see whether our value of 3.079 is large or small, we compare it with the percentiles of the t_3 distribution as given in Table T.2. The 95, 97.5 and 99.5 percentiles of this two-sided distribution are 2.353, 3.182 and 5.841 respectively. Our value of 3.079 is larger than the 95 percentile of 2.353, so as above we have found some (weak) evidence to suggest that our populations of adults and babies differ in terms of their mean selenium levels.

Our second test statistic $(\mathbf{y}\cdot\mathbf{U_3})^2/\{[(\mathbf{y}\cdot\mathbf{U_4})^2+(\mathbf{y}\cdot\mathbf{U_5})^2+(\mathbf{y}\cdot\mathbf{U_6})^2]/3\}$ can also be rewritten in the form $A^2/(B^2/3)$. By taking square roots we can rewrite this test statistic as

$$t'' = \frac{\mathbf{y}\cdot\mathbf{U_3}}{\sqrt{[(\mathbf{y}\cdot\mathbf{U_4})^2+(\mathbf{y}\cdot\mathbf{U_5})^2+(\mathbf{y}\cdot\mathbf{U_6})^2]/3}}$$

$$= \frac{A}{B/\sqrt{3}} = \frac{7}{\sqrt{(512+0+50)/3}} = 0.511$$

where $A/(B/\sqrt{3})$ is again in the form given in Chapter 1.

As stated in the last subsection, if $\mu_2 = \mu_3$ the projection lengths $\mathbf{y}\cdot\mathbf{U_3}$, $\mathbf{y}\cdot\mathbf{U_4}$, $\mathbf{y}\cdot\mathbf{U_5}$ and $\mathbf{y}\cdot\mathbf{U_6}$ all come from an $N(0,\sigma^2)$ distribution. Hence the test statistic $\mathbf{y}\cdot\mathbf{U_3}\,/\sqrt{[(\mathbf{y}\cdot\mathbf{U_4})^2+(\mathbf{y}\cdot\mathbf{U_5})^2+(\mathbf{y}\cdot\mathbf{U_6})^2]/3}$ comes from the t_3 distribution. Our observed value of 0.511 is not large since it lies within the 90% "normal range" bounded by the 5 and 95 percentiles of −2.353 and 2.353 respectively. Hence as above we have found no evidence to suggest that the populations of full-term and premature babies differ in terms of their mean selenium levels.

Comments

In the above analysis the direction $\mathbf{U_1}$ is used to estimate the mean of the three population means, μ, the directions $\mathbf{U_2}$ and $\mathbf{U_3}$ are used to estimate the contrasts $[\mu_1 - (\mu_2 + \mu_3)/2]$ and $(\mu_3 - \mu_2)$, and the directions $\mathbf{U_4}$, $\mathbf{U_5}$ and $\mathbf{U_6}$ are used to estimate the population variance σ^2. In the long run the projection length $\mathbf{y}\cdot\mathbf{U_1} = \sqrt{6}\bar{y}$ averages to $\sqrt{6}\mu$, so $\bar{y} = 43.667$ serves as our estimate of μ. Similarly, the projection length $\mathbf{y}\cdot\mathbf{U_2} = (2\bar{y}_1 - \bar{y}_2 - \bar{y}_3)/\sqrt{3}$ averages to $(2\mu_1 - \mu_2 - \mu_3)/\sqrt{3}$, so $(2\bar{y}_1 - \bar{y}_2 - \bar{y}_3) = 2 \times 68 - 28 - 35 = 73$ serves as our estimate of $(2\mu_1 - \mu_2 - \mu_3)$; thus the difference in blood plasma selenium level between adults and babies $[\mu_1 - (\mu_2 + \mu_3)/2]$ is estimated to be $73/2 = 36.5\,\mu$g/L. Likewise, the projection length $\mathbf{y}\cdot\mathbf{U_3} = (\bar{y}_3 - \bar{y}_2)$ averages to $(\mu_3 - \mu_2)$, so $(\bar{y}_3 - \bar{y}_2) = 35 - 28 = 7$ serves as our estimate of $(\mu_3 - \mu_2)$, the difference in selenium level between full-term and premature babies. Lastly, the squared projection lengths $(\mathbf{y}\cdot\mathbf{U_4})^2$, $(\mathbf{y}\cdot\mathbf{U_5})^2$ and $(\mathbf{y}\cdot\mathbf{U_6})^2$ all average to σ^2, so $s^2 = [(\mathbf{y}\cdot\mathbf{U_4})^2 + (\mathbf{y}\cdot\mathbf{U_5})^2 + (\mathbf{y}\cdot\mathbf{U_6})^2]/3 = (512 + 0 + 50)/3 = 187.3$ serves as our estimate of σ^2.

In addition, it can be shown that in the long run the squared projection length $(\mathbf{y}\cdot\mathbf{U_1})^2$ averages to $6\mu^2 + \sigma^2$, while $(\mathbf{y}\cdot\mathbf{U_2})^2$ averages to $(2\mu_1 - \mu_2 - \mu_3)^2/3 + \sigma^2$ and $(\mathbf{y}\cdot\mathbf{U_3})^2$ averages to $(\mu_3 - \mu_2)^2 + \sigma^2$. Hence if $\mu_1 \neq (\mu_2 + \mu_3)/2$ the quantity $(\mathbf{y}\cdot\mathbf{U_2})^2$ is inflated, and our first test statistic $(\mathbf{y}\cdot\mathbf{U_2})^2/s^2$ can be thought of as a potentially inflated estimate of σ^2 divided by an unbiased estimate of σ^2. Similarly, if $\mu_2 \neq \mu_3$ the quantity $(\mathbf{y}\cdot\mathbf{U_3})^2$ is inflated, and our second test statistic $(\mathbf{y}\cdot\mathbf{U_3})^2/s^2$ can be thought of as a potentially inflated estimate of σ^2 divided by an unbiased estimate of σ^2.

4.3 Full Data Set

We now rework the previous section using the full data set, consisting of three samples of size 24, as given in Table 4.1. The corresponding *observation vector* is

$$\mathbf{y} = \begin{bmatrix} 60 \\ \vdots \\ 24 \\ \vdots \\ 31 \\ \vdots \end{bmatrix}$$

The Basic Idea

The basic idea is the same as in §4.2. Four cases can occur. Firstly, if there are no differences in mean selenium levels between adults, full-term babies and premature babies ($\mu_1 = \mu_2 = \mu_3 = \mu$), then our observation vector will be part of a scatter centered on the point $(\mu, \ldots, \mu)$ in 72-dimensional space (c.f. Figure 4.2(a)).

Secondly, if adults and babies differ in their mean selenium levels ($\mu_1 \neq (\mu_2 + \mu_3)/2$), but full-term and premature babies have the same mean selenium levels ($\mu_2 = \mu_3$), then our observation vector will be part of a scatter centered on the point $(\mu_1, \ldots, \mu_2, \ldots, \mu_2, \ldots)$ in 72-dimensional space (c.f. Figure 4.2(b)). In this case the scatter is moved away from the equiangular line in the direction $[2, \ldots, -1, \ldots, -1, \ldots]^T$ in 72-dimensional space (c.f. details in §4.2).

Thirdly, if adults and babies have the same mean selenium level ($\mu_1 = (\mu_2+\mu_3)/2$), but full-term and premature babies differ in mean selenium level ($\mu_2 \neq \mu_3$), then our observation vector will be part of a scatter centered on the point $((\mu_2 + \mu_3)/2, \ldots, \mu_2, \ldots, \mu_3, \ldots)$ (c.f. Figure 4.2(c)). In this case the scatter is moved away from the equiangular line in the direction $[0, \ldots, -1, \ldots, 1, \ldots]^T$ (c.f. details in §4.2).

Lastly, if adults and babies differ in mean selenium level ($\mu_1 \neq (\mu_2 + \mu_3)/2$), and full-term and premature babies also differ in mean selenium level ($\mu_2 \neq \mu_3$), then our observation vector will be part of a scatter centered on the point $(\mu_1, \ldots, \mu_2, \ldots, \mu_3, \ldots)$ (c.f. Figure 4.2(d)). In this case the scatter is moved away from the equiangular line in a direction which has components in both of the directions $[2, \ldots, -1, \ldots, -1, \ldots]^T$ and $[0, \ldots, -1, \ldots, 1, \ldots]^T$.

An Appropriate Coordinate System

For our purposes, then, an appropriate set of coordinate axes for 72-space is

$$
\begin{array}{ccccccccc}
\mathbf{U}_1 & \mathbf{U}_2 & \mathbf{U}_3 & \mathbf{U}_4 & \cdots & \mathbf{U}_{27} & \cdots & \mathbf{U}_{50} & \cdots \\
\frac{\begin{bmatrix} 1 \\ 1 \\ 1 \\ \vdots \\ 1 \\ 1 \\ 1 \\ \vdots \\ 1 \\ 1 \\ 1 \\ \vdots \end{bmatrix}}{\sqrt{72}} &
\frac{\begin{bmatrix} 2 \\ 2 \\ 2 \\ \vdots \\ -1 \\ -1 \\ -1 \\ \vdots \\ -1 \\ -1 \\ -1 \\ \vdots \end{bmatrix}}{\sqrt{144}} &
\frac{\begin{bmatrix} 0 \\ 0 \\ 0 \\ \vdots \\ -1 \\ -1 \\ -1 \\ \vdots \\ 1 \\ 1 \\ 1 \\ \vdots \end{bmatrix}}{\sqrt{48}} &
\frac{\begin{bmatrix} -1 \\ 1 \\ 0 \\ \vdots \\ 0 \\ 0 \\ 0 \\ \vdots \\ 0 \\ 0 \\ 0 \\ \vdots \end{bmatrix}}{\sqrt{2}} &
\cdots &
\frac{\begin{bmatrix} 0 \\ 0 \\ 0 \\ \vdots \\ -1 \\ 1 \\ 0 \\ \vdots \\ 0 \\ 0 \\ 0 \\ \vdots \end{bmatrix}}{\sqrt{2}} &
\cdots &
\frac{\begin{bmatrix} 0 \\ 0 \\ 0 \\ \vdots \\ 0 \\ 0 \\ 0 \\ \vdots \\ -1 \\ 1 \\ 0 \\ \vdots \end{bmatrix}}{\sqrt{2}} &
\cdots
\end{array}
$$

where the unlisted coordinate axes lie in the error space and follow the pattern established on page 51. As in §4.2, $\mathbf{U_1}$ is the direction associated with the overall mean μ, $\mathbf{U_2}$ is the direction associated with the comparison of *adults with babies* $[\mu_1 - (\mu_2 + \mu_3)/2]$, and $\mathbf{U_3}$ is the direction associated with the comparison of *full-term with premature babies* $(\mu_3 - \mu_2)$.

The Orthogonal Decomposition

We now project our observation vector $\mathbf{y} = [60, \ldots, 24, \ldots, 31, \ldots]^T$ onto each of our coordinate axes. The resulting orthogonal decomposition is

$$\mathbf{y} = (\mathbf{y}\cdot\mathbf{U_1})\mathbf{U_1} + (\mathbf{y}\cdot\mathbf{U_2})\mathbf{U_2} + (\mathbf{y}\cdot\mathbf{U_3})\mathbf{U_3} + (\mathbf{y}\cdot\mathbf{U_4})\mathbf{U_4} + \cdots + (\mathbf{y}\cdot\mathbf{U_{72}})\mathbf{U_{72}}$$
$$\mathbf{y} = 378.07\mathbf{U_1} + 161.17\mathbf{U_2} - 28.29\mathbf{U_3} - 21.21\mathbf{U_4} + \cdots + 5.06\mathbf{U_{72}}$$

In vectors this is

$$\begin{bmatrix} 60 \\ \vdots \\ 24 \\ \vdots \\ 31 \\ \vdots \end{bmatrix} = \begin{bmatrix} 44.556 \\ \vdots \\ 44.556 \\ \vdots \\ 44.556 \\ \vdots \end{bmatrix} + \begin{bmatrix} 26.861 \\ \vdots \\ -13.431 \\ \vdots \\ -13.431 \\ \vdots \end{bmatrix} + \begin{bmatrix} 0 \\ \vdots \\ 4.083 \\ \vdots \\ -4.083 \\ \vdots \end{bmatrix} + \begin{bmatrix} -11.417 \\ \vdots \\ -11.208 \\ \vdots \\ -3.958 \\ \vdots \end{bmatrix}$$

observation vector   overall mean vector   contrast vectors   error vector

where we have collapsed the last 69 projection vectors into a single "error" vector. If we so desire, this can be simplified further to the fitted model

$$\begin{bmatrix} 60 \\ \vdots \\ 24 \\ \vdots \\ 31 \\ \vdots \end{bmatrix} = \begin{bmatrix} 44.556 \\ \vdots \\ 44.556 \\ \vdots \\ 44.556 \\ \vdots \end{bmatrix} + \begin{bmatrix} 26.861 \\ \vdots \\ -9.347 \\ \vdots \\ -17.514 \\ \vdots \end{bmatrix} + \begin{bmatrix} -11.417 \\ \vdots \\ -11.208 \\ \vdots \\ -3.958 \\ \vdots \end{bmatrix}$$
$$\mathbf{y} = \bar{\mathbf{y}} + (\bar{\mathbf{y}}_i - \bar{\mathbf{y}}) + (\mathbf{y} - \bar{\mathbf{y}}_i)$$

observation vector   overall mean vector   treatment vector   error vector

which in turn can be simplified by adding the overall mean vector to the treatment vector, yielding the vector of treatment means. This allows us to satisfy our first study objective by yielding least squares estimates for μ_1, μ_2 and μ_3 of $\bar{y}_1 = 71.4$ (adults), $\bar{y}_2 = 35.2$ (full-term babies) and $\bar{y}_3 = 27.0$ (premature babies).

To summarize, we have broken our observation vector $\mathbf{y}$ into three orthogonal components, an *overall mean vector*, two *contrast vectors*, and an *error vector*, as shown in Figure 4.5.

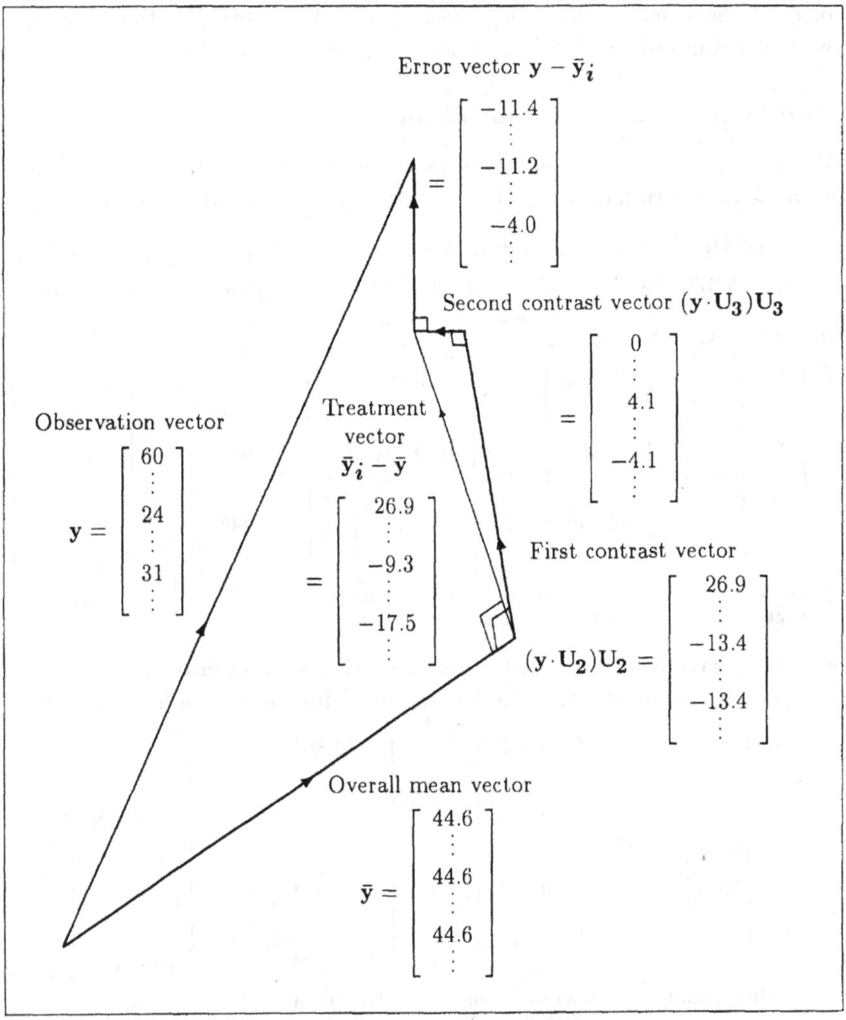

Figure 4.5: Orthogonal decomposition of the observation vector for the case of independent samples of size 24 from three study populations. Vectors have been rounded to one decimal place. Please excuse us for using the same picture as in the $n = 2$ case; in particular, the second contrast vector should have been redrawn to be negative!

Distributions of the Projection Lengths

Proceeding as in §4.2, we find that $\mathbf{y}\cdot\mathbf{U}_1$ comes from an $N(\sqrt{72}\mu, \sigma^2)$ distribution, $\mathbf{y}\cdot\mathbf{U}_2$ comes from an $N\left[\sqrt{24}(2\mu_1 - \mu_2 - \mu_3)/\sqrt{6}, \sigma^2\right]$ distribution, $\mathbf{y}\cdot\mathbf{U}_3$ comes from an $N\left[\sqrt{24}(\mu_3 - \mu_2)/\sqrt{2}, \sigma^2\right]$ distribution, and $\mathbf{y}\cdot\mathbf{U}_4, \ldots, \mathbf{y}\cdot\mathbf{U}_{72}$ all come from an $N(0, \sigma^2)$ distribution.

That is, the distributions of $\mathbf{y}\cdot\mathbf{U}_1$, $\mathbf{y}\cdot\mathbf{U}_2$ and $\mathbf{y}\cdot\mathbf{U}_3$ are centered on potentially nonzero quantities, whereas the distributions of $\mathbf{y}\cdot\mathbf{U}_4, \ldots, \mathbf{y}\cdot\mathbf{U}_{72}$ are always centered on zero.

Testing the Hypotheses

We now investigate our second and third study objectives, which involve testing:

- whether selenium levels differ between adults and babies;
- whether selenium levels differ between full-term and premature babies.

This involves testing whether our two contrasts differ from zero. For each test we check whether the appropriate squared projection length $[(\mathbf{y}\cdot\mathbf{U}_2)^2$ or $(\mathbf{y}\cdot\mathbf{U}_3)^2]$ is similar to, or considerably greater than, the average of the squared projection lengths for the "error" directions, $[(\mathbf{y}\cdot\mathbf{U}_4)^2 + \cdots + (\mathbf{y}\cdot\mathbf{U}_{72})^2]/69$.

The relevant *Pythagorean breakup* is

$$\begin{aligned}
\|\mathbf{y}\|^2 &= (\mathbf{y}\cdot\mathbf{U}_1)^2 + (\mathbf{y}\cdot\mathbf{U}_2)^2 + (\mathbf{y}\cdot\mathbf{U}_3)^2 + (\mathbf{y}\cdot\mathbf{U}_4)^2 + \cdots + (\mathbf{y}\cdot\mathbf{U}_{72})^2 \\
\|\mathbf{y}\|^2 &= \|\bar{\mathbf{y}}\|^2 + (\mathbf{y}\cdot\mathbf{U}_2)^2 + (\mathbf{y}\cdot\mathbf{U}_3)^2 + \|\mathbf{y} - \bar{\mathbf{y}}_i\|^2 \\
182994 &= 142934.2 + 25974.7 + 800.3 + 13284.7
\end{aligned}$$

as illustrated in Figure 4.6.

For testing whether $\mu_1 = (\mu_2 + \mu_3)/2$ the test statistic is

$$F' = \frac{(\mathbf{y}\cdot\mathbf{U}_2)^2}{[(\mathbf{y}\cdot\mathbf{U}_4)^2 + \cdots + (\mathbf{y}\cdot\mathbf{U}_{72})^2]/69} = \frac{(\mathbf{y}\cdot\mathbf{U}_2)^2}{\|\mathbf{y} - \bar{\mathbf{y}}_i\|^2/69} = \frac{25974.7}{13284.7/69} = 134.91$$

If $\mu_1 = (\mu_2 + \mu_3)/2$ this comes from the $F_{1,69}$ distribution. Since the observed F value (134.91) is larger than the 99 percentile of the $F_{1,69}$ distribution (7.02), we reject the hypothesis that $\mu_1 = (\mu_2 + \mu_3)/2$, and conclude that there is very strong evidence to suggest that the mean blood plasma selenium level was higher for adults than for babies (in 1989–1990). Note that we are very much more certain of our conclusion than we were in §4.2. This illustrates the increase in the power of the test associated with increasing the sample size from 2 to 24.

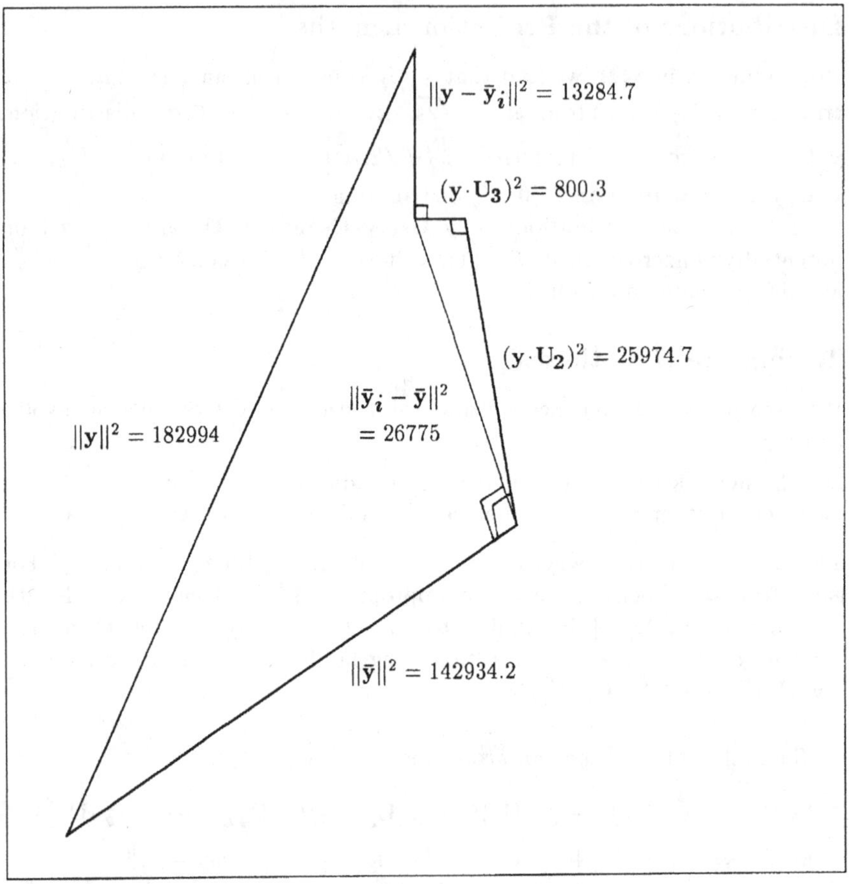

Figure 4.6: The Pythagorean breakup of the squared length of the observation vector for the case of three independent samples.

For testing whether $\mu_2 = \mu_3$ the test statistic is

$$F'' = \frac{(\mathbf{y} \cdot \mathbf{U_3})^2}{[(\mathbf{y} \cdot \mathbf{U_4})^2 + \cdots + (\mathbf{y} \cdot \mathbf{U_{72}})^2]/69} = \frac{(\mathbf{y} \cdot \mathbf{U_3})^2}{||\mathbf{y} - \bar{\mathbf{y}}_i||^2/69} = \frac{800.3}{13284.7/69} = 4.16$$

If $\mu_2 = \mu_3$ this comes from the $F_{1,69}$ distribution. Since the observed F value (4.16) is larger than the 95 percentile of the $F_{1,69}$ distribution (3.98), we reject the hypothesis that $\mu_2 = \mu_3$, and conclude that there is evidence to suggest that the mean blood plasma selenium level was higher for full-term babies than for premature neonatal babies (in 1989–1990). Note that in §4.2 we had no evidence of such a difference, and in fact our sample means were in the reverse order. This again illustrates the increase in the power of the test associated with increasing the sample size from 2 to 24.

Analysis of Variance Table

The calculations of the last subsection are traditionally set out in an "analysis of variance" table. This table is based on a rearrangement of our orthogonal decompositions, obtained by taking the overall mean vector over to the left-hand sides of the equations as follows:

$$\begin{array}{rcl} \mathbf{y} - (\mathbf{y}\cdot\mathbf{U}_1)\mathbf{U}_1 &=& (\mathbf{y}\cdot\mathbf{U}_2)\mathbf{U}_2 + (\mathbf{y}\cdot\mathbf{U}_3)\mathbf{U}_3 \;+\; (\mathbf{y} - \bar{\mathbf{y}}_i) \\ \mathbf{y} - \bar{\mathbf{y}} &=& (\bar{\mathbf{y}}_i - \bar{\mathbf{y}}) \;+\; (\mathbf{y} - \bar{\mathbf{y}}_i) \end{array}$$

These rearranged decompositions lead to the following two rearrangements of our Pythagorean breakups:

$$\begin{array}{rcl} \|\mathbf{y}\|^2 - (\mathbf{y}\cdot\mathbf{U}_1)^2 &=& (\mathbf{y}\cdot\mathbf{U}_2)^2 + (\mathbf{y}\cdot\mathbf{U}_3)^2 + \|\mathbf{y} - \bar{\mathbf{y}}_i\|^2 \\ 182994 - 142934.2 &=& 25974.7 + 800.3 \;+\; 13284.7 \end{array}$$

and

$$\begin{array}{rcl} \|\mathbf{y} - \bar{\mathbf{y}}\|^2 &=& \|\bar{\mathbf{y}}_i - \bar{\mathbf{y}}\|^2 + \|\mathbf{y} - \bar{\mathbf{y}}_i\|^2 \\ 40059.8 &=& 26775 \;+\; 13284.7 \end{array}$$

These two Pythagorean breakups are summarized in the third column of the analysis of variance table, as shown in Table 4.3. This column is traditionally headed "Sums of Squares" (SS), in reference to the squaring and summing involved in the calculation of these sums of squared projection lengths. The other columns of the table are as follows. The first column is traditionally headed "Source of Variation." The second column is headed "degrees of freedom" (df), which means the number of orthogonal directions onto which the observation vector has been projected (or equivalently, the dimension of the appropriate subspace). The fourth column is headed "Mean Square" (MS), since it contains the mean of the squared projection lengths (the sum of the squared projection lengths divided by the number of them). Lastly, the fifth column is headed "F" since it contains our two F test statistics.

Source of Variation	df	SS	MS	F
Treatments	2	26775		
Adults − babies	1	25974.7	25974.7	134.91
Full-term − premature	1	800.3	800.3	4.16
Error	69	13284.7	192.53	
Total	71	40059.8		

Table 4.3: Analysis of variance table.

Contrast t Tests

Note that our first test statistic, $(\mathbf{y}\cdot\mathbf{U_2})^2/\{[(\mathbf{y}\cdot\mathbf{U_4})^2+\cdots+(\mathbf{y}\cdot\mathbf{U_{72}})^2]/69\} = (\mathbf{y}\cdot\mathbf{U_2})^2/(\|\mathbf{y}-\bar{\mathbf{y}}_i\|^2/69)$, can be transformed by taking square roots to

$$\frac{\mathbf{y}\cdot\mathbf{U_2}}{\sqrt{[(\mathbf{y}\cdot\mathbf{U_4})^2+\cdots+(\mathbf{y}\cdot\mathbf{U_{72}})^2]/69}} = \frac{\mathbf{y}\cdot\mathbf{U_2}}{\|\mathbf{y}-\bar{\mathbf{y}}_i\|/\sqrt{69}} = \frac{1934\sqrt{144}}{\sqrt{13284.7/69}} = 11.615$$

If $\mu_1 = (\mu_2+\mu_3)/2$ this comes from the t_{69} distribution. As previously, the size of our observed value constitutes very strong evidence against the notion that the adults and babies had the same mean blood plasma selenium levels.

Note that the algebraic expression for our calculated t value is

$$t' = \frac{\mathbf{y}\cdot\mathbf{U_2}}{\|\mathbf{y}-\bar{\mathbf{y}}_i\|/\sqrt{3(n-1)}} = \frac{\sqrt{n}(2\bar{y}_1-\bar{y}_2-\bar{y}_3)/\sqrt{(-2)^2+1^2+1^2}}{\sqrt{\sum_{i=1}^{3}\sum_{j=1}^{n}(y_{ij}-\bar{y}_i)^2/[3(n-1)]}}$$

$$= \frac{2\bar{y}_1-\bar{y}_2-\bar{y}_3}{\sqrt{6}s/\sqrt{n}}$$

where s^2 is the pooled variance estimate and where $n = 24$ in this example.

Our second test statistic, $(\mathbf{y}\cdot\mathbf{U_3})^2/\{[(\mathbf{y}\cdot\mathbf{U_4})^2+\cdots+(\mathbf{y}\cdot\mathbf{U_{72}})^2]/69\} = (\mathbf{y}\cdot\mathbf{U_3})^2/(\|\mathbf{y}-\bar{\mathbf{y}}_i\|^2/69)$, can be transformed by taking square roots to

$$\frac{\mathbf{y}\cdot\mathbf{U_3}}{\sqrt{[(\mathbf{y}\cdot\mathbf{U_4})^2+\cdots+(\mathbf{y}\cdot\mathbf{U_{72}})^2]/69}} = \frac{\mathbf{y}\cdot\mathbf{U_3}}{\|\mathbf{y}-\bar{\mathbf{y}}_i\|/\sqrt{69}} = \frac{-196\sqrt{48}}{\sqrt{13284.7/69}} = -2.039$$

If $\mu_2 = \mu_3$ this comes from the t_{69} distribution. As previously, the size of our observed value constitutes evidence against the notion that premature and full-term babies had the same mean blood plasma selenium levels.

The algebraic expression for our second calculated t value is

$$t'' = \frac{\mathbf{y}\cdot\mathbf{U_3}}{\|\mathbf{y}-\bar{\mathbf{y}}_i\|/\sqrt{3(n-1)}} = \frac{\sqrt{n}(\bar{y}_3-\bar{y}_2)/\sqrt{(-1)^2+1^2}}{\sqrt{\sum_{i=1}^{3}\sum_{j=1}^{n}(y_{ij}-\bar{y}_i)^2/[3(n-1)]}}$$

$$= \frac{\bar{y}_3-\bar{y}_2}{\sqrt{2}s/\sqrt{n}}$$

where s^2 is the pooled variance estimate and where $n = 24$.

Comment

As in §4.2, the "model space" directions $\mathbf{U_1}$, $\mathbf{U_2}$ and $\mathbf{U_3}$ are used to estimate the population means μ_1, μ_2 and μ_3 (or, equivalently, their mean and two contrasts between them), while the "error space" directions $\mathbf{U_3}, \ldots, \mathbf{U_{72}}$ are used to estimate the common population variance σ^2.

Assumption Checking

In the last sentence of §4.1 we commented upon the fact that we have a problem with one of our basic assumptions, that of equality of variance. In fact, the estimated variances for the three populations, using the data in Table 4.1, work out to be $s_1^2 = 431.9$, $s_2^2 = 50.6$ and $s_3^2 = 95.1$. This is in agreement with our expectation that the adult population would have a higher variability than the two baby populations.

In this situation, where the population with the largest sample mean also has the largest variability, a logarithm transformation appears to be appropriate. If our data is reanalyzed following a logarithm transformation of each data value (using any base of logarithm), our two F test statistics become 107.57 and 11.63 respectively. That is, the evidence pointing to a difference between adults and babies is decreased (though it is still overwhelming!), while the evidence pointing to a difference between full-term and premature babies is increased.

Interpretation of Results

In interpreting the results obtained above, the medical researchers had to take into account other information which became available. Two main factors had to be considered. The first was that, unbeknown to the researchers, the blood plasma selenium levels of the study populations were rising during the period of the study (1989–1990) due to an increasing usage of imported, high-selenium wheat made possible by the deregulation of the New Zealand economy (Winterbourn et al. 1992). The percentage rise for adults was estimated to be about a 10–30% rise during the period 1989–1990, using data in Winterbourn et al. (1992). In the current study the three samples presented in Table 4.1 differed in the average time at which they were taken (within the 1989–1990 period), so this may account for some of the apparent differences between study populations.

The second factor was that for the full-term babies the blood samples were taken from the umbilical cord at the time of birth, whereas for the premature babies the samples were taken at various times from 0 to 7 days of age. This is important since a separate study has shown that blood plasma selenium levels drop substantially (by about 13%) during the first week of a baby's life (Sluis et al. 1992). This could account for some or all of the difference in selenium levels between full-term and premature babies.

The net results, considering these two factors, are as follows. There appears to be little doubt that the observed difference in mean selenium level between adults and babies is a real difference, since the difference is too large to be accounted for by any difference in sampling time (during the 1989–1990 period). However, there is reasonable doubt about whether the observed difference between the means for full-term and premature babies is a real effect, since the difference is small enough to be accounted for by the difference in sampling time (at birth or within the first week) between the two baby categories.

4.4 General Case

In §4.3 we dealt with the case of three independent samples of size n from three study populations. The more general case is that of k independent samples of size n from k study populations. In this case the observation vector is of the form

$$\mathbf{y} = [y_{11}, \ldots, y_{21}, \ldots, y_{k1}, \ldots]^T$$

where y_{ij} refers to the jth observation from the ith study population.

When we had three study populations we were able to specify two orthogonal contrasts between the population means. With k study populations we are able to specify $(k-1)$ orthogonal contrasts between the population means. These contrasts are chosen to correspond to comparisons of particular interest, and are very much tied to the nature of the populations or experimental treatments. Contrasts generally fall into four categories:

1. Class comparisons (as in this chapter).
2. Factorial comparisons.
3. Polynomial contrasts.
4. Pairwise comparisons.

The creation of a meaningful set of orthogonal contrasts (of one or more of the above categories) is an art which we have no space to teach; however, the interested reader can refer to Chapters 7–11 of Saville and Wood (1991) for a full account.

In general, the contrast $c = c_1\mu_1 + \cdots + c_k\mu_k$ is associated with the direction

$$\mathbf{U}_c = \frac{1}{\sqrt{n\sum_{i=1}^{k} c_i^2}} \begin{bmatrix} c_1 \\ \vdots \\ c_2 \\ \vdots \\ c_k \\ \vdots \end{bmatrix}$$

For example, in §4.2 the contrast $(2\mu_1 - \mu_2 - \mu_3)$ was associated with the direction $[2, 2, -1, -1, -1, -1]^T/\sqrt{12}$.

If a complete set of $(k-1)$ orthogonal contrasts is specified, we have $(k-1)$ associated orthogonal directions $\mathbf{U}_2, \ldots, \mathbf{U}_k$ which form the beginnings of an appropriate orthogonal coordinate system for kn-space. The other axes consist of $\mathbf{U}_1$, the unit vector in the equiangular direction, and $k(n-1)$ unit vectors which span the error space, $\mathbf{U}_{k+1}, \ldots, \mathbf{U}_{kn}$. The latter follow the pattern established on page 51 of §3.3.

The associated orthogonal decomposition of the observation vector $\mathbf{y}$ is

$$\begin{aligned}\mathbf{y} &= (\mathbf{y}\cdot\mathbf{U}_1)\mathbf{U}_1 + (\mathbf{y}\cdot\mathbf{U}_2)\mathbf{U}_2 + \cdots + (\mathbf{y}\cdot\mathbf{U}_k)\mathbf{U}_k + (\mathbf{y}\cdot\mathbf{U}_{k+1})\mathbf{U}_{k+1} + \cdots + (\mathbf{y}\cdot\mathbf{U}_{kn})\mathbf{U}_{kn} \\ &= \underset{\text{mean vector}}{\bar{\mathbf{y}}} + \underset{\text{contrast vectors}}{(\mathbf{y}\cdot\mathbf{U}_2)\mathbf{U}_2 + \cdots + (\mathbf{y}\cdot\mathbf{U}_k)\mathbf{U}_k} + \underset{\text{error vector}}{(\mathbf{y} - \bar{\mathbf{y}}_i)}\end{aligned}$$

as shown in Figure 4.7.

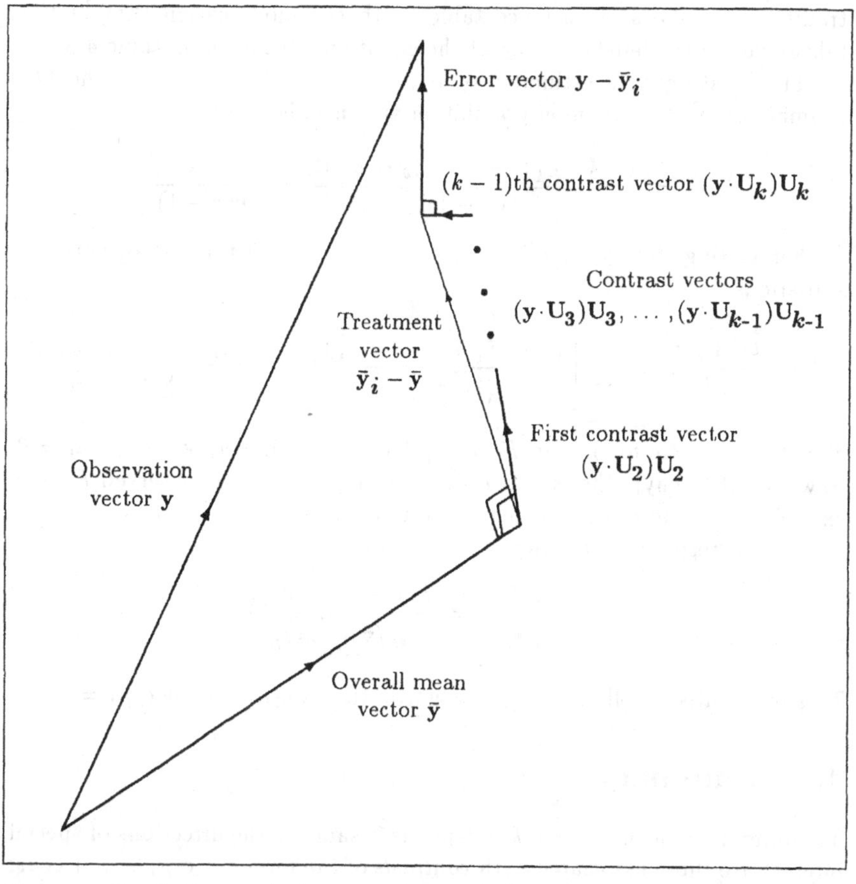

Figure 4.7: Orthogonal decomposition of the observation vector for the case of k independent samples of size n. Try to imagine a further $(k-3)$ dimensions where we have put the dots!

This orthogonal decomposition can also be rewritten as

$$\begin{aligned}\mathbf{y} &= \bar{\mathbf{y}} + (\bar{\mathbf{y}}_i - \bar{\mathbf{y}}) + (\mathbf{y} - \bar{\mathbf{y}}_i) \\ &= \text{mean vector} + \text{treatment vector} + \text{error vector}\end{aligned}$$

as also shown in Figure 4.7.

The corresponding Pythagorean breakups are

$$\begin{aligned}
\|\mathbf{y}\|^2 &= (\mathbf{y}\cdot\mathbf{U}_1)^2 + (\mathbf{y}\cdot\mathbf{U}_2)^2 + \cdots + (\mathbf{y}\cdot\mathbf{U}_k)^2 + (\mathbf{y}\cdot\mathbf{U}_{k+1})^2 + \cdots + (\mathbf{y}\cdot\mathbf{U}_{kn})^2 \\
&= \|\bar{\mathbf{y}}\|^2 + (\mathbf{y}\cdot\mathbf{U}_2)^2 + \cdots + (\mathbf{y}\cdot\mathbf{U}_k)^2 + \|\mathbf{y} - \bar{\mathbf{y}}_i\|^2 \\
&= \|\bar{\mathbf{y}}\|^2 + \|\bar{\mathbf{y}}_i - \bar{\mathbf{y}}\|^2 + \|\mathbf{y} - \bar{\mathbf{y}}_i\|^2
\end{aligned}$$

As an aside, the last breakups form the sums of squares column of the traditional analysis of variance table (with the modification that $\|\bar{\mathbf{y}}\|^2$ is taken over to the left-hand side of the equation) as shown in Table 4.3.

The least squares estimates of $\mu_1, \ldots, \mu_k$ are $\bar{y}_1, \ldots, \bar{y}_k$ and the best estimate of σ^2, the common population variance, is

$$s^2 = \frac{(\mathbf{y}\cdot\mathbf{U}_{k+1})^2 + \cdots + (\mathbf{y}\cdot\mathbf{U}_{kn})^2}{k(n-1)} = \frac{\|\mathbf{y} - \bar{\mathbf{y}}_i\|^2}{k(n-1)}$$

For testing the hypothesis $c_1\mu_1 + \cdots + c_k\mu_k = 0$ the appropriate test statistic is

$$F = \frac{(\mathbf{y}\cdot\mathbf{U}_c)^2}{s^2} = \frac{1}{s^2}\left[\frac{\sqrt{n}(c_1\bar{y}_1 + \cdots + c_k\bar{y}_k)}{\sqrt{\sum_{i=1}^k c_i^2}}\right]^2 = \frac{(c_1\bar{y}_1 + \cdots + c_k\bar{y}_k)^2}{s^2\sum_{i=1}^k c_i^2/n}$$

This test statistic follows an $F_{1,k(n-1)}$ distribution if $c_1\mu_1 + \cdots + c_k\mu_k = 0$, so we reject the hypothesis that $c_1\mu_1 + \cdots + c_k\mu_k = 0$ if the observed F value exceeds the 95 percentile of this distribution.

The corresponding t value is

$$t = \frac{\mathbf{y}\cdot\mathbf{U}_c}{s} = \frac{c_1\bar{y}_1 + \cdots + c_k\bar{y}_k}{s\sqrt{\sum_{i=1}^k c_i^2/n}}$$

This test statistic follows a $t_{k(n-1)}$ distribution if $c_1\mu_1 + \cdots + c_k\mu_k = 0$.

4.5 Summary

To summarize, in the case of k independent samples the directions of special interest are those associated with *contrasts* of the form $c = c_1\mu_1 + \ldots + c_k\mu_k$, where the coefficients $c_1, \ldots, c_k$ sum to zero. If a particular contrast c is nonzero, then the length of the projection of the vector of observations onto the direction $\mathbf{U}_c = [c_1, \cdots, c_2, \ldots, c_k, \ldots]^T/\sqrt{n\sum_{i=1}^k c_i^2}$ will be inflated relative to the projection lengths for the "error" directions. Thus to distinguish between the cases $c = 0$ and $c \neq 0$ we divide the squared projection length for the $\mathbf{U}_c$ direction by the average of the squared projection lengths for the $k(n-1)$ orthogonal directions in the "error space"; if the resulting ratio is small we decide that c could be equal to zero, while if the ratio is large we decide that c could not be equal to zero.

If a complete set of $(k-1)$ orthogonal contrasts is specified, corresponding to the directions $\mathbf{U}_2, \ldots, \mathbf{U}_k$, then the orthogonal decomposition of the observation vector $\mathbf{y}$ is

$$\begin{aligned}\mathbf{y} &= (\mathbf{y}\cdot\mathbf{U}_1)\mathbf{U}_1 + (\mathbf{y}\cdot\mathbf{U}_2)\mathbf{U}_2 + \cdots + (\mathbf{y}\cdot\mathbf{U}_k)\mathbf{U}_k + (\mathbf{y}\cdot\mathbf{U}_{k+1})\mathbf{U}_{k+1} + \cdots + (\mathbf{y}\cdot\mathbf{U}_{kn})\mathbf{U}_{kn} \\ &= \underset{\text{mean vector}}{\bar{\mathbf{y}}} + \underset{\text{contrast vectors}}{(\mathbf{y}\cdot\mathbf{U}_2)\mathbf{U}_2 + \cdots + (\mathbf{y}\cdot\mathbf{U}_k)\mathbf{U}_k} + \underset{\text{error vector}}{(\mathbf{y} - \bar{\mathbf{y}}_i)}\end{aligned}$$

as shown in Figure 4.7.

The corresponding Pythagorean breakups are

$$\begin{aligned}\|\mathbf{y}\|^2 &= (\mathbf{y}\cdot\mathbf{U}_1)^2 + (\mathbf{y}\cdot\mathbf{U}_2)^2 + \cdots + (\mathbf{y}\cdot\mathbf{U}_k)^2 + (\mathbf{y}\cdot\mathbf{U}_{k+1})^2 + \cdots + (\mathbf{y}\cdot\mathbf{U}_{kn})^2 \\ &= \|\bar{\mathbf{y}}\|^2 + (\mathbf{y}\cdot\mathbf{U}_2)^2 + \cdots + (\mathbf{y}\cdot\mathbf{U}_k)^2 + \|\mathbf{y} - \bar{\mathbf{y}}_i\|^2\end{aligned}$$

For testing the hypothesis $c = c_1\mu_1 + \cdots + c_k\mu_k = 0$ the test statistic is

$$F = \frac{(\mathbf{y}\cdot\mathbf{U}_c)^2}{\|\mathbf{y} - \bar{\mathbf{y}}_i\|^2/[k(n-1)]} = \frac{(\mathbf{y}\cdot\mathbf{U}_c)^2}{s^2} = \frac{(c_1\bar{y}_1 + \cdots + c_k\bar{y}_k)^2}{s^2\sum_{i=1}^k c_i^2/n}$$

This test statistic follows an $F_{1,k(n-1)}$ distribution if $c_1\mu_1 + \cdots + c_k\mu_k = 0$.

The equivalent t test is

$$t = \frac{\mathbf{y}\cdot\mathbf{U}_c}{s} = \frac{c_1\bar{y}_1 + \cdots + c_k\bar{y}_k}{s\sqrt{\sum_{i=1}^k c_i^2/n}}$$

This test statistic follows a $t_{k(n-1)}$ distribution if $c_1\mu_1 + \cdots + c_k\mu_k = 0$.

For an analysis of the examples in this chapter using the angle θ as the test statistic, the reader is referred to Appendix D. For computer methods refer to Appendix C. For the estimation of the 95% confidence interval for a contrast $c_1\mu_1 + \cdots + c_k\mu_k$ refer to pp. 164, 173, 174, 175, 538, 539 and 540 of Saville and Wood (1991).

For further examples the reader can refer to the exercises at the end of this chapter, or to Chapters 7–11 of Saville and Wood (1991).

For the unequal replications case the reader is referred to Appendix B of Saville and Wood (1991).

Class Exercise

In this class exercise we shall simulate an experiment in which six lambs are allocated in a completely random manner to three experimental treatments. The experiment is designed to determine:

(A) Whether it is necessary to drench lambs for worm control.

(B) Whether a single drench at 3 months of age is adequate, compared with an additional drench 1 month later.

The experimental treatments are as follows:

(1) Control: undrenched lambs.

(2) One drench: lambs drenched at 3 months of age.

(3) Two drenches: lambs drenched at 3 and 4 months of age.

The success of the drench program is judged by the weight gain of the lambs from 3 months of age to 6 months of age. During this 3 month experimental period the reader can imagine the six lambs living in a single group and feeding on identical pasture. Any possible recontamination of the drenched lambs by the undrenched lambs will be ignored.

(a) For the class exercise each student is asked to use the random numbers in Table T.1 to select two lambs from each of the populations in Table 4.4. The six corresponding weight gains form the observation vector $\mathbf{y}$.

These six observations will be treated as independent samples of size two from three normally distributed populations with means μ_1, μ_2 and μ_3 and a common variance σ^2. The questions of interest can be phrased in terms of independent hypothesis tests as follows:

(A) Is $\mu_1 = \frac{\mu_2 + \mu_3}{2}$? (Or, is $2\mu_1 - \mu_2 - \mu_3 = 0$?)

(B) Is $\mu_2 = \mu_3$? (Or, is $\mu_2 - \mu_3 = 0$?)

An appropriate coordinate system for 6-space is

$$\mathbf{U}_1 = \frac{\begin{bmatrix}1\\1\\1\\1\\1\\1\end{bmatrix}}{\sqrt{6}} \quad \mathbf{U}_2 = \frac{\begin{bmatrix}2\\2\\-1\\-1\\-1\\-1\end{bmatrix}}{\sqrt{12}} \quad \mathbf{U}_3 = \frac{\begin{bmatrix}0\\0\\1\\1\\-1\\-1\end{bmatrix}}{\sqrt{4}} \quad \mathbf{U}_4 = \frac{\begin{bmatrix}1\\-1\\0\\0\\0\\0\end{bmatrix}}{\sqrt{2}} \quad \mathbf{U}_5 = \frac{\begin{bmatrix}0\\0\\1\\-1\\0\\0\end{bmatrix}}{\sqrt{2}} \quad \mathbf{U}_6 = \frac{\begin{bmatrix}0\\0\\0\\0\\1\\-1\end{bmatrix}}{\sqrt{2}}$$

Here $\mathbf{U}_1$ spans the overall mean space, $\mathbf{U}_2$ and $\mathbf{U}_3$ are the directions associated with the contrasts $(2\mu_1 - \mu_2 - \mu_3)$ and $(\mu_2 - \mu_3)$, and $\mathbf{U}_4$, $\mathbf{U}_5$ and $\mathbf{U}_6$ span the error space.

(b) Each class member is now asked to calculate the scalars $\mathbf{y}\cdot\mathbf{U}_1$, $\mathbf{y}\cdot\mathbf{U}_2$, $\mathbf{y}\cdot\mathbf{U}_3$, $\mathbf{y}\cdot\mathbf{U}_4$, $\mathbf{y}\cdot\mathbf{U}_5$ and $\mathbf{y}\cdot\mathbf{U}_6$.

The teacher is then asked to plot a histogram of the class results for $\mathbf{y}\cdot\mathbf{U}_1$. This will approximate the distribution of the projection length $\mathbf{y}\cdot\mathbf{U}_1$. What is the theoretical mean of this distribution if you assume in god-like manner that $\mu_1 = 10$, $\mu_2 = 15$ and $\mu_3 = 20$? Does your theoretical answer match with the histogram?

Population one: Control

Lamb no.	Wt. gain	Lamb no.	Wt. gain	Lamb no.	Wt. gain	Lamb no.	Wt. gain	Lamb no.	Wt. gain
1	11	11	7	21	12	31	10	41	10
2	10	12	9	22	8	32	10	42	12
3	8	13	9	23	9	33	8	43	9
4	11	14	7	24	8	34	11	44	10
5	11	15	9	25	11	35	11	45	11
6	10	16	10	26	9	36	12	46	7
7	12	17	7	27	9	37	10	47	9
8	12	18	11	28	10	38	10	48	10
9	12	19	10	29	7	39	9	49	10
10	9	20	10	30	9	40	9	50	17

Population two: One drench

1	18	11	15	21	15	31	17	41	14
2	15	12	14	22	15	32	17	42	16
3	10	13	16	23	16	33	16	43	17
4	16	14	12	24	13	34	14	44	15
5	16	15	18	25	13	35	14	45	14
6	15	16	14	26	17	36	16	46	18
7	14	17	17	27	15	37	16	47	14
8	12	18	12	28	16	38	15	48	14
9	15	19	17	29	13	39	14	49	14
10	15	20	14	30	11	40	16	50	15

Population three: Two drenches

1	20	11	22	21	19	31	23	41	17
2	19	12	18	22	19	32	20	42	21
3	18	13	15	23	21	33	20	43	21
4	18	14	27	24	23	34	24	44	19
5	22	15	18	25	21	35	20	45	21
6	19	16	23	26	21	36	15	46	19
7	18	17	17	27	20	37	20	47	17
8	18	18	18	28	20	38	15	48	17
9	19	19	25	29	21	39	23	49	20
10	18	20	20	30	21	40	20	50	23

Table 4.4: Populations of lamb weight gains, in kilograms, for simulation exercise.

The teacher is also asked to plot five more histograms, on the same axes, one for each of $\mathbf{y}\cdot\mathbf{U}_2$, $\mathbf{y}\cdot\mathbf{U}_3$, $\mathbf{y}\cdot\mathbf{U}_4$, $\mathbf{y}\cdot\mathbf{U}_5$ and $\mathbf{y}\cdot\mathbf{U}_6$. Again, what are the theoretical means for these histograms?

(c) The above histograms all have a true variance of σ^2, assuming that $\sigma_1^2 = \sigma_2^2 = \sigma_3^2$. Each class member is now asked to estimate σ^2 by substituting their own values into the formula

$$s^2 = \frac{(\mathbf{y}\cdot\mathbf{U}_4)^2 + (\mathbf{y}\cdot\mathbf{U}_5)^2 + (\mathbf{y}\cdot\mathbf{U}_6)^2}{3}$$

The teacher is asked to draw a histogram of the class results. How variable are these estimates of σ^2?

(d) Each class member is now asked to test the hypothesis $\mu_1 = (\mu_2 + \mu_3)/2$ by calculating the test statistic $F = (\mathbf{y}\cdot\mathbf{U}_2)^2/s^2$. If the hypothesis is true this comes from the $F_{1,3}$ distribution which has a 95 percentile of 10.13. Reject the hypothesis if your calculated F is greater than 10.13 and accept it if your calculated F is less than 10.13.

The teacher is now asked to draw a histogram of the test statistics F. What percentage of the class rejected the null hypothesis?

(e) Each class member can now repeat this last section for the null hypothesis $\mu_2 = \mu_3$ by calculating the test statistic $F = (\mathbf{y}\cdot\mathbf{U}_3)^2/s^2$.

The teacher is again asked to draw a histogram of the test statistics F. What percentage of the class rejected the null hypothesis? How does this compare with the results from the last section (d)?

(f) If time permits, each class member can calculate the orthogonal decomposition

$$\begin{bmatrix} y_{11} \\ y_{12} \\ y_{21} \\ y_{22} \\ y_{31} \\ y_{32} \end{bmatrix} = \begin{bmatrix} \bar{y} \\ \bar{y} \\ \bar{y} \\ \bar{y} \\ \bar{y} \\ \bar{y} \end{bmatrix} + \begin{bmatrix} \bar{y}_1 - \bar{y} \\ \bar{y}_1 - \bar{y} \\ \bar{y}_2 - \bar{y} \\ \bar{y}_2 - \bar{y} \\ \bar{y}_3 - \bar{y} \\ \bar{y}_3 - \bar{y} \end{bmatrix} + \begin{bmatrix} y_{11} - \bar{y}_1 \\ y_{12} - \bar{y}_1 \\ y_{21} - \bar{y}_2 \\ y_{22} - \bar{y}_2 \\ y_{31} - \bar{y}_3 \\ y_{32} - \bar{y}_3 \end{bmatrix}$$

and confirm that s^2 can also be calculated using the formula

$$s^2 = \frac{\|\mathbf{y} - \bar{\mathbf{y}}_i\|^2}{3} = \frac{(y_{11} - \bar{y}_1) + \cdots + (y_{32} - \bar{y}_3)^2}{3}$$

Exercises

Note that solutions to exercises marked with an * are given in Appendix E.

(4.1)* An experiment was carried out on a property with a blackberry weed problem to see whether blackberry is controlled more effectively by goats or sheep at equivalent stocking rates. A secondary objective was to determine whether some breeds of sheep or goats are more effective at controlling blackberry than other breeds.

Eight fields with similar amounts of blackberry were allocated in a completely random manner to eight mobs of animals, being two replicates of the treatments listed in the table. At yearly intervals an aerial photograph was taken of the trial, and the percentage of blackberry cover assessed for each field. The data which will be analyzed is the reduction in percentage of blackberry cover between the initial photograph and the photograph taken 1 year later. For example, 44% = 76% − 32%, where 76% is the initial cover and 32% the cover after 1 year. These data are given in the table.

Treatments	Reduction in cover	
1. Romney sheep	44	40
2. Merino sheep	42	50
3. Angora goats	51	55
4. Feral goats	58	52

Questions of interest are:

(A) On average, do the two breeds of sheep differ from the two breeds of goat in their effectiveness as blackberry control agents?

(B) Do Romney sheep differ from Merino sheep?

(C) Do angora goats differ from feral goats?

(a) Write down a set of three orthogonal contrasts corresponding to these questions of interest.

(b) Write down the associated unit vectors, $\mathbf{U}_2$, $\mathbf{U}_3$ and $\mathbf{U}_4$.

(c) Confirm that these unit vectors are orthogonal to one another. They will serve as an orthogonal coordinate system for the "treatment space."

(d) Complete the orthogonal coordinate system for 8-space by writing down a coordinate axis, $\mathbf{U}_1$, for the overall mean space and four orthogonal axes, $\mathbf{U}_5$, $\mathbf{U}_6$, $\mathbf{U}_7$ and $\mathbf{U}_8$, for the error space.

(e) Calculate the signed projection lengths $\mathbf{y}\cdot\mathbf{U}_1, \ldots, \mathbf{y}\cdot\mathbf{U}_8$ in the orthogonal decomposition $\mathbf{y} = (\mathbf{y}\cdot\mathbf{U}_1)\mathbf{U}_1 + \cdots + (\mathbf{y}\cdot\mathbf{U}_8)\mathbf{U}_8$.

(f) Write out the corresponding Pythagorean breakup.

(g) Test the hypothesis corresponding to question (A) by calculating the test statistic

$$F = \frac{(\mathbf{y}\cdot\mathbf{U}_2)^2}{[(\mathbf{y}\cdot\mathbf{U}_5)^2 + \cdots + (\mathbf{y}\cdot\mathbf{U}_8)^2]/4}$$

What is your conclusion?

(h) Test hypotheses corresponding to questions (B) and (C) similarly, stating conclusions.

(4.2) An experiment is conducted to compare the honeydew yield of three races of light honey bee and two races of dark honey bee. Fifteen hives, three of each variety, were placed in a single apiary. The table shows the results of the experiment.

Treatment		Honey Yield (in kg)		
Light	Italian	52.6	58.4	49.2
	Caucasian	51.4	56.3	53.1
	Caucasian (new variety)	52.7	53.6	51.2
Dark	German	60.5	58.6	61.3
	African "killer" bee	71.2	76.3	78.4

(a) Write down the contrasts corresponding to:

- light bees versus dark bees;
- Caucasian (both races) versus Italian;
- Caucasian versus new race of Caucasian;
- German versus African "killer" bee.

(b) Write down the associated unit vectors, $\mathbf{U_2}$, $\mathbf{U_3}$, $\mathbf{U_4}$ and $\mathbf{U_5}$.

(c) Are these contrasts mutually orthogonal? Give reasons.

(d) Calculate the contrast sums of squares, $(\mathbf{y}\cdot\mathbf{U_2})^2$, $(\mathbf{y}\cdot\mathbf{U_3})^2$, $(\mathbf{y}\cdot\mathbf{U_4})^2$ and $(\mathbf{y}\cdot\mathbf{U_5})^2$.

(e) Write down the Pythagorean breakup in the form

$$||\mathbf{y}||^2 = (\mathbf{y}\cdot\mathbf{U_1})^2 + \cdots + (\mathbf{y}\cdot\mathbf{U_5})^2 + ||\text{error vector}||^2$$

(f) Test the four hypotheses of the form H_0: contrast $= 0$ which correspond to the contrasts you wrote down in (a). What are your conclusions?

(4.3) A survey of household cats was carried out to determine:

(i) whether, on average, male cats differ from females in weight;

(ii) whether, on average, Siamese cats differ in weight from ordinary cats;

(iii) whether the sex weight difference is the same for both Siamese and ordinary cats.

The local veterinary association was approached for lists of all known Siamese and ordinary fully grown cats in the area. From each of these lists three males and three females were chosen at random. The populations of study and the resulting cat weights were as follows:

Population	Cat weights (kg)		
1. Siamese males	4.1	5.2	3.6
2. Ordinary males	6.3	7.5	8.7
3. Siamese females	3.0	4.7	3.4
4. Ordinary females	5.7	4.7	6.1

(a) Write down the observation vector $\mathbf{y}$.

(b) Rewritten in terms of population means our questions of interest are:

(i) Is $\mu_1 + \mu_2 = \mu_3 + \mu_4$?

(ii) Is $\mu_1 + \mu_3 = \mu_2 + \mu_4$?

(iii) Is $\mu_1 - \mu_3 = \mu_2 - \mu_4$?

Write down a complete set of orthogonal contrasts which corresponds to this set of questions. Also write down the corresponding unit vectors $\mathbf{U}_i$.

(c) Calculate the projection lengths $\mathbf{y}\cdot\mathbf{U}_i$ corresponding to the questions of interest.

(d) Write out the error vector $\mathbf{y} - \bar{\mathbf{y}}_i$ and calculate s^2.

(e) Test the hypotheses of interest. What are your conclusions?

Note: In this study the populations have a 2×2 *factorial* structure, with factors sex (male or female) and type of cat (Siamese or ordinary). For further discussion of this type of design the interested reader can refer to §9.1 and §9.2 of Saville and Wood (1991).

(4.4) An experiment was conducted to compare four crops which can be grown as green manure for improving the fertility and structure of the soil. The site chosen for the experiment was one of the less fertile fields on an experiment station. The experimental treatments listed below were assigned in a completely random manner to twelve field plots.

1. Peas (cultivar Onward) sown at 300 kg/ha.
2. Barley (cultivar Zephyr) sown at 150 kg/ha.
3. Lupins (old standard cultivar) sown at 200 kg/ha.
4. Lupins (new cultivar, Uniharvest) sown at 200 kg/ha.

All plots were sown on the first day of autumn. On the first day of winter, 3 months later, a 2 m by 8 m area was harvested from the center of each plot. Plants were cut off at ground level so that only the above ground portions were included in the harvest samples. These samples were then dried and weighed. The data given in the table are the dry weights of the harvest samples converted to units of t/ha.

Treatment	Dry matter of "tops"			Mean
Peas	4.7	4.9	4.5	4.7
Barley	3.4	3.9	3.2	3.5
Lupins (old)	5.2	5.1	6.2	5.5
Lupins (new)	5.8	5.0	5.1	5.3

Three out of the four crops are *legumes*, plants which fix nitrogen from the atmosphere. These are the peas and the two lupin cultivars. The fourth crop, barley, is nonleguminous.

Questions of interest are:

- Did the three legume crops produce on average more or less dry matter in the tops than the nonlegume crop?
- Did one legume — peas — produce more or less than the average of the cultivars of the other legume — lupins?
- Was there any difference between the two lupin cultivars?

(a) Write down a complete set of three orthogonal contrasts corresponding to these questions of interest.

(b) Write down the corresponding unit vectors, $\mathbf{U_2}$, $\mathbf{U_3}$ and $\mathbf{U_4}$ which span the treatment space.

(c) Calculate the squared lengths of the projections of $\mathbf{y}$ onto these directions. These are $(\mathbf{y}\cdot\mathbf{U_2})^2$, $(\mathbf{y}\cdot\mathbf{U_3})^2$ and $(\mathbf{y}\cdot\mathbf{U_4})^2$.

(d) Write down the fitted model in the form $\mathbf{y} = \bar{\mathbf{y}}_i + (\mathbf{y} - \bar{\mathbf{y}}_i)$.

(e) Estimate σ^2 using the formula $s^2 = \|\mathbf{y} - \bar{\mathbf{y}}_i\|^2/8$.

(f) Write down and test hypotheses relating to the questions of interest. What are your conclusions?

(4.5)* The following experiment was carried out to determine the best method of applying a fertilizer mix to corn. Three treatments, given below, were assigned completely at random to nine plots. The weight of corn, in kilograms, yielded by each plot was as follows:

1. Control (no fertilizer)	45.1	46.7	47.4
2. 300 kg/ha plowed under	56.7	57.3	54.6
3. 300 kg/ha broadcast	53.3	55.0	54.7

The two questions of interest are:

- Did fertilizer increase corn yield?
- Was there a difference due to method of application?

Set up contrasts and unit vectors relevant to this experiment and calculate F tests to allow these two questions to be answered. What are your conclusions? Exhibit your working by listing test hypotheses and a relevant Pythagorean breakup.

(4.6) An experiment is conducted to test the tolerance of wheat to herbicides used in the control of yarrow. A weed free area is divided into twelve plots and six treatments, namely, five herbicides and a control, are assigned in a completely random manner to the plots. The treatments and wheat grain yields in t/ha are:

Control (no herbicide)	6.5	6.1
Systemic herbicide A	5.2	5.0
Systemic herbicide B	4.9	5.3
Contact herbicide P, formulation 1	5.8	5.7
Contact herbicide P, formulation 2	5.4	5.6
Contact herbicide Q, formulation 3	5.0	5.3

Questions of interest are:

- Do the herbicides, on average, affect the grain yield?
- Do systemic herbicides differ from contact herbicides?
- Does "systemic A" differ from "systemic B"?
- Does "contact P" differ from "contact Q"?
- Do the formulations of "contact P" differ?

Set up a system of five orthogonal contrasts relating to these questions and write down the unit vectors $\mathbf{U_2}$, $\mathbf{U_3}$, $\mathbf{U_4}$, $\mathbf{U_5}$, and $\mathbf{U_6}$ corresponding to these contrasts. Calculate $(\mathbf{y}\cdot\mathbf{U_2})^2, \ldots, (\mathbf{y}\cdot\mathbf{U_6})^2$, obtain an estimate, s^2, for σ^2, and perform the five F tests. What are your conclusions?

(4.7)* Write down an appropriate set of orthogonal contrasts for each of the following treatment lists. (Assume each treatment is equally replicated.)

(a) Treatments:

1. Superphosphate applied in the autumn.
2. Superphosphate applied in the spring.
3. Control (no superphosphate).

(b) Treatments:

1. Protein based diet, brand A.
2. Starch based diet, brand B.
3. Protein based diet, brand C.
4. Starch based diet, brand D.

(c) Treatments:

1. Field peas (legume).
2. Beans (legume).
3. Oats (nonlegume).
4. Mustard (nonlegume).
5. Garden peas (legume).

Chapter 5

Simple Regression

Over a century ago Sir Francis Galton investigated the manner in which a son's height depended on that of the father. He showed that the heights of tall fathers' sons were distributed about a value somewhat less than that of their father. The paper was entitled "Regression towards mediocrity in hereditary stature," and the word regression has stayed with the subject.

In this book we deal with the simplest relationship between two variables x and y, the straight line relationship $y = \alpha + \beta x$. Our main interest will be in testing whether the slope β is zero.

In §5.1 of this chapter we shall describe a data set consisting of air pollution levels y and "inversion" effect levels x for a set of winter evenings in Christchurch, New Zealand. The method for fitting a straight line is introduced by analyzing a subset of this data set in §5.2. This is followed by an analysis of the full data set in §5.3. The chapter closes with a summary in §5.4, a class exercise and general exercises.

5.1 Air Pollution

During the winter of 1992 a shortage of water in the hydroelectricity storage lakes in the South Island of New Zealand caused a higher than normal usage of solid-fuel burners in the South Island city of Christchurch. The resulting air pollution was at its highest level in the late evening, and was thought to be most severe on evenings when an "inversion effect" occurred (this is when the air temperature is higher above the ground than at ground level, the opposite to the normal situation). On a normal evening the "hot air" rises, dissipating any pollution; however, when an inversion effect occurs, the air does not rise, so the pollution builds up throughout the evening.

To allow us to investigate this relationship, we present in Table 5.1 four key variables for each of the 30 days of June, 1992, which is winter in the Southern Hemisphere! The first variable is the air pollution level, or more precisely the level of suspended particulate in micrograms per cubic meter,

(June) Day	Air pollution	Temperatures (°C) Above ground	Ground	Inversion effect
1	*	*	*	*
2	290	6.0	5.1	0.9
3	79	8.9	7.8	1.1
4	33	5.7	6.3	−0.6
5	27	6.6	7.0	−0.4
6	59	6.6	7.1	−0.5
7	26	7.6	8.1	−0.5
8	33	8.5	9.0	−0.5
9	23	9.5	9.8	−0.3
10	118	4.7	4.8	−0.1
11	33	10.0	10.1	−0.1
12	394	5.6	3.0	2.6
13	73	6.2	6.4	−0.2
14	381	2.6	1.2	1.4
15	41	4.8	5.0	−0.2
16	51	8.3	7.3	1.0
17	42	5.5	5.7	−0.2
18	136	4.1	4.0	0.1
19	232	1.0	0.6	0.4
20	546	1.2	−0.8	2.0
21	*	*	*	*
22	*	*	*	*
23	*	*	*	*
24	33	4.2	4.6	−0.4
25	236	1.7	1.1	0.6
26	154	2.8	2.1	0.7
27	388	2.1	−0.5	2.6
28	151	3.4	3.2	0.2
29	71	2.7	3.1	−0.4
30	207	1.6	−0.8	2.4

Table 5.1: Air pollution levels, temperatures at and above ground level, and their difference (the "inversion effect") for St. Albans, Christchurch at 11.10 pm for each evening of June, 1992. Data by courtesy of Mr Bob Ayrey, Canterbury Regional Council. Missing values are denoted by *.

recorded at 11.10 pm each evening at a monitoring station in the suburb of St. Albans, Christchurch. The second and third variables are the air temperatures at ground level and 3 m above the ground, respectively, recorded simultaneously with the air pollution level. The fourth variable is a measure of the inversion effect, the difference between the two temperatures (above ground minus ground). For a pictorial view of the relationship we graph the air pollution level against the inversion effect in Figure 5.1.

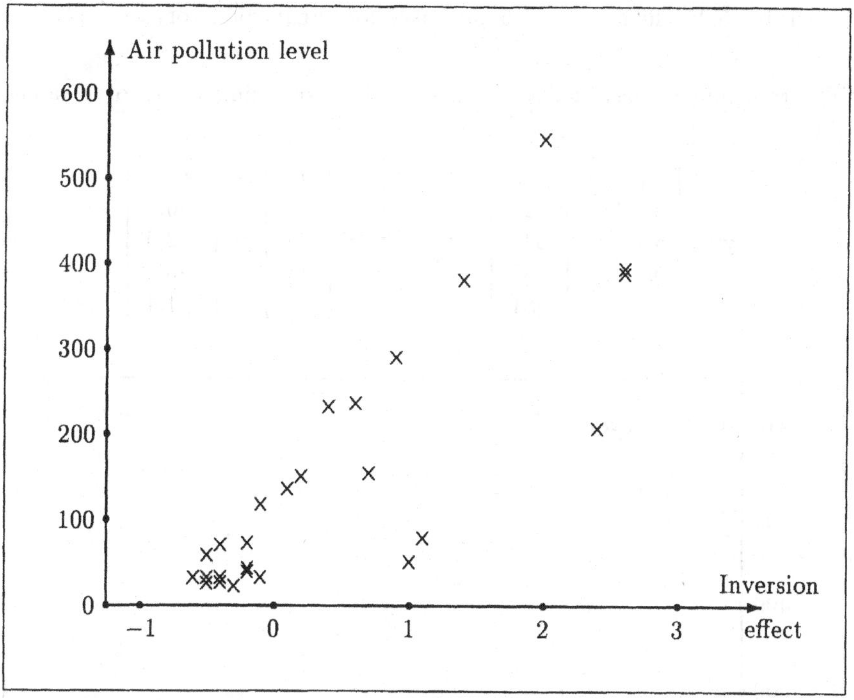

Figure 5.1: Scattergram of air pollution level against the inversion effect for the winter month of June, 1992. At 11.10 pm on a "normal" evening the ground temperature is slightly higher than the above ground temperature, so the x values are negative, as seen in the bottom left corner of the graph.

5.2 Sample of Size Five

We now apply our geometric method to the problem of fitting a straight line to such data. For ease of explanation we prefer initially to work with a low sample size. We therefore pick five days from Table 5.1: the days 13, 5, 20, 8 and 14 June, 1992. The reduced data set is shown in Table 5.2, and the reduced scattergram in Figure 5.2.

(June) Day	Air pollution	Temperatures (°C) Above ground	Ground	Inversion effect
13	73	6.2	6.4	−0.2
5	27	6.6	7.0	−0.4
20	546	1.2	−0.8	2.0
8	33	8.5	9.0	−0.5
14	381	2.6	1.2	1.4

Table 5.2: The sample of 5 days used for our introductory example.

The resulting *observation vector* and the corresponding vector of x values are

$$\mathbf{y} = \begin{bmatrix} y_1 \\ y_2 \\ y_3 \\ y_4 \\ y_5 \end{bmatrix} = \begin{bmatrix} 73 \\ 27 \\ 546 \\ 33 \\ 381 \end{bmatrix} \quad \text{and} \quad \mathbf{x} = \begin{bmatrix} x_1 \\ x_2 \\ x_3 \\ x_4 \\ x_5 \end{bmatrix} = \begin{bmatrix} -0.2 \\ -0.4 \\ 2.0 \\ -0.5 \\ 1.4 \end{bmatrix}$$

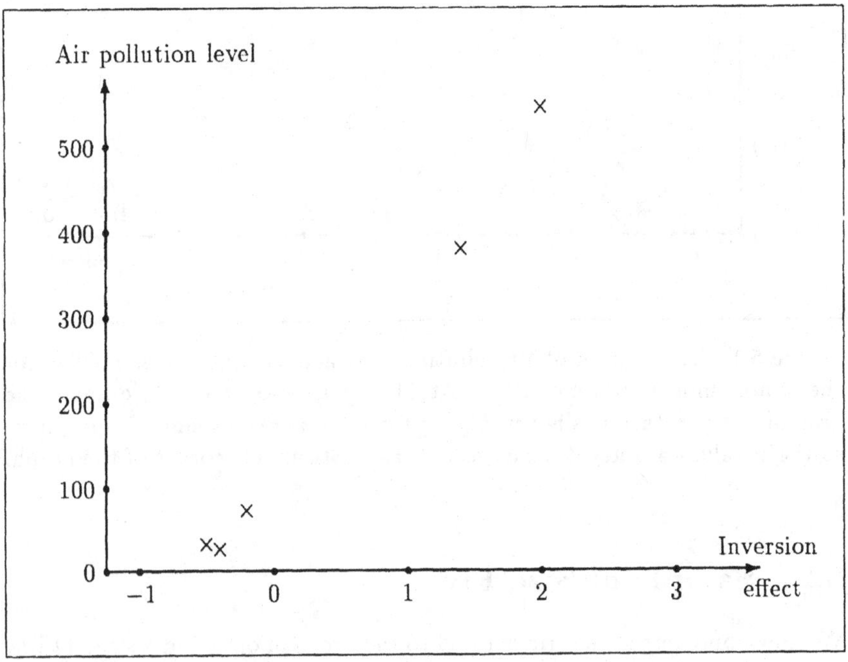

Figure 5.2: Scattergram of air pollution level against the inversion effect for our sample of 5 days.

Objective

The study objective is as follows:

> To investigate whether air pollution level is related to the inversion effect.

Model

We digress briefly to consider the assumptions underlying the analysis we are about to carry out.

The simple (linear) regression model firstly assumes that the relationship between the x and y variables is a straight line, $y = \alpha + \beta x$, where α is the intercept and β is the slope. Secondly, it assumes that the x values are measured without error. Thirdly, it assumes that for a given value of x, the possible values for y are normally distributed about the point on the line, $y = \alpha + \beta x$, with a variance σ^2 which is the same for all x values. Fourthly, it assumes that the observed y values are independently drawn. The situation is illustrated in Figure 5.3.

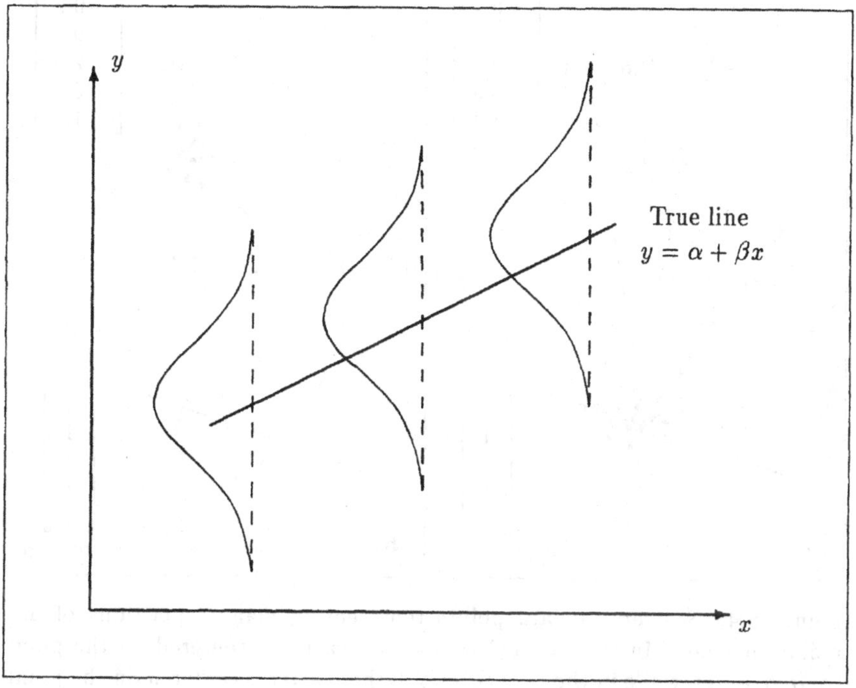

Figure 5.3: The assumptions of the simple regression model.

The Basic Idea

Figure 5.4 explains the basic idea for our latest situation. If the pollution level and the inversion effect are unrelated, then $\beta = 0$ and our observation vector will be part of a scatter centered on the point $(\alpha, \alpha, \alpha, \alpha, \alpha)$ as shown in Figure 5.4(a). However, if they are related, then $\beta \neq 0$ and our observation vector will be part of a scatter centered on the tip of the vector

$$\begin{bmatrix} \alpha + \beta x_1 \\ \alpha + \beta x_2 \\ \alpha + \beta x_3 \\ \alpha + \beta x_4 \\ \alpha + \beta x_5 \end{bmatrix} = \begin{bmatrix} \alpha - 0.2\beta \\ \alpha - 0.4\beta \\ \alpha + 2.0\beta \\ \alpha - 0.5\beta \\ \alpha + 1.4\beta \end{bmatrix} = \alpha \begin{bmatrix} 1 \\ 1 \\ 1 \\ 1 \\ 1 \end{bmatrix} + \beta \begin{bmatrix} -0.2 \\ -0.4 \\ 2.0 \\ -0.5 \\ 1.4 \end{bmatrix}$$

as shown in Figure 5.4(b). In other words, if $\beta = 0$ the scatter is centered on the equiangular line, whereas if $\beta \neq 0$ the scatter is moved away from the equiangular line in the plane defined by the vectors $[1, 1, 1, 1, 1]^T$ and $[-0.2, -0.4, 2.0, -0.5, 1.4]^T$.

Note that when thinking of many repetitions of our study, we are thinking of studies with the *same five x values*. That is, in Figure 5.4 the variation

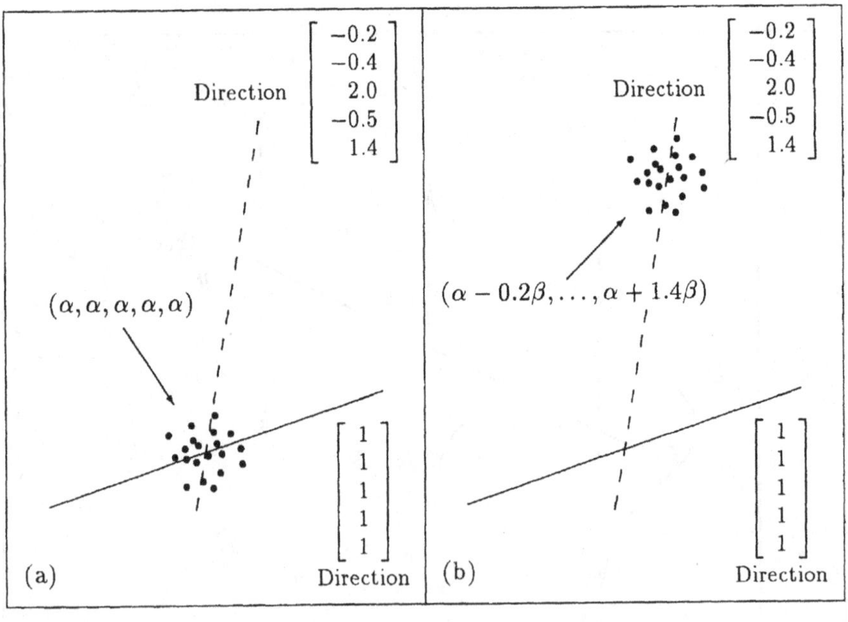

Figure 5.4: Scatters of data points representing many repetitions of our pollution study. In the case (a) $\beta = 0$ the scatter is centered on the point $(\alpha, \alpha, \alpha, \alpha, \alpha)$, while in the case (b) $\beta \neq 0$ the scatter is centered on the point $(\alpha - 0.2\beta, \alpha - 0.4\beta, \alpha + 2.0\beta, \alpha - 0.5\beta, \alpha + 1.4\beta)$.

in the sample points is solely due to the likely variation in the y values corresponding to five fixed values for x.

An Appropriate Coordinate System

In order to be able to fit the model in our standard manner we require a set of orthogonal coordinate axes for 5-space which includes two orthogonal directions which lie within the plane defined by the vectors $[1, 1, 1, 1, 1]^T$ and $[-0.2, -0.4, 2.0, -0.5, 1.4]^T$.

To start with, we can take $\mathbf{U_1} = [1, 1, 1, 1, 1]^T/\sqrt{5}$. This means we require a second unit vector, $\mathbf{U_2}$, lying in the above plane, and perpendicular to $\mathbf{U_1}$. How do we find such a vector?

Figure 5.5 provides the clues we require. In it we first draw the direction $\mathbf{U_1} = [1, 1, 1, 1, 1]^T/\sqrt{5}$ and the vector $\mathbf{x} = [-0.2, -0.4, 2.0, -0.5, 1.4]^T$. These two determine the plane in which we are interested. We then project

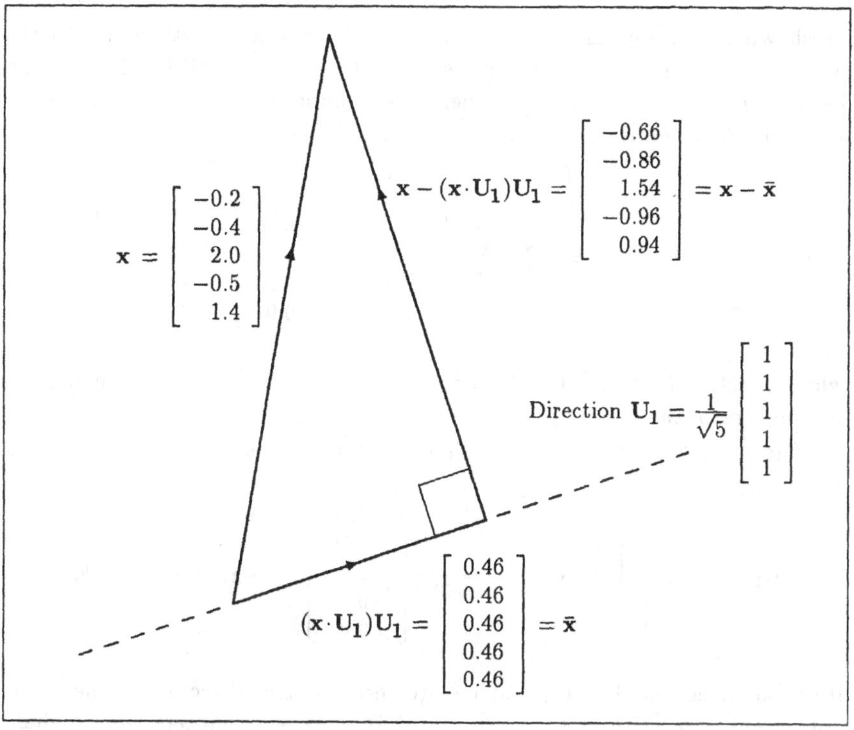

Figure 5.5: To obtain a vector perpendicular to the equiangular direction $(\mathbf{U_1})$ in the plane defined by $\mathbf{U_1}$ and the vector of x values, we subtract from the vector $\mathbf{x}$ its projection onto $\mathbf{U_1}$, namely $(\mathbf{x} \cdot \mathbf{U_1})\mathbf{U_1} = \bar{\mathbf{x}}$. This yields the required direction, $(\mathbf{x} - \bar{\mathbf{x}})$.

the vector $\mathbf{x}$ onto the direction $\mathbf{U_1}$, yielding the projection vector

$$(\mathbf{x}\cdot\mathbf{U_1})\mathbf{U_1} = \left(\begin{bmatrix} -0.2 \\ -0.4 \\ 2.0 \\ -0.5 \\ 1.4 \end{bmatrix} \cdot \frac{1}{\sqrt{5}}\begin{bmatrix} 1 \\ 1 \\ 1 \\ 1 \\ 1 \end{bmatrix}\right) \frac{1}{\sqrt{5}}\begin{bmatrix} 1 \\ 1 \\ 1 \\ 1 \\ 1 \end{bmatrix} = 0.46\begin{bmatrix} 1 \\ 1 \\ 1 \\ 1 \\ 1 \end{bmatrix} = \bar{\mathbf{x}}$$

We then subtract this projection vector $\bar{\mathbf{x}}$ from the vector of x values, yielding

$$\mathbf{x} - (\mathbf{x}\cdot\mathbf{U_1})\mathbf{U_1} = \begin{bmatrix} -0.2 \\ -0.4 \\ 2.0 \\ -0.5 \\ 1.4 \end{bmatrix} - \begin{bmatrix} 0.46 \\ 0.46 \\ 0.46 \\ 0.46 \\ 0.46 \end{bmatrix} = \begin{bmatrix} -0.66 \\ -0.86 \\ 1.54 \\ -0.96 \\ 0.94 \end{bmatrix} = \mathbf{x} - \bar{\mathbf{x}}$$

as shown in Figure 5.5. The resulting vector, $\mathbf{x} - \bar{\mathbf{x}}$, is orthogonal to $\mathbf{U_1}$ and lies in the plane determined by the vectors $[1, 1, 1, 1, 1]^{\mathrm{T}}$ and $[-0.2, -0.4, 2.0, -0.5, 1.4]^{\mathrm{T}}$. It therefore remains only to make it a unit vector by dividing it by its length, $\|\mathbf{x} - \bar{\mathbf{x}}\|$. This yields

$$\mathbf{U_2} = \frac{\mathbf{x} - \bar{\mathbf{x}}}{\|\mathbf{x} - \bar{\mathbf{x}}\|} = \frac{1}{\sqrt{5.352}}\begin{bmatrix} -0.66 \\ -0.86 \\ 1.54 \\ -0.96 \\ 0.94 \end{bmatrix}$$

where $5.352 = (-0.66)^2 + (-0.86)^2 + 1.54^2 + (-0.96)^2 + 0.94^2$, the sum of the squares of the $(x_i - \bar{x})$ values.

The resulting orthogonal coordinate system for 5-space is

$$\mathbf{U_1} = \frac{1}{\sqrt{5}}\begin{bmatrix} 1 \\ 1 \\ 1 \\ 1 \\ 1 \end{bmatrix}, \quad \mathbf{U_2} = \frac{1}{\sqrt{5.352}}\begin{bmatrix} -0.66 \\ -0.86 \\ 1.54 \\ -0.96 \\ 0.94 \end{bmatrix}, \quad \mathbf{U_3},\ \mathbf{U_4} \text{ and } \mathbf{U_5}$$

Here the directions $\mathbf{U_3}$, $\mathbf{U_4}$ and $\mathbf{U_5}$ are "error space" directions. These are not as easy to write down explicitly as in previous chapters, so we shall not do so, since this is not essential to our development. Let the reader be assured that such directions do exist! (For example, we could choose three directions at random and make them orthogonal to $\mathbf{U_1}$, $\mathbf{U_2}$ and to one another by extending the "Gram–Schmidt" process of orthogonalization introduced in Figure 5.5.)

The Orthogonal Decomposition

We now project our observation vector $\mathbf{y} = [73, 27, 546, 33, 381]^T$ onto our first two coordinate axes, $\mathbf{U_1}$ and $\mathbf{U_2}$. The first projection vector is

$$(\mathbf{y}\cdot\mathbf{U_1})\mathbf{U_1} = \left(\begin{bmatrix}73\\27\\546\\33\\381\end{bmatrix}\cdot\frac{1}{\sqrt{5}}\begin{bmatrix}1\\1\\1\\1\\1\end{bmatrix}\right)\frac{1}{\sqrt{5}}\begin{bmatrix}1\\1\\1\\1\\1\end{bmatrix} = \begin{bmatrix}212\\212\\212\\212\\212\end{bmatrix} = \bar{\mathbf{y}}$$

The second projection vector is $(\mathbf{y}\cdot\mathbf{U_2})\mathbf{U_2}$

$$= \left(\begin{bmatrix}73\\27\\546\\33\\381\end{bmatrix}\cdot\frac{1}{\sqrt{5.352}}\begin{bmatrix}-0.66\\-0.86\\1.54\\-0.96\\0.94\end{bmatrix}\right)\frac{1}{\sqrt{5.352}}\begin{bmatrix}-0.66\\-0.86\\1.54\\-0.96\\0.94\end{bmatrix}$$

$$= \frac{73\times(-0.66)+27\times(-0.86)+546\times 1.54+33\times(-0.96)+381\times 0.94}{\sqrt{5.352}\times\sqrt{5.352}}\begin{bmatrix}-0.66\\-0.86\\1.54\\-0.96\\0.94\end{bmatrix}$$

$$= \frac{1095.9}{5.352}\begin{bmatrix}-0.66\\-0.86\\1.54\\-0.96\\0.94\end{bmatrix} = 204.765\begin{bmatrix}-0.66\\-0.86\\1.54\\-0.96\\0.94\end{bmatrix} = b(\mathbf{x}-\bar{\mathbf{x}}) = \begin{bmatrix}-135.145\\-176.098\\315.338\\-196.574\\192.479\end{bmatrix}$$

Here 204.765 is the estimated slope b. By looking back through the above calculations we can see that the algebraic formula for the estimated slope is

$$b = \frac{y_1\times(x_1-\bar{x})+\cdots+y_5\times(x_5-\bar{x})}{(x_1-\bar{x})^2+\cdots+(x_5-\bar{x})^2} = \frac{1095.9}{5.352} = 204.765$$

The resulting orthogonal decomposition of the observation vector is

$$\begin{array}{ccccccc}
\mathbf{y} & = & (\mathbf{y}\cdot\mathbf{U_1})\mathbf{U_1} & + & (\mathbf{y}\cdot\mathbf{U_2})\mathbf{U_2} & + & (\mathbf{y}\cdot\mathbf{U_3})\mathbf{U_3}+(\mathbf{y}\cdot\mathbf{U_4})\mathbf{U_4}+(\mathbf{y}\cdot\mathbf{U_5})\mathbf{U_5} \\
\begin{bmatrix}73\\27\\546\\33\\381\end{bmatrix} & = & \begin{bmatrix}212\\212\\212\\212\\212\end{bmatrix} & + & \begin{bmatrix}-135.1\\-176.1\\315.3\\-196.6\\192.5\end{bmatrix} & + & \begin{bmatrix}-3.9\\-8.9\\18.7\\17.6\\-23.5\end{bmatrix} \\
\mathbf{y} & = & \bar{\mathbf{y}} & + & b(\mathbf{x}-\bar{\mathbf{x}}) & + & [\mathbf{y}-\bar{\mathbf{y}}-b(\mathbf{x}-\bar{\mathbf{x}})] \\
\text{observation vector} & & \text{mean vector} & & \text{slope vector} & & \text{error vector}
\end{array}$$

where the error vector is obtained by subtraction.

In summary, we have broken our observation vector **y** into three orthogonal components, a *mean vector*, a *slope vector*, and an *error vector*, as shown in Figure 5.6.

The equation of the line which we have fitted is

$$y = 212 + 204.8(x - 0.46)$$

The resulting fitted line is displayed, with the scattergram, in Figure 5.7.

Note that the equation is in the general "orthogonal" form

$$y = \bar{y} + b(x - \bar{x})$$

We may convert the equation back to our original "nonorthogonal" form by

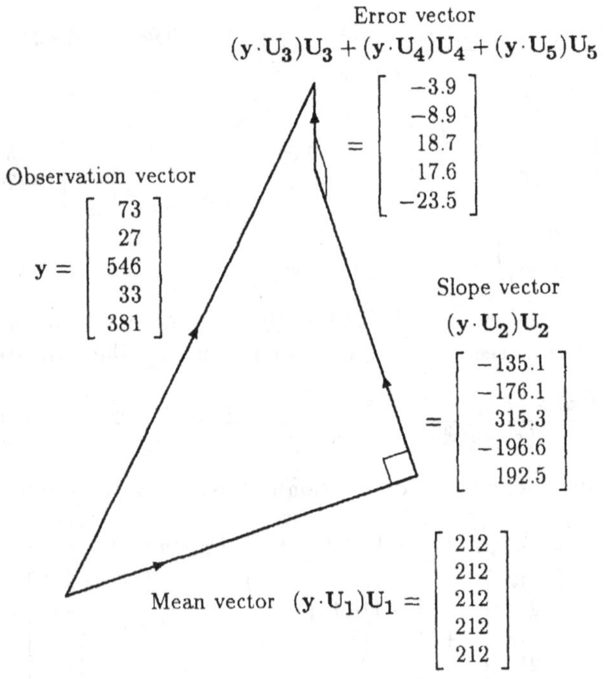

Figure 5.6: Orthogonal decomposition of the observation vector for the simple regression case. The figure is approximately to scale, with the major exception being that the error vector has been tripled in length to make it more visible.

expanding the bracket and collecting terms:

$$\begin{aligned} y &= 212 + 204.8x - 204.8 \times 0.46 \\ &= (212 - 204.8 \times 0.46) + 204.8x \end{aligned}$$

or

$$y = 117.8 + 204.8x$$

In general, the nonorthogonal form is

$$y = (\bar{y} - b\bar{x}) + bx$$

Hence, in general, the intercept α is estimated by $a = \bar{y} - b\bar{x}$ (= 117.8 here), and the slope β is estimated by

$$b = \frac{y_1 \times (x_1 - \bar{x}) + \cdots + y_n \times (x_n - \bar{x})}{(x_1 - \bar{x})^2 + \cdots + (x_n - \bar{x})^2}$$

where n is the sample size.

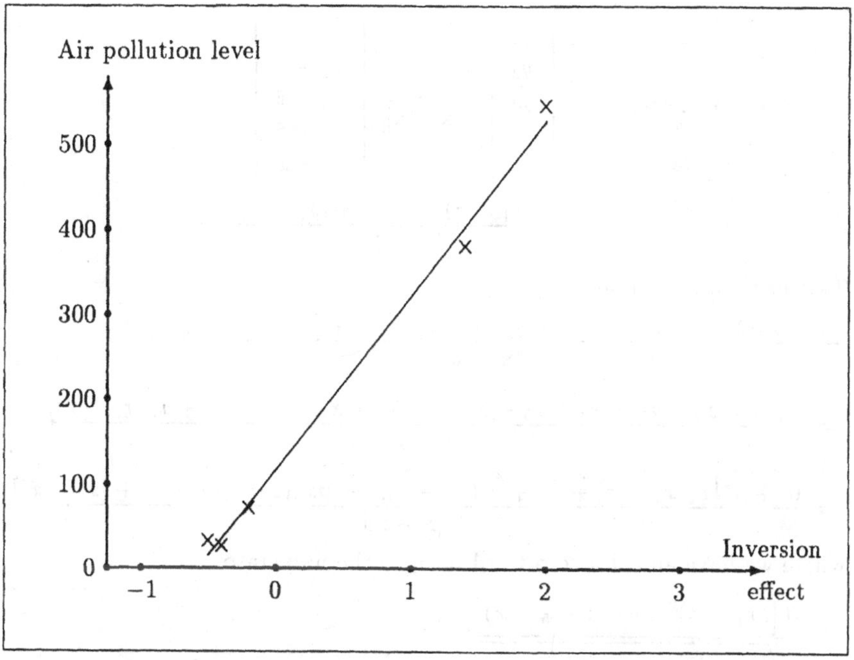

Figure 5.7: Fitted regression line superimposed on the scattergram of air pollution level against the inversion effect.

Distributions of the Projection Lengths

To prepare for testing whether the true slope β could plausibly be zero, we now work out the means of the distributions of the projection lengths $\mathbf{y}\cdot\mathbf{U_1}$, $\mathbf{y}\cdot\mathbf{U_2}$, $\mathbf{y}\cdot\mathbf{U_3}$, $\mathbf{y}\cdot\mathbf{U_4}$ and $\mathbf{y}\cdot\mathbf{U_5}$. To save having to redo this exercise later we deal in fairly general terms, using algebraic symbols such as y_1, y_2 and x_1, x_2 instead of using the particular x values in our example. Readers who have an aversion to algebraic manipulation should now skip to the summary paragraph at the end of this subsection.

$$\text{Firstly,}\quad \mathbf{y}\cdot\mathbf{U_1} = \begin{bmatrix} y_1 \\ y_2 \\ y_3 \\ y_4 \\ y_5 \end{bmatrix} \cdot \frac{1}{\sqrt{5}} \begin{bmatrix} 1 \\ 1 \\ 1 \\ 1 \\ 1 \end{bmatrix} = \frac{y_1 + y_2 + y_3 + y_4 + y_5}{\sqrt{5}}$$

This has a mean value of

$$\frac{(\alpha + \beta x_1) + (\alpha + \beta x_2) + (\alpha + \beta x_3) + (\alpha + \beta x_4) + (\alpha + \beta x_5)}{\sqrt{5}}$$

$$= \frac{5\alpha + \beta(x_1 + x_2 + x_3 + x_4 + x_5)}{\sqrt{5}} = \frac{5\alpha + 5\beta\bar{x}}{\sqrt{5}} = \sqrt{5}(\alpha + \beta\bar{x})$$

$$\text{Secondly,}\quad \mathbf{y}\cdot\mathbf{U_2} = \begin{bmatrix} y_1 \\ y_2 \\ y_3 \\ y_4 \\ y_5 \end{bmatrix} \cdot \frac{1}{\|\mathbf{x} - \bar{\mathbf{x}}\|} \begin{bmatrix} x_1 - \bar{x} \\ x_2 - \bar{x} \\ x_3 - \bar{x} \\ x_4 - \bar{x} \\ x_5 - \bar{x} \end{bmatrix}$$

$$= \frac{y_1(x_1 - \bar{x}) + \cdots + y_5(x_5 - \bar{x})}{\|\mathbf{x} - \bar{\mathbf{x}}\|}$$

This has a mean value of

$$\frac{(\alpha + \beta x_1)(x_1 - \bar{x}) + (\alpha + \beta x_2)(x_2 - \bar{x}) + \cdots + (\alpha + \beta x_5)(x_5 - \bar{x})}{\|\mathbf{x} - \bar{\mathbf{x}}\|}$$

$$= \frac{\alpha(x_1 + x_2 + x_3 + x_4 + x_5 - 5\bar{x}) \;+\; \beta\left[x_1(x_1 - \bar{x}) + \cdots + x_5(x_5 - \bar{x})\right]}{\|\mathbf{x} - \bar{\mathbf{x}}\|}$$

$$= \frac{0 \;+\; \beta\left[x_1(x_1 - \bar{x}) + \cdots + x_5(x_5 - \bar{x})\right] \;-\; \beta\bar{x}\left[(x_1 - \bar{x}) + \cdots + (x_5 - \bar{x})\right]}{\|\mathbf{x} - \bar{\mathbf{x}}\|}$$

where we have added a zero third term to the numerator

$$= \frac{\beta\left[(x_1 - \bar{x})^2 + \cdots + (x_5 - \bar{x})^2\right]}{\sqrt{(x_1 - \bar{x})^2 + \cdots + (x_5 - \bar{x})^2}}$$

$$= \beta\sqrt{(x_1 - \bar{x})^2 + \cdots + (x_5 - \bar{x})^2} = \beta\|\mathbf{x} - \bar{\mathbf{x}}\|$$

Lastly, we need to show that for each of the unspecified coordinate axis directions in the error space, $\mathbf{U_3}$, $\mathbf{U_4}$ and $\mathbf{U_5}$, the mean of the projection length is zero. We do this for a general such direction, $\mathbf{U_e} = [e_1, e_2, e_3, e_4, e_5]^T$.

$$\text{Now } \mathbf{y}\cdot\mathbf{U_e} = \begin{bmatrix} y_1 \\ y_2 \\ y_3 \\ y_4 \\ y_5 \end{bmatrix} \cdot \begin{bmatrix} e_1 \\ e_2 \\ e_3 \\ e_4 \\ e_5 \end{bmatrix} = y_1e_1 + y_2e_2 + y_3e_3 + y_4e_4 + y_5e_5$$

This has a mean value of

$$[\alpha + \beta x_1]\, e_1 + [\alpha + \beta x_2]\, e_2 + [\alpha + \beta x_3]\, e_3 + [\alpha + \beta x_4]\, e_4 + [\alpha + \beta x_5]\, e_5$$

$$= \;[(\alpha + \beta\bar{x}) + \beta(x_1 - \bar{x})]\, e_1 \;+\; \cdots \;+\; [(\alpha + \beta\bar{x}) + \beta(x_5 - \bar{x})]\, e_5$$

where we have added and subtracted $\beta\bar{x}$ in each term

$$= \;(\alpha + \beta\bar{x})(e_1 + \cdots + e_5) \;+\; \beta\,[(x_1 - \bar{x})e_1 + \cdots + (x_5 - \bar{x})e_5]$$

$$= \;(\alpha + \beta\bar{x}) \begin{bmatrix} 1 \\ 1 \\ 1 \\ 1 \\ 1 \end{bmatrix} \cdot \begin{bmatrix} e_1 \\ e_2 \\ e_3 \\ e_4 \\ e_5 \end{bmatrix} + \beta \begin{bmatrix} x_1 - \bar{x} \\ x_2 - \bar{x} \\ x_3 - \bar{x} \\ x_4 - \bar{x} \\ x_5 - \bar{x} \end{bmatrix} \cdot \begin{bmatrix} e_1 \\ e_2 \\ e_3 \\ e_4 \\ e_5 \end{bmatrix}$$

$= \;0\;$ since $\mathbf{U_e}$ is orthogonal to both $\mathbf{U_1}$ and $\mathbf{U_2}$.

In addition, it can readily be shown that all five projection lengths have variance σ^2. Thus $\mathbf{y}\cdot\mathbf{U_1}$ comes from an $N(\sqrt{5}(\alpha + \beta\bar{x}), \sigma^2)$ distribution, $\mathbf{y}\cdot\mathbf{U_2}$ comes from an $N(\beta\|\mathbf{x} - \bar{\mathbf{x}}\|, \sigma^2)$ distribution, and $\mathbf{y}\cdot\mathbf{U_3}$, $\mathbf{y}\cdot\mathbf{U_4}$ and $\mathbf{y}\cdot\mathbf{U_5}$ all come from an $N(0, \sigma^2)$ distribution.

The point to note is that the distributions of $\mathbf{y}\cdot\mathbf{U_1}$ and $\mathbf{y}\cdot\mathbf{U_2}$ are centered on potentially nonzero quantities whereas the distributions of $\mathbf{y}\cdot\mathbf{U_3}$, $\mathbf{y}\cdot\mathbf{U_4}$ and $\mathbf{y}\cdot\mathbf{U_5}$ are always centered on zero. Furthermore, for deciding whether β is zero or nonzero, the direction $\mathbf{U_2}$ *is the direction of interest*, since the projection length $\mathbf{y}\cdot\mathbf{U_2}$ averages to zero if $\beta = 0$, and averages to a nonzero quantity if $\beta \neq 0$.

Testing the Hypothesis

We now proceed to investigate the study objective: Is air pollution level related to the inversion effect? In terms of our simple regression model this translates into the question: Is the true slope $\beta = 0$?

The relevant *Pythagorean breakup* is

$$\|\mathbf{y}\|^2 = (\mathbf{y}\cdot\mathbf{U_1})^2 + (\mathbf{y}\cdot\mathbf{U_2})^2 + (\mathbf{y}\cdot\mathbf{U_3})^2 + (\mathbf{y}\cdot\mathbf{U_4})^2 + (\mathbf{y}\cdot\mathbf{U_5})^2$$

$$\text{or,} \quad 450424 = 224720 + 224402 + 1302$$

as illustrated in Figure 5.8. Note that we have calculated the squared lengths of the vectors using the vectors worked out on page 111. For the last two squared lengths we needed three decimal places to allow reasonable accuracy. We therefore used the more accurate form of $(\mathbf{y}\cdot\mathbf{U}_2)\mathbf{U}_2$ given on page 111, together with a corresponding, more accurate error vector. In spite of this, our sums of squares are less accurate in the last digit than those obtained using a computer. This slight inaccuracy is unimportant in our example, but is mentioned in case it causes confusion for the reader.

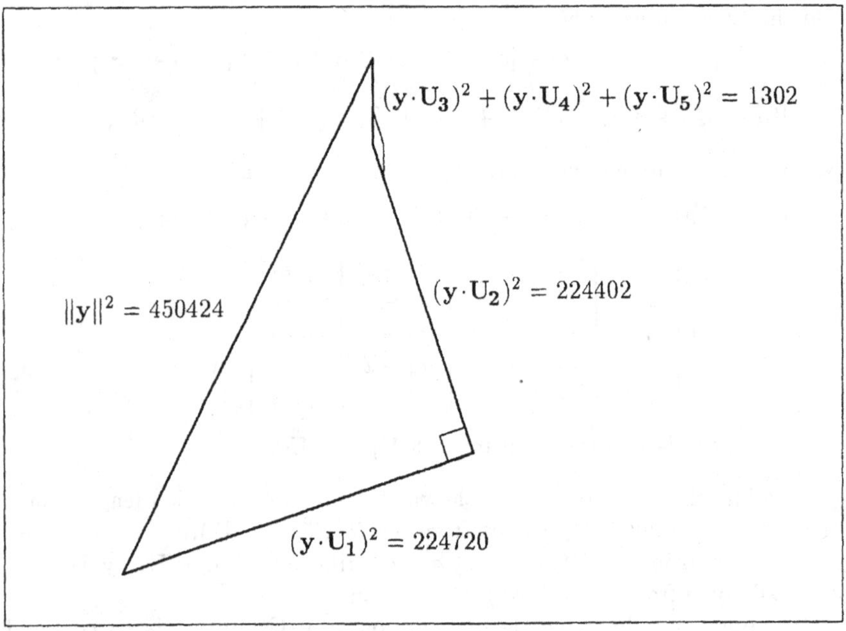

Figure 5.8: The Pythagorean breakup of the squared length of the observation vector as the sum of the squared lengths of the five orthogonal projections, $(\mathbf{y}\cdot\mathbf{U}_i)^2$, for our simple regression example.

To test the hypothesis that $\beta = 0$ we check whether the squared projection length $(\mathbf{y}\cdot\mathbf{U}_2)^2$ is similar to, or considerably greater than, the average of the squared projection lengths $(\mathbf{y}\cdot\mathbf{U}_3)^2$, $(\mathbf{y}\cdot\mathbf{U}_4)^2$ and $(\mathbf{y}\cdot\mathbf{U}_5)^2$. The resulting test statistic is

$$F = \frac{(\mathbf{y}\cdot\mathbf{U}_2)^2}{[(\mathbf{y}\cdot\mathbf{U}_3)^2 + (\mathbf{y}\cdot\mathbf{U}_4)^2 + (\mathbf{y}\cdot\mathbf{U}_5)^2]/3} = \frac{224402}{1302/3} = \frac{224402}{434} = 517$$

Is this observed F value large or small? Now if $\beta = 0$ the projection lengths $\mathbf{y}\cdot\mathbf{U}_2$, $\mathbf{y}\cdot\mathbf{U}_3$, $\mathbf{y}\cdot\mathbf{U}_4$ and $\mathbf{y}\cdot\mathbf{U}_5$ all come from an $N(0, \sigma^2)$ distribution, so the test statistic $F = (\mathbf{y}\cdot\mathbf{U}_2)^2/\{[(\mathbf{y}\cdot\mathbf{U}_3)^2 + (\mathbf{y}\cdot\mathbf{U}_4)^2 + (\mathbf{y}\cdot\mathbf{U}_5)^2]/3\}$

comes from the $F_{1,3}$ distribution defined in Appendix B. Hence to see whether our observed value of 517 is large or small, we compare it with the percentiles of the $F_{1,3}$ distribution given in Table T.2. The 90, 95 and 99 percentiles are 5.54, 10.13 and 34.12, respectively, so our observed value of 517 is clearly large, being much greater than the 99 percentile of 34.12. That is, we have found very strong evidence to support the idea that the air pollution level is related to the inversion effect.

Equivalent t Test

Our test statistic $F = (\mathbf{y}\cdot\mathbf{U}_2)^2/\{[(\mathbf{y}\cdot\mathbf{U}_3)^2 + (\mathbf{y}\cdot\mathbf{U}_4)^2 + (\mathbf{y}\cdot\mathbf{U}_5)^2]/3\}$ can be rewritten as $A^2/(B^2/3)$, where B is the length of the error vector. By taking the square root, this leads to the form given in Chapter 1,

$$t = \frac{\mathbf{y}\cdot\mathbf{U}_2}{\sqrt{[(\mathbf{y}\cdot\mathbf{U}_3)^2 + (\mathbf{y}\cdot\mathbf{U}_4)^2 + (\mathbf{y}\cdot\mathbf{U}_5)^2]/3}} = \frac{A}{B/\sqrt{3}} = \frac{\sqrt{224402}}{\sqrt{1302/3}} = 22.74$$

This t value is then compared with the percentiles of the t_3 distribution given in Table T.2, yielding an identical conclusion to that obtained above.

Comments

In the above analysis the direction $\mathbf{U}_1$ is used to estimate the constant term $(\alpha + \beta\bar{x})$, the direction $\mathbf{U}_2$ is used to estimate the true slope β, and the directions $\mathbf{U}_3$, $\mathbf{U}_4$ and $\mathbf{U}_5$ are used to estimate the variance σ^2. In the long run the projection length $\mathbf{y}\cdot\mathbf{U}_2 = b\|\mathbf{x} - \bar{\mathbf{x}}\|$ averages to $\beta\|\mathbf{x} - \bar{\mathbf{x}}\|$, so $b = 204.8$ serves as our estimate of β. Also, the squared projection lengths $(\mathbf{y}\cdot\mathbf{U}_3)^2$, $(\mathbf{y}\cdot\mathbf{U}_4)^2$ and $(\mathbf{y}\cdot\mathbf{U}_5)^2$ all average to σ^2, so $s^2 = [(\mathbf{y}\cdot\mathbf{U}_3)^2 + (\mathbf{y}\cdot\mathbf{U}_4)^2 + (\mathbf{y}\cdot\mathbf{U}_5)^2]/3 = 1302/3 = 434$ serves as our estimate of σ^2.

In addition, it can be shown that in the long run the squared projection length $(\mathbf{y}\cdot\mathbf{U}_2)^2$ averages to $\beta^2\|\mathbf{x} - \bar{\mathbf{x}}\|^2 + \sigma^2$. Hence, if $\beta \neq 0$ the quantity $(\mathbf{y}\cdot\mathbf{U}_2)^2$ is inflated, and our test statistic $(\mathbf{y}\cdot\mathbf{U}_2)^2/s^2$ can be thought of as an inflated estimate of σ^2 divided by an unbiased estimate of σ^2.

5.3 General Case

To illustrate the general case we now pick out the essential steps in our geometric approach and apply them to the full data set of 26 "non-missing" days given in Table 5.1. The resulting *observation vector* and the corresponding vector of x values are

$$\mathbf{y} = \begin{bmatrix} y_1 \\ \vdots \\ y_{26} \end{bmatrix} = \begin{bmatrix} 290 \\ \vdots \\ 207 \end{bmatrix} \quad \text{and} \quad \mathbf{x} = \begin{bmatrix} x_1 \\ \vdots \\ x_{26} \end{bmatrix} = \begin{bmatrix} 0.9 \\ \vdots \\ 2.4 \end{bmatrix}$$

The Basic Idea

The basic idea is the same as in §5.2. If the air pollution level and the inversion effect are unrelated, then $\beta = 0$ and our observation vector will be part of a scatter centered on the point $(\alpha, \ldots, \alpha)$ as shown in Figure 5.9(a). If they are related however, then $\beta \neq 0$ and our observation vector will be part of a scatter centered on the tip of the vector

$$\begin{bmatrix} \alpha + \beta x_1 \\ \vdots \\ \alpha + \beta x_{26} \end{bmatrix} = \begin{bmatrix} \alpha + 0.9\beta \\ \vdots \\ \alpha + 2.4\beta \end{bmatrix} = \alpha \begin{bmatrix} 1 \\ \vdots \\ 1 \end{bmatrix} + \beta \begin{bmatrix} 0.9 \\ \vdots \\ 2.4 \end{bmatrix}$$

as shown in Figure 5.9(b). In other words, if $\beta = 0$ the scatter is centered on the equiangular line, whereas if $\beta \neq 0$ the scatter is moved away from the equiangular line in the plane defined by the vectors $[1, \ldots, 1]^T$ and $[0.9, \ldots, 2.4]^T$.

As in §5.2, note that when thinking of many repetitions of our study, we are thinking of studies with the same 26 x values. That is, in Figure 5.9 the variation in the sample points is solely due to the likely variation in the y values corresponding to 26 fixed values for x.

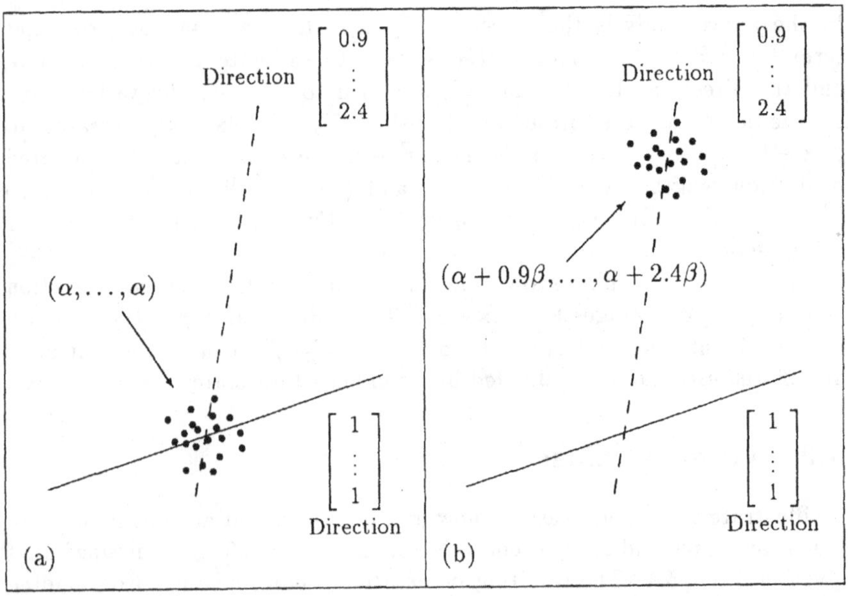

Figure 5.9: Scatters of data points representing many repetitions of our pollution study. In the case (a) $\beta = 0$ the scatter is centered on the point $(\alpha, \ldots, \alpha)$, while in the case (b) $\beta \neq 0$ the scatter is centered on the point $(\alpha + 0.9\beta, \ldots, \alpha + 2.4\beta)$.

An Appropriate Coordinate System

An appropriate orthogonal coordinate system for 26-space is

$$\mathbf{U_1} = \frac{1}{\sqrt{26}}\begin{bmatrix}1\\ \vdots\\ 1\end{bmatrix}, \quad \mathbf{U_2} = \frac{1}{\sqrt{25.9646}}\begin{bmatrix}0.454\\ \vdots\\ 1.946\end{bmatrix}, \quad \mathbf{U_3}, \ \ldots, \ \mathbf{U_{26}}$$

Here $\mathbf{U_1}$ is the usual equiangular direction and $\mathbf{U_2}$ is the vector of deviations $(x_i - \bar{x})$ of the x values about their mean $\bar{x} = 0.446$, divided by its length $\|\mathbf{x} - \bar{\mathbf{x}}\| = \sqrt{25.9646}$. The directions $\mathbf{U_3}, \ldots, \mathbf{U_{26}}$ are 24 orthogonal "error space" directions.

The Orthogonal Decomposition

We now project our observation vector $\mathbf{y} = [290, \ldots, 207]^{\mathrm{T}}$ onto our first two coordinate axes, $\mathbf{U_1}$ and $\mathbf{U_2}$. The first projection vector is

$$(\mathbf{y}\cdot\mathbf{U_1})\mathbf{U_1} = \left(\begin{bmatrix}290\\ \vdots\\ 207\end{bmatrix}\cdot\frac{1}{\sqrt{26}}\begin{bmatrix}1\\ \vdots\\ 1\end{bmatrix}\right)\frac{1}{\sqrt{26}}\begin{bmatrix}1\\ \vdots\\ 1\end{bmatrix} = \begin{bmatrix}148.3\\ \vdots\\ 148.3\end{bmatrix} = \bar{\mathbf{y}}$$

The second projection vector is $(\mathbf{y}\cdot\mathbf{U_2})\mathbf{U_2}$

$$= \left(\begin{bmatrix}290\\ \vdots\\ 207\end{bmatrix}\cdot\frac{1}{\sqrt{25.9646}}\begin{bmatrix}0.454\\ \vdots\\ 1.954\end{bmatrix}\right)\frac{1}{\sqrt{25.9646}}\begin{bmatrix}0.454\\ \vdots\\ 1.954\end{bmatrix}$$

$$= \frac{290 \times 0.454 + \cdots + 207 \times 1.954}{\sqrt{25.9646} \times \sqrt{25.9646}}\begin{bmatrix}0.454\\ \vdots\\ 1.954\end{bmatrix}$$

$$= \frac{3035.1}{25.9646}\begin{bmatrix}0.454\\ \vdots\\ 1.954\end{bmatrix} = 116.9\begin{bmatrix}0.454\\ \vdots\\ 1.954\end{bmatrix} = b(\mathbf{x} - \bar{\mathbf{x}}) = \begin{bmatrix}53.1\\ \vdots\\ 228.4\end{bmatrix}$$

Here the estimated slope is $b = 116.9$. The resulting orthogonal decomposition of the observation vector is

$$\begin{array}{ccccccc}
\mathbf{y} & = & (\mathbf{y}\cdot\mathbf{U_1})\mathbf{U_1} & + & (\mathbf{y}\cdot\mathbf{U_2})\mathbf{U_2} & + & (\mathbf{y}\cdot\mathbf{U_3})\mathbf{U_3} + \cdots + (\mathbf{y}\cdot\mathbf{U_{26}})\mathbf{U_{26}} \\
\begin{bmatrix}290\\ \vdots\\ 207\end{bmatrix} & = & \begin{bmatrix}148.3\\ \vdots\\ 148.3\end{bmatrix} & + & \begin{bmatrix}53.1\\ \vdots\\ 228.4\end{bmatrix} & + & \begin{bmatrix}88.6\\ \vdots\\ -169.7\end{bmatrix} \\
\mathbf{y} & = & \bar{\mathbf{y}} & + & b(\mathbf{x} - \bar{\mathbf{x}}) & + & [\mathbf{y} - \bar{\mathbf{y}} - b(\mathbf{x} - \bar{\mathbf{x}})] \\
\text{observation vector} & & \text{mean vector} & & \text{slope vector} & & \text{error vector}
\end{array}$$

where the error vector is obtained by subtraction.

As in §5.2, we have broken our observation vector $\mathbf{y}$ into three orthogonal components, a *mean vector*, a *slope vector*, and an *error vector*, as shown in Figure 5.10.

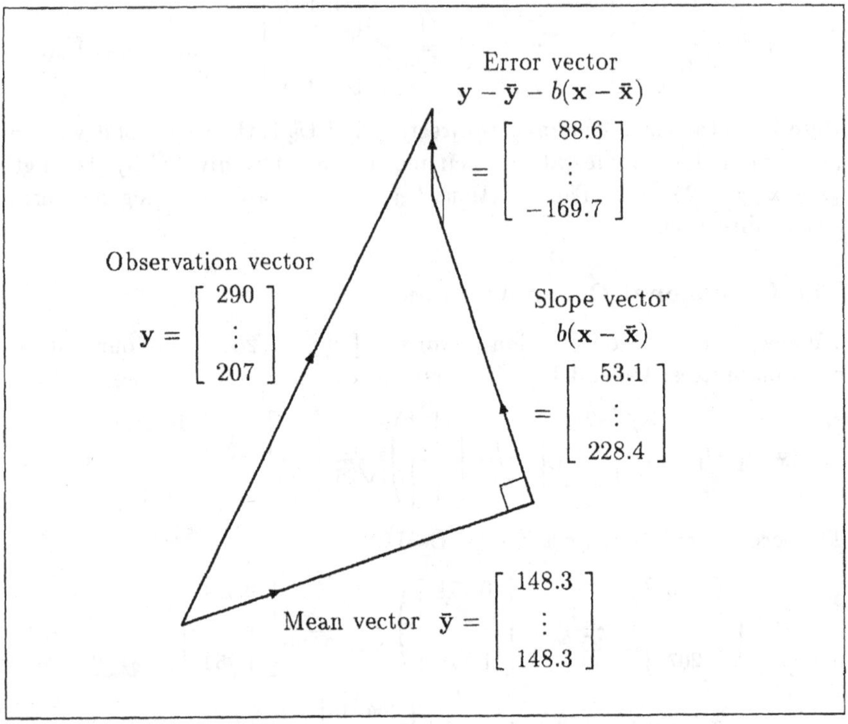

Figure 5.10: Orthogonal decomposition of the observation vector for the simple regression case, using the full data set from Table 5.1.

The equation of the line which we have fitted is, in "orthogonal" form:

$$y = 148.3 + 116.9(x - 0.446)$$

The resulting fitted line is displayed, with the scattergram, in Figure 5.11.

When converted to "nonorthogonal" form the equation becomes:

$$\begin{aligned} y &= 148.3 + 116.9x - 116.9 \times 0.446 \\ &= (148.3 - 116.9 \times 0.446) + 116.9x \end{aligned}$$

or,

$$y = 96.2 + 116.9x$$

As noted in §5.2, the intercept α is estimated by $a = \bar{y} - b\bar{x} = 96.2$, and the slope β is estimated by

$$b = \frac{\mathbf{y} \cdot (\mathbf{x} - \bar{\mathbf{x}})}{\|\mathbf{x} - \bar{\mathbf{x}}\|^2} = \frac{y_1 \times (x_1 - \bar{x}) + \cdots + y_n \times (x_n - \bar{x})}{(x_1 - \bar{x})^2 + \cdots + (x_n - \bar{x})^2} = 116.9$$

where n is the sample size (26 here).

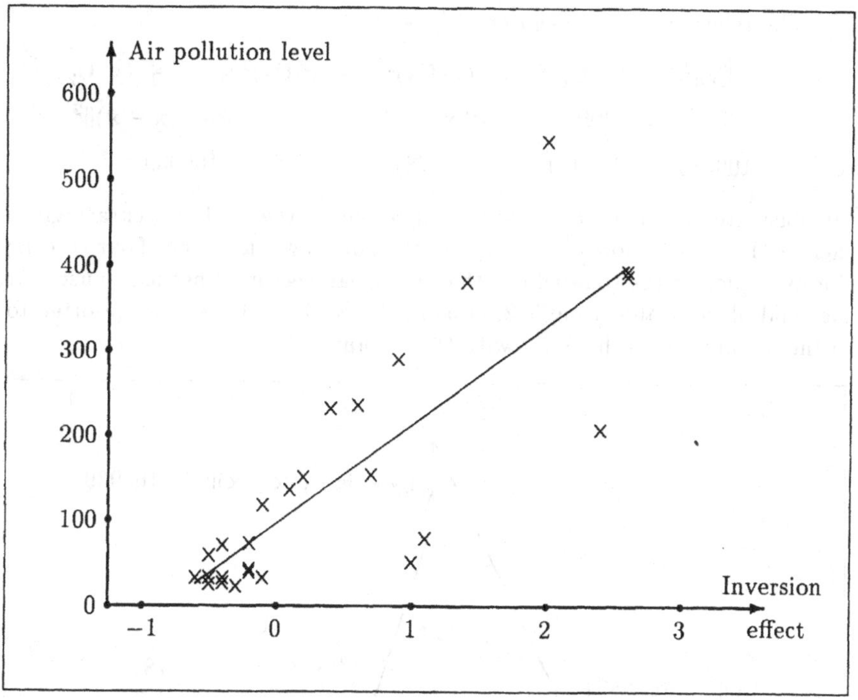

Figure 5.11: Fitted regression line superimposed on the scattergram of air pollution level against the inversion effect for the winter month of June, 1992.

Distributions of the Projection Lengths

Proceeding as in §5.2, we find that $\mathbf{y}\cdot\mathbf{U}_1$ comes from an $N(\sqrt{26}(\alpha+\beta\bar{x}),\sigma^2)$ distribution, $\mathbf{y}\cdot\mathbf{U}_2$ comes from an $N(\beta\|\mathbf{x}-\bar{\mathbf{x}}\|,\sigma^2)$ distribution, and $\mathbf{y}\cdot\mathbf{U}_3, \ldots, \mathbf{y}\cdot\mathbf{U}_{26}$ all come from an $N(0,\sigma^2)$ distribution.

As before, the point to note is that the distributions of $\mathbf{y}\cdot\mathbf{U}_1$ and $\mathbf{y}\cdot\mathbf{U}_2$ are centered on potentially nonzero quantities whereas the distributions of $\mathbf{y}\cdot\mathbf{U}_3, \ldots, \mathbf{y}\cdot\mathbf{U}_{26}$ are always centered on zero. Furthermore, for deciding whether β is zero or nonzero, the direction $\mathbf{U}_2$ *is the direction of interest*, since the projection length $\mathbf{y}\cdot\mathbf{U}_2$ averages to zero if $\beta = 0$, and averages to a nonzero quantity if $\beta \neq 0$.

Testing the Hypothesis

We now proceed to investigate the study objective: Is air pollution level related to the inversion effect? That is: Is the true slope $\beta = 0$?

The relevant *Pythagorean breakup* is

$$\begin{aligned} \|\mathbf{y}\|^2 &= (\mathbf{y}\cdot\mathbf{U}_1)^2 + (\mathbf{y}\cdot\mathbf{U}_2)^2 + (\mathbf{y}\cdot\mathbf{U}_3)^2 + \cdots + (\mathbf{y}\cdot\mathbf{U}_{26})^2 \\ \text{or,}\quad \|\mathbf{y}\|^2 &= \|\bar{\mathbf{y}}\|^2 + b^2\|\mathbf{x}-\bar{\mathbf{x}}\|^2 + \|\mathbf{y}-\bar{\mathbf{y}}-b(\mathbf{x}-\bar{\mathbf{x}})\|^2 \\ \text{or,}\quad 1096891 &= 572171 + 354781 + 169939 \end{aligned}$$

as illustrated in Figure 5.12. Note that to obtain the level of accuracy given here we have used more decimal places than displayed for some of our vectors; for example, for the squared length of the mean vector we needed to use four decimal places, calculating 572171 as 26×148.3462^2. In fact, we resorted to using a computer to help out with the arithmetic!

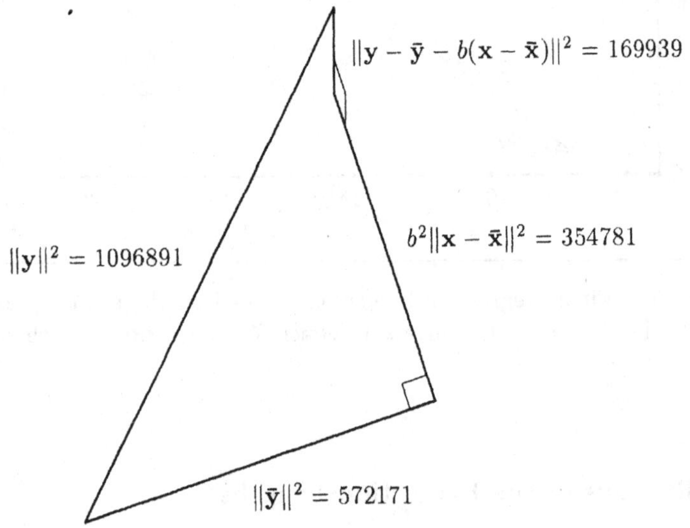

Figure 5.12: The Pythagorean breakup of the squared length of the observation vector for the general simple regression case, using our full data set as an example.

In our Pythagorean breakup the squared projection length $(\mathbf{y}\cdot\mathbf{U}_2)^2$ is often referred to as the *regression sum of squares*. Expanded algebraically it becomes

$$(\mathbf{y}\cdot\mathbf{U}_2)^2 = \left[\mathbf{y}\cdot\frac{(\mathbf{x}-\bar{\mathbf{x}})}{\|\mathbf{x}-\bar{\mathbf{x}}\|}\right]^2 = \frac{[\sum_{i=1}^n y_i(x_i-\bar{x})]^2}{\sum_{i=1}^n (x_i-\bar{x})^2} = \frac{[\sum_{i=1}^n (y_i-\bar{y})(x_i-\bar{x})]^2}{\sum_{i=1}^n (x_i-\bar{x})^2}$$

where we have used the equality $\sum_{i=1}^n \bar{y}(x_i-\bar{x}) = 0$ to rearrange the numerator. The last form of the numerator is often preferred since it is symmetric in x and y.

To test the hypothesis that $\beta = 0$ we check whether the squared projection length $(\mathbf{y}\cdot\mathbf{U_2})^2$ is similar to, or considerably greater than, the average of the squared projection lengths for the "error" directions, $\left[(\mathbf{y}\cdot\mathbf{U_3})^2 + \cdots + (\mathbf{y}\cdot\mathbf{U_{26}})^2\right]/24$. The resulting test statistic is

$$\begin{aligned} F &= \frac{(\mathbf{y}\cdot\mathbf{U_2})^2}{\left[(\mathbf{y}\cdot\mathbf{U_3})^2 + \cdots + (\mathbf{y}\cdot\mathbf{U_{26}})^2\right]/24} = \frac{b^2\|\mathbf{x}-\bar{\mathbf{x}}\|^2}{\|\mathbf{y}-\bar{\mathbf{y}}-b(\mathbf{x}-\bar{\mathbf{x}})\|^2/24} \\ &= \frac{354781}{169939/24} = \frac{354781}{7081} = 50.10 \end{aligned}$$

If $\beta = 0$ this comes from the $F_{1,24}$ distribution. Since the observed F value (50.10) is larger than the 99 percentile of the $F_{1,24}$ distribution we reject the hypothesis that $\beta = 0$, and conclude that there is very strong evidence to suggest that air pollution levels and the inversion effect are related.

Equivalent t Test

Note that our test statistic $F = (\mathbf{y}\cdot\mathbf{U_2})^2/\{\left[(\mathbf{y}\cdot\mathbf{U_3})^2 + \cdots + (\mathbf{y}\cdot\mathbf{U_{26}})^2\right]/24\} = b^2\|\mathbf{x}-\bar{\mathbf{x}}\|^2/\left[\|\mathbf{y}-\bar{\mathbf{y}}-b(\mathbf{x}-\bar{\mathbf{x}})\|^2/24\right] = 50.10$ can be transformed to

$$\begin{aligned} t &= \frac{\mathbf{y}\cdot\mathbf{U_2}}{\sqrt{\left[(\mathbf{y}\cdot\mathbf{U_3})^2 + \cdots + (\mathbf{y}\cdot\mathbf{U_{26}})^2\right]/24}} = \frac{b\|\mathbf{x}-\bar{\mathbf{x}}\|}{\|\mathbf{y}-\bar{\mathbf{y}}-b(\mathbf{x}-\bar{\mathbf{x}})\|/\sqrt{24}} \\ &= \frac{\sqrt{354781}}{\sqrt{169939/24}} = 7.078 \end{aligned}$$

If $\beta = 0$ this comes from the t_{24} distribution. As previously, the size of our observed t value constitutes very strong evidence against the idea that air pollution and the inversion effect are unrelated.

Note that an algebraic expression for t is

$$t = \frac{b\|\mathbf{x}-\bar{\mathbf{x}}\|}{s} = \frac{b\sqrt{\sum_{i=1}^{n}(x_i-\bar{x})^2}}{s} = \frac{b}{s/\sqrt{\sum_{i=1}^{n}(x_i-\bar{x})^2}}$$

where the denominator in the last expression may be familiar as the "standard error of the slope."

Correlation Coefficient

There is a third equivalent way of testing whether the slope β is zero. This is to calculate the *correlation coefficient*, r, and compare the observed value with the percentiles of the distribution of r under the hypothesis $\beta = 0$ (given in Table T.2).

The correlation coefficient can be defined as the cosine of the angle θ between the vectors $(\mathbf{y}-\bar{\mathbf{y}})$ and $(\mathbf{x}-\bar{\mathbf{x}})$, as shown in Figure 5.13 (note that b is conveniently positive). In our example it works out to be

$$r = \cos\theta = \frac{\mathbf{x}-\bar{\mathbf{x}}}{\|\mathbf{x}-\bar{\mathbf{x}}\|}\cdot\frac{\mathbf{y}-\bar{\mathbf{y}}}{\|\mathbf{y}-\bar{\mathbf{y}}\|} = \frac{\sum_{i=1}^{n}(x_i-\bar{x})(y_i-\bar{y})}{\sqrt{\sum_{i=1}^{n}(x_i-\bar{x})^2}\sqrt{\sum_{i=1}^{n}(y_i-\bar{y})^2}}$$

$$= \frac{3035.1}{\sqrt{25.9646}\sqrt{524720}} = 0.822$$

This value is much greater than the 99.5 percentile for the null distribution of the correlation coefficient with 24 "degrees of freedom," 0.496 in Table T.2, so we again conclude there is very strong evidence against the hypothesis $\beta = 0$.

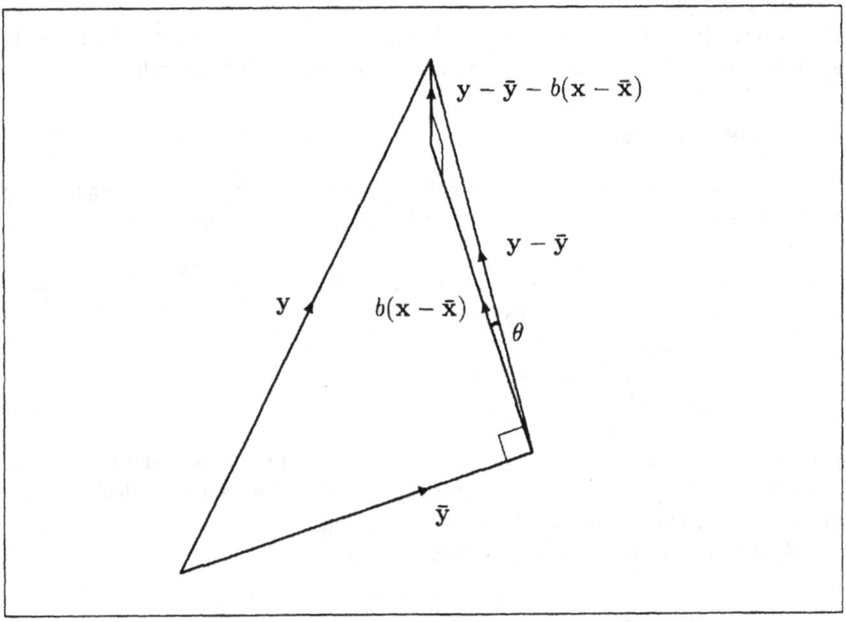

Figure 5.13: The orthogonal decomposition, showing also the "corrected" observation vector $(\mathbf{y}-\bar{\mathbf{y}})$, and the angle θ whose cosine is the correlation coefficient, r.

Note that the t value can also be expressed as a trigonometric function of the angle θ, since $t = A/(B/\sqrt{n-2}) = \sqrt{n-2}\cot\theta$. Interested readers can derive a formula linking t to r, and convince themselves that the two tests are equivalent. (See also Exercise 5.13.)

In Appendix D we discuss a fourth equivalent way of testing whether the slope β is zero. This is to use the angle θ itself as the test statistic.

Comments

As in §5.2, the "model space" directions $\mathbf{U_1}$ and $\mathbf{U_2}$ are used to estimate the intercept and slope (or, equivalently, a constant and the slope), while the "error space" directions $\mathbf{U_3}, \ldots, \mathbf{U_{26}}$ are used to estimate the variance σ^2.

The reader may have noticed that, whilst our sample size has increased from 5 to 26, our F value has actually decreased, from about 517 down to about 50. This is the opposite of what we would anticipate in the long run, and has arisen simply because we were lucky in our choice of a sample of size five, with our five points lying exceptionally close to the fitted line.

Note that we have not proven that the inversion effect *causes* high air pollution levels. There could be a third related variable which could equally be the guilty party, or one of the guilty parties. In our case ground temperature is such a variable, since the colder the temperature, the greater the number of people who light their fires, and the greater the air pollution. Also, the colder the temperature, the greater the chance of an inversion effect, and the greater the air pollution. That is, in our data set coldness and the inversion effect are closely related, or *confounded*, to use statistical language.

Lastly, the reader may have noticed that the third assumption in Figure 5.3 has been violated, since the variance of our observed y values increases with increasing x value. A remedy which seems appropriate in our situation is to take the logarithm of the y values. The resulting scattergram and fitted line are displayed in Figure 5.14. Note that the fitted line becomes

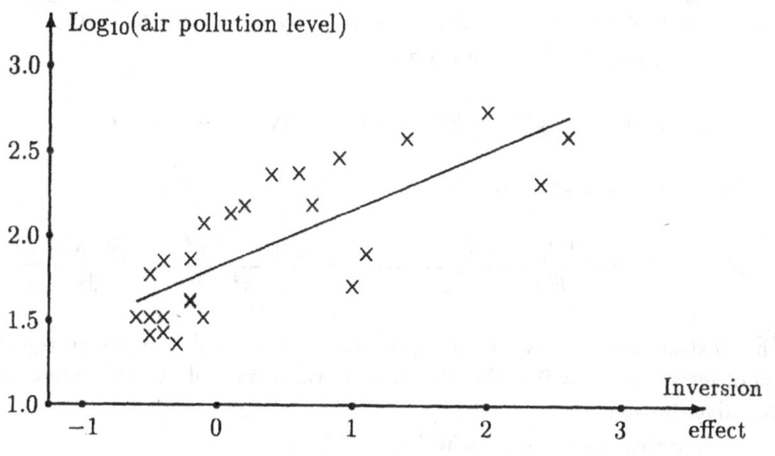

Figure 5.14: Figure 5.11 redone with air pollution level logarithm (base 10) transformed. The equation of the fitted line is $\log y = 1.8168 + 0.3417x$, with associated variance estimate $s^2 = 0.07016$.

an exponential curve when this picture is back transformed to the original scale. We shall not pursue this any further; for a full treatment the interested reader can refer to Chapter 15 of Saville and Wood (1991), where an identical situation arose.

5.4 Summary

To summarize, in the simple regression case the direction of special interest is the direction $\mathbf{U_2} = (\mathbf{x} - \bar{\mathbf{x}})/||\mathbf{x} - \bar{\mathbf{x}}||$. If the slope β is nonzero, then this is the direction for which the length of the projection is increased. More fully, if $\beta \neq 0$ the length of the projection of the vector of observations onto the direction $\mathbf{U_2}$ will be inflated relative to the projection lengths for the "error" directions $\mathbf{U_3}, \ldots, \mathbf{U_n}$. Thus to distinguish between the cases $\beta = 0$ and $\beta \neq 0$ we divide the squared projection length for the $\mathbf{U_2}$ direction by the average of the squared projection lengths for the $(n-1)$ perpendicular directions in the "error space"; if the resulting ratio is small we decide that β could be equal to zero, while if the ratio is large we decide that β could not be equal to zero.

In general, the projection of the vector of observations, $\mathbf{y} = [y_1, \ldots, y_n]^T$, onto the direction $\mathbf{U_2} = (\mathbf{x} - \bar{\mathbf{x}})/||\mathbf{x} - \bar{\mathbf{x}}||$ is the vector $b(\mathbf{x} - \bar{\mathbf{x}})$, where $b = \mathbf{y} \cdot (\mathbf{x} - \bar{\mathbf{x}})/||\mathbf{x} - \bar{\mathbf{x}}||^2$. This leads to the orthogonal decomposition $\mathbf{y} = \bar{\mathbf{y}} + b(\mathbf{x} - \bar{\mathbf{x}}) + [\mathbf{y} - \bar{\mathbf{y}} - b(\mathbf{x} - \bar{\mathbf{x}})]$ shown in Figure 5.15. Here the "error vector" $[\mathbf{y} - \bar{\mathbf{y}} - b(\mathbf{x} - \bar{\mathbf{x}})]$ is a sum of $(n-2)$ orthogonal projection vectors, one for each of the orthogonal coordinate axes spanning the error space.

The resulting Pythagorean breakup is

$$||\mathbf{y}||^2 = ||\bar{\mathbf{y}}||^2 + b^2||\mathbf{x} - \bar{\mathbf{x}}||^2 + ||\mathbf{y} - \bar{\mathbf{y}} - b(\mathbf{x} - \bar{\mathbf{x}})||^2$$

This leads to the test statistic

$$F = \frac{b^2||\mathbf{x} - \bar{\mathbf{x}}||^2}{||\mathbf{y} - \bar{\mathbf{y}} - b(\mathbf{x} - \bar{\mathbf{x}})||^2/(n-2)} = \frac{b^2||\mathbf{x} - \bar{\mathbf{x}}||^2}{s^2} = \frac{[\mathbf{y} \cdot (\mathbf{x} - \bar{\mathbf{x}})]^2}{s^2||\mathbf{x} - \bar{\mathbf{x}}||^2}$$

This test statistic follows an $F_{1,(n-2)}$ distribution if $\beta = 0$, so we reject the hypothesis that $\beta = 0$ if the observed F value exceeds the 95 percentile of this distribution.

The equivalent t test of whether $\beta = 0$ is

$$t = \frac{b||\mathbf{x} - \bar{\mathbf{x}}||}{s} = \frac{\mathbf{y} \cdot (\mathbf{x} - \bar{\mathbf{x}})}{s||\mathbf{x} - \bar{\mathbf{x}}||}$$

This test statistic follows a $t_{(n-2)}$ distribution if $\beta = 0$.

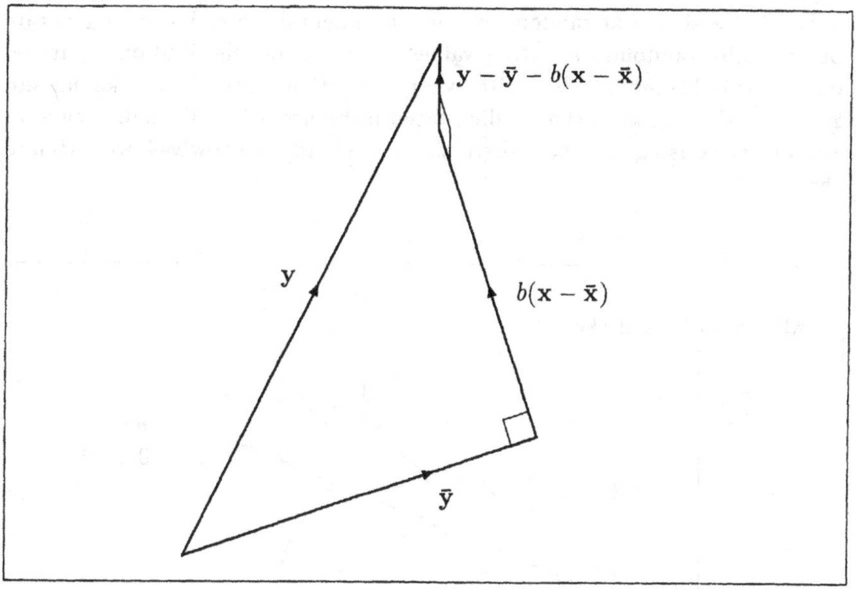

Figure 5.15: The orthogonal decomposition of the observation vector for the simple regression case.

Another equivalent test of whether $\beta = 0$ is provided by comparing the correlation coefficient r with tabulated percentiles of its distribution under the hypothesis $\beta = 0$. Here

$$r = \cos\theta = \frac{\mathbf{x}-\bar{\mathbf{x}}}{\|\mathbf{x}-\bar{\mathbf{x}}\|} \cdot \frac{\mathbf{y}-\bar{\mathbf{y}}}{\|\mathbf{y}-\bar{\mathbf{y}}\|} = \frac{\sum_{i=1}^{n}(x_i-\bar{x})(y_i-\bar{y})}{\sqrt{\sum_{i=1}^{n}(x_i-\bar{x})^2}\sqrt{\sum_{i=1}^{n}(y_i-\bar{y})^2}}$$

where θ is the angle between the "corrected" **x** and **y** vectors, as shown in Figure 5.13.

Yet another equivalent method for testing whether $\beta = 0$ is to use the angle θ itself as the test statistic. This is covered in Appendix D. For computer methods refer to Appendix C. For estimation of the 95% confidence interval for the true slope β, points on the line and predicted y values refer to pp. 402–408, 420–421 and 538–540 of Saville and Wood (1991).

For additional reading on the geometry of the simple regression case, the reader can refer to Chapter 15 and Appendix D of Saville and Wood (1991).

Class Exercise

In this class exercise we shall simulate an experiment which examines the influence of boron fertilizer on the seed yield of alfalfa. Experimental treatments will consist of borax applications at rates of 2, 4, 6, 8 and 10 kg borax/ha, assigned at random to five experimental plots. We shall generate our data by randomly selecting values from normal distributions centered on the straight line $y = 80 + 10x$, where y = alfalfa seed yield in kg/ha and x = rate of borax in kg/ha, as illustrated in Figure 5.16. We shall then analyze our data using the linear regression model, and see how well we estimate the true line.

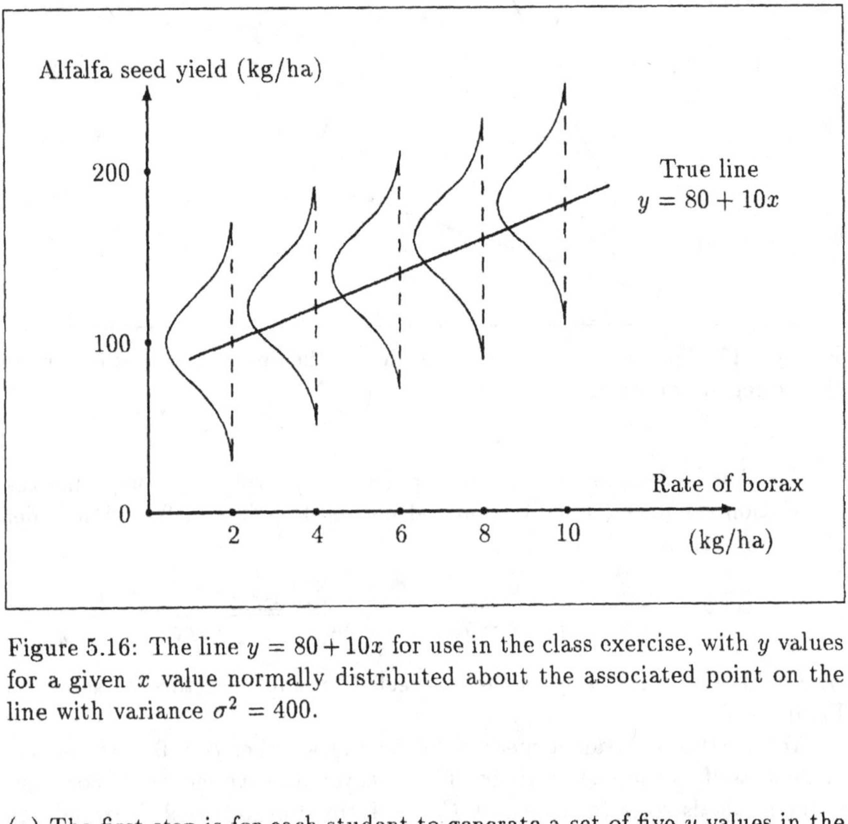

Figure 5.16: The line $y = 80 + 10x$ for use in the class exercise, with y values for a given x value normally distributed about the associated point on the line with variance $\sigma^2 = 400$.

(a) The first step is for each student to generate a set of five y values in the following manner. Pick a random number in the range 1 to 20 from Table T.1, then go to the column for $x = 2$ kg borax/ha in Table 5.3 and find the corresponding y value. This will be your alfalfa seed yield in kg/ha for the plot receiving boron at a rate of 2 kg borax/ha. Repeat this process for $x = 4$, 6, 8 and 10 kg borax/ha using the other four columns of Table 5.3. Don't worry if you strike the same random number more than once.

Write down the resulting observation vector $\mathbf{y}$ and the vector of corresponding x values, $\mathbf{x}$.

(b) Using graph paper, draw a scattergram of your data.

Random number	Alfalfa seed yields, y, for x values of:				
	2 kg/ha	4 kg/ha	6 kg/ha	8 kg/ha	10 kg/ha
1	60	78	102	120	140
2	72	90	108	132	152
3	78	96	114	138	158
4	84	102	120	144	164
5	84	102	126	144	164
6	84	108	126	150	170
7	90	108	132	150	170
8	96	114	138	150	170
9	96	114	138	156	176
10	96	120	138	156	176
11	102	120	144	162	182
12	102	126	144	162	182
13	102	126	150	168	182
14	108	132	150	168	188
15	114	132	150	168	194
16	114	138	156	174	194
17	120	138	156	180	200
18	126	144	162	186	206
19	132	150	168	192	212
20	138	162	180	198	218

Table 5.3: Samples of twenty y values normally distributed with mean $y = 80 + 10x$ and standard deviation $\sigma = 20$ kg/ha, for five rates of applied borax, $x = 2, 4, 6, 8$ and 10 kg/ha.

An appropriate coordinate system for 5-space is

$$\mathbf{U_1} = \frac{\begin{bmatrix} 1 \\ 1 \\ 1 \\ 1 \\ 1 \end{bmatrix}}{\sqrt{5}} \quad \mathbf{U_2} = \frac{\begin{bmatrix} -2 \\ -1 \\ 0 \\ 1 \\ 2 \end{bmatrix}}{\sqrt{10}} \quad \mathbf{U_3} = \frac{\begin{bmatrix} 2 \\ -1 \\ -2 \\ -1 \\ 2 \end{bmatrix}}{\sqrt{14}} \quad \mathbf{U_4} = \frac{\begin{bmatrix} -1 \\ 2 \\ 0 \\ -2 \\ 1 \end{bmatrix}}{\sqrt{10}} \quad \mathbf{U_5} = \frac{\begin{bmatrix} 1 \\ -4 \\ 6 \\ -4 \\ 1 \end{bmatrix}}{\sqrt{70}}$$

Here $\mathbf{U_1}$ is associated with the overall size of the observations, $\mathbf{U_2}$ is associated with the true slope β, and $\mathbf{U_3}, \ldots, \mathbf{U_5}$ are associated with the random variation about the true line.

(c) Recalculate $\mathbf{U_2}$ in the form $(\mathbf{x} - \bar{\mathbf{x}})/||\mathbf{x} - \bar{\mathbf{x}}||$ and confirm that it reduces to the vector $\mathbf{U_2}$ written above.

(d) Each class member is asked to calculate the scalars $\mathbf{y} \cdot \mathbf{U_1}$, $\mathbf{y} \cdot \mathbf{U_2}$, $\mathbf{y} \cdot \mathbf{U_3}$, $\mathbf{y} \cdot \mathbf{U_4}$ and $\mathbf{y} \cdot \mathbf{U_5}$ using his or her own observation vector, confirming as a check that $||\mathbf{y}||^2 = (\mathbf{y}\cdot\mathbf{U}_1)^2 + \cdots + (\mathbf{y}\cdot\mathbf{U}_5)^2$. The teacher can then plot the five scalars in five histograms (with a common scale). These histograms will approximate the distributions of the projection lengths $\mathbf{y}\cdot\mathbf{U}_1, \ldots, \mathbf{y}\cdot\mathbf{U}_5$.

Questions:

1. Which of the projection lengths appear to have mean zero?
2. Is the variability similar between the histograms?

(e) Each class member is now asked to calculate the test statistic

$$F = \frac{(\mathbf{y}\cdot\mathbf{U}_2)^2}{\left[(\mathbf{y}\cdot\mathbf{U}_3)^2 + (\mathbf{y}\cdot\mathbf{U}_4)^2 + (\mathbf{y}\cdot\mathbf{U}_5)^2\right]/3}$$

and compare his or her answer with the 95 percentile of the $F_{1,3}$ distribution.

The teacher can then plot the class results in the form of a histogram. In what percentage of simulated experiments was the hypothesis "slope $\beta = 0$" rejected at the 5% level of significance? This percentage is an approximation to the power of the 5% level test for our assumed values $\beta = 10$ and $\sigma^2 = 400$ and our experimental design.

(f) Each class member is now asked to write out the fitted model in the form

$$\mathbf{y} = (\mathbf{y}\cdot\mathbf{U}_1)\mathbf{U}_1 + (\mathbf{y}\cdot\mathbf{U}_2)\mathbf{U}_2 + \text{error vector}$$

writing out each vector in 5-space in full. Sketch this orthogonal decomposition as a vector diagram, labeling each vector appropriately.

(g) Rewrite the slope vector $(\mathbf{y}\cdot\mathbf{U}_2)\mathbf{U}_2$ in the form $b(\mathbf{x} - \bar{\mathbf{x}})$. What is your least squares estimate b of the slope β?

(h) Add your regression line $y = \bar{y} + b(x - \bar{x})$ to your scattergram. Does it look correct? Mark in the fitted values and error values.

At this point the teacher is asked to draw up a histogram of the estimated slopes b. Is this histogram centered approximately on the true value $\beta = 10$?

(i) Each class member is now asked to use their results from (f) to calculate the Pythagorean breakup

$$||\mathbf{y}||^2 = (\mathbf{y}\cdot\mathbf{U}_1)^2 + (\mathbf{y}\cdot\mathbf{U}_2)^2 + ||\text{error vector}||^2$$

and recalculate F as $(\mathbf{y}\cdot\mathbf{U}_2)^2/\left[||\text{error vector}||^2/3\right]$. Check that your answer agrees with your answer in (e).

(j) The variance σ^2 can now be estimated using either

$$s^2 = \left[(\mathbf{y}\cdot\mathbf{U}_3)^2 + (\mathbf{y}\cdot\mathbf{U}_4)^2 + (\mathbf{y}\cdot\mathbf{U}_5)^2\right]/3 \quad \text{or} \quad s^2 = ||\text{error vector}||^2/3$$

At this point the teacher is asked to draw a histogram of the s^2 values. These estimates of σ^2 will in the long run average to $\sigma^2 = 400$, though individual estimates may not be very close to this value. Does this long run average seem plausible?

(k) Lastly, the correlation coefficient r can be calculated by each class member using the formula

$$r = \cos\theta = \frac{(\mathbf{x}-\bar{\mathbf{x}})}{||\mathbf{x}-\bar{\mathbf{x}}||} \cdot \frac{(\mathbf{y}-\bar{\mathbf{y}})}{||\mathbf{y}-\bar{\mathbf{y}}||}$$

The hypothesis $\beta = 0$ can then be tested in an alternative, equivalent manner using the percentage points of the correlation coefficient as given in Table T.2. The significance of the result should agree with that obtained in (e) above, since the F and r tests are mathematically equivalent.

Exercises

Note that solutions to exercises marked with an * are given in Appendix E.

(5.1)* (a) In §5.2, rewrite the orthogonal decomposition

$$\mathbf{y} = \bar{\mathbf{y}} + b(\mathbf{x}-\bar{\mathbf{x}}) + [\mathbf{y}-\bar{\mathbf{y}}-b(\mathbf{x}-\bar{\mathbf{x}})]$$

in the alternative form

$$(\mathbf{y}-\bar{\mathbf{y}}) = b(\mathbf{x}-\bar{\mathbf{x}}) + [\mathbf{y}-\bar{\mathbf{y}}-b(\mathbf{x}-\bar{\mathbf{x}})]$$

by subtracting the overall mean vector from both sides of the equation. Redraw Figure 5.6, adding the vector $(\mathbf{y}-\bar{\mathbf{y}})$ and highlighting the triangle which represents our new decomposition.

(b) Calculate the associated Pythagorean breakup

$$||\mathbf{y}-\bar{\mathbf{y}}||^2 = b^2||\mathbf{x}-\bar{\mathbf{x}}||^2 + ||\mathbf{y}-\bar{\mathbf{y}}-b(\mathbf{x}-\bar{\mathbf{x}})||^2$$

and recalculate the test statistic using the formula

$$F = \frac{b^2||\mathbf{x}-\bar{\mathbf{x}}||^2}{||\mathbf{y}-\bar{\mathbf{y}}-b(\mathbf{x}-\bar{\mathbf{x}})||^2/3}$$

You should get the same answer as in §5.2.

Note that this is the Pythagorean breakup which forms the basis of the "analysis of variance" table given in standard statistical textbooks.

(5.2) From Table 5.1 take the air pollution levels (y) and inversion effects (x) for the four days: June 9, 12, 15 and 18. Our aims are to estimate the relationship $y = \alpha + \beta x$ and to test whether $\beta = 0$.

(a) Draw a scattergram of air pollution level versus the inversion effect.

(b) Write down the observation vector, $\mathbf{y}$.

(c) Write down the vector of x values, and the unit vectors $\mathbf{U}_1$ (in the equiangular direction) and $\mathbf{U}_2 = (\mathbf{x} - \bar{\mathbf{x}})/\|\mathbf{x} - \bar{\mathbf{x}}\|$.

(d) Calculate the projection lengths $\mathbf{y} \cdot \mathbf{U}_1$ and $\mathbf{y} \cdot \mathbf{U}_2$.

(e) Work out the projection vectors $(\mathbf{y} \cdot \mathbf{U}_1)\mathbf{U}_1$ and $(\mathbf{y} \cdot \mathbf{U}_2)\mathbf{U}_2$, and write down the orthogonal decomposition

$$\mathbf{y} = (\mathbf{y} \cdot \mathbf{U}_1)\mathbf{U}_1 + (\mathbf{y} \cdot \mathbf{U}_2)\mathbf{U}_2 + \text{error vector}$$

Sketch this decomposition as a vector diagram, labeling each vector appropriately.

(f) Work out the estimate of the slope β, and write down the equation of the fitted line in the form $y = \bar{y} + b(x - \bar{x})$. Add this line to your scattergram. Does it look correct?

(g) Write down the appropriate Pythagorean breakup.

(h) Test the hypothesis "slope $\beta = 0$" by calculating

$$F = (\mathbf{y} \cdot \mathbf{U}_2)^2/(\|\text{error vector}\|^2/2)$$

and comparing it with an appropriate reference distribution. Say what this distribution is. What is your conclusion?

(5.3) Redo the example in §5.2 using the *temperature at ground level* as the x variable for the same 5 days as summarized in Table 5.2.

(a) Draw a scattergram of y versus x.

(b) Work out your new unit vector $\mathbf{U}_2 = (\mathbf{x} - \bar{\mathbf{x}})/\|\mathbf{x} - \bar{\mathbf{x}}\|$.

(c) Write down the new orthogonal decomposition

$$\mathbf{y} = (\mathbf{y} \cdot \mathbf{U}_1)\mathbf{U}_1 + (\mathbf{y} \cdot \mathbf{U}_2)\mathbf{U}_2 + \text{error vector}$$

(d) Write down the equation of the fitted line, and add the line to the scattergram. Is the line a reasonable fit to the data?

(e) Test the hypothesis "slope $\beta = 0$" using the test statistic

$$F = (\mathbf{y} \cdot \mathbf{U}_2)^2/(\|\text{error vector}\|^2/3)$$

What is your conclusion?

This exercise demonstrates the fact that air pollution level is related to ground temperature as well as to the inversion effect.

(5.4)* An experiment was carried out to investigate the bee-pollination requirements of kiwifruit orchards. Before flowering, a large orchard was divided into six equally sized and environmentally similar areas. These were allocated, two per treatment, in a completely random manner to the three levels of beehive density listed in the table. After the fruit had "set" on the vines the number of fruit were counted in 20 one-meter lengths of row, randomly positioned within each area of the kiwifruit orchard. The average number of fruit per meter in each area is given in the table.

Treatment	Number of kiwifruit set/meter	
8 hives per hectare	33.4	30.6
12 hives per hectare	44.7	46.3
16 hives per hectare	63.7	61.3

(a) Draw a scattergram of these data, treating the number of kiwifruit set per meter as the y variable, and the number of hives per hectare as the x variable.

(b) Write down the unit vector $\mathbf{U_2} = (\mathbf{x} - \bar{\mathbf{x}})/||\mathbf{x} - \bar{\mathbf{x}}||$ and reduce it to its simplest terms.

(c) Fit the linear regression model. Write down the equation of the line and add it to the scattergram.

(d) Write down the usual Pythagorean breakup and work out your estimate of σ^2. Test the hypothesis that the number of kiwifruit set per meter is not related to the number of hives per hectare.

(e) For the linear regression model in our simple exercise, a coordinate system for the error space is given by the directions

$$\mathbf{U_3} = \frac{1}{\sqrt{12}}\begin{bmatrix} 1 \\ 1 \\ -2 \\ -2 \\ 1 \\ 1 \end{bmatrix}, \; \mathbf{U_4} = \frac{1}{\sqrt{2}}\begin{bmatrix} -1 \\ 1 \\ 0 \\ 0 \\ 0 \\ 0 \end{bmatrix}, \; \mathbf{U_5} = \frac{1}{\sqrt{2}}\begin{bmatrix} 0 \\ 0 \\ -1 \\ 1 \\ 0 \\ 0 \end{bmatrix}, \; \mathbf{U_6} = \frac{1}{\sqrt{2}}\begin{bmatrix} 0 \\ 0 \\ 0 \\ 0 \\ -1 \\ 1 \end{bmatrix}$$

assuming the data order is $\mathbf{y} = [33.4, 30.6, 44.7, \ldots]^T$. Calculate the projection lengths $\mathbf{y}\cdot\mathbf{U_3}$, $\mathbf{y}\cdot\mathbf{U_4}$, $\mathbf{y}\cdot\mathbf{U_5}$ and $\mathbf{y}\cdot\mathbf{U_6}$. Show that the average of the squares of these lengths is s^2, the estimate of σ^2 found in (d).

The reader who is interested in the link between regression and "analysis of variance" may care to look up Exercise 15.3 in Saville and Wood (1991), where we develop the link by carrying this exercise on further.

(5.5) Jessica, the daughter of one of the authors, was bottle fed as a baby. Her parents became concerned whenever she drank less than anticipated, being uncertain as to what variation in milk intake was normal. They therefore recorded her milk intake at each feed, subsequently bulking it up to daily totals and entering it on graph paper taped to the refrigerator. (Daily milk intake (y) was plotted against age (x).) This was started when Jessica was 19 days old, and carried on until she was 27 weeks old, at which time solids were introduced into her diet. The graph served as a *quality control chart*, providing Jessica's parents with a good idea as to whether a particular daily intake was abnormally low. A tidy version of the resulting graph is displayed in Figure 5.17.

In general terms, Jessica's milk intake increased in a fairly linear fashion over the 25 week period, increasing from about 450–650 ml/day to about 700–900 ml/day. With one exception, no abnormal deviations in milk intake occurred, to the relief of her parents.

Times of special concern were the dates of Jessica's injections, which are marked on Figure 5.17 with arrows. At exactly 6 weeks (42 days) of age Jessica had her first "triple vaccine injection" (for whooping cough, diphtheria and tetanus). A week later she had her first Hepatitis B vaccine

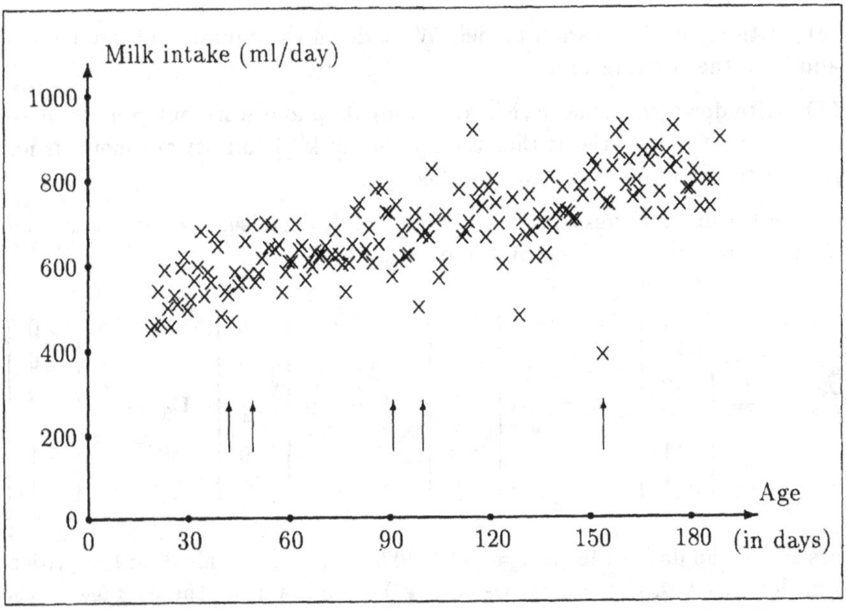

Figure 5.17: Quality control chart of Jessica's daily milk intake (y) plotted against days of age (x). Medical injections, as described in the text, occurred on the days marked with arrows.

injection. At 13 weeks of age she had her second triple vaccine injection plus her first polio sip, followed by her second Hepatitis B injection 9 days later. Lastly, at 22 weeks of age she had her third triple vaccine injection plus her second polio sip. None of these events affected Jessica's milk intake with the exception of the last event, when her milk intake was halved on the day of the injection plus sip.

As a sample of the data, the table gives Jessica's milk intakes for six regularly spaced days.

Age (days):	30	60	90	120	150	180
Milk intake (ml):	495	610	720	780	810	775

(a) Draw a scattergram of the data in the table.

(b) Write down the unit vectors $\mathbf{U_1}$ and $\mathbf{U_2}$ which are appropriate for this data set.

(c) Work out the resulting orthogonal decomposition.

(d) Work out the equation of the fitted line and add the line to the scattergram. What is your estimate of the *rate of increase* of milk intake in milliliters per day?

(e) Draw, by freehand, a line through the data presented in Figure 5.17. Use this approximate line to estimate the rate of increase in milk intake for the full data set. Was your result in (d) an overestimate or an underestimate?

(f) Does the data in the table provide evidence of an increase in milk intake over the period 30–180 days of age? Quote the appropriate test statistic and the associated reference distribution.

(5.6) In the table we present the New Zealand kiwifruit production, in thousands of tonnes, for the years 1977–1983.

Year	1977	1978	1979	1980	1981	1982	1983
Production (thousands of tonnes)	7.97	9.58	18.65	17.97	29.79	25.35	35.30

(a) Fit a simple regression line to the data, write down its equation, and draw it onto a scattergram of the data.

(b) Imagine you are the marketing manager for the New Zealand Kiwifruit Authority. Use the regression line to predict the production for 1984. How confident would you feel about your prediction?

(5.7)* One of the authors acquired a kitten in late November, 1986. The kitten had been born on October 12, 1986, and was given the name "Paw Paw." Since Paw Paw was the first kitten ever owned by the author, he decided to plot a graph of its increasing weight. Paw Paw was initially weighed on the kitchen scales, but when she became too heavy the bathroom scales were used. The resulting weights are given in the table.

Age (days):	55	65	73	86	109	153	171	210
Weight (kg):	0.9	1.1	1.4	1.6	2.1	3.2	3.2	3.9
Age (days):	230	248	270	293	313	393	459	
Weight (kg):	4.3	4.5	4.5	4.3	4.5	5.0	5.0	

(a) Draw a scattergram of these data, treating Paw Paw's weight, in kilograms, as the dependent variable y and age, in days, as the independent variable x. Paw Paw was spayed when she was 155 days old, causing a check in her growth. Also, her weight reached a plateau when she was about 8 months old.

(b) Fit a regression line to the portion of the growth curve prior to the plateau. Include the first ten values, covering the period 55–248 days of age inclusive. Write down the equation of the line and add the line to the relevant portion of the scattergram. Does the straight line appear to be a reasonable fit?

(c) Calculate the rate of increase of Paw Paw's weight, in grams per day, for the period 55–248 days of age.

(5.8) An experiment is conducted on ten cars to investigate the relationship between the amount of a petrol additive and the reduction in the quantity of nitrogen oxides in the exhaust. The data are shown in the table.

Amount of additive (x)	1	1	2	3	4	4	5	6	6	7
Reduction in nitrogen oxides (y)	2.1	2.5	3.1	3.0	3.8	3.2	4.3	3.9	4.4	4.8

(a) On graph paper, draw a scattergram of y against x.

(b) Fit a regression line to the data. Write down the regression equation and add the fitted line to the scattergram.

(c) Write down the appropriate Pythagorean breakup and use it to test the null hypothesis $H_0 : \beta = 0$ against the alternative hypothesis $H_1 : \beta \neq 0$. What is your conclusion?

(5.9) Symptoms of lead poisoning were detected in two users of an indoor small bore rifle range (George et al., 1993). Data were therefore collected on number of rounds fired per week and red cell lead levels (μmol/L) at the end of the 6 month (winter) indoor shooting season. The full data set for 49 individuals is displayed in Figure 5.18, and the data for 10 randomly selected individuals is given in the table below.

Rounds per week (x)	60	60	60	40	90	100	40	26	60	15
Red cell lead level (y)	2.93	2.54	1.84	1.42	2.54	4.33	1.39	1.98	1.95	1.30

(a) Use graph paper to draw a scattergram of this subset of the data.

(b) Write down $\mathbf{U_2} = (\mathbf{x} - \bar{\mathbf{x}})/||\mathbf{x} - \bar{\mathbf{x}}||$, calculate the projection vector $(\mathbf{y}\cdot\mathbf{U_2})\mathbf{U_2}$, and estimate the regression slope β. Work out the equation of the fitted line and add it to your scattergram.

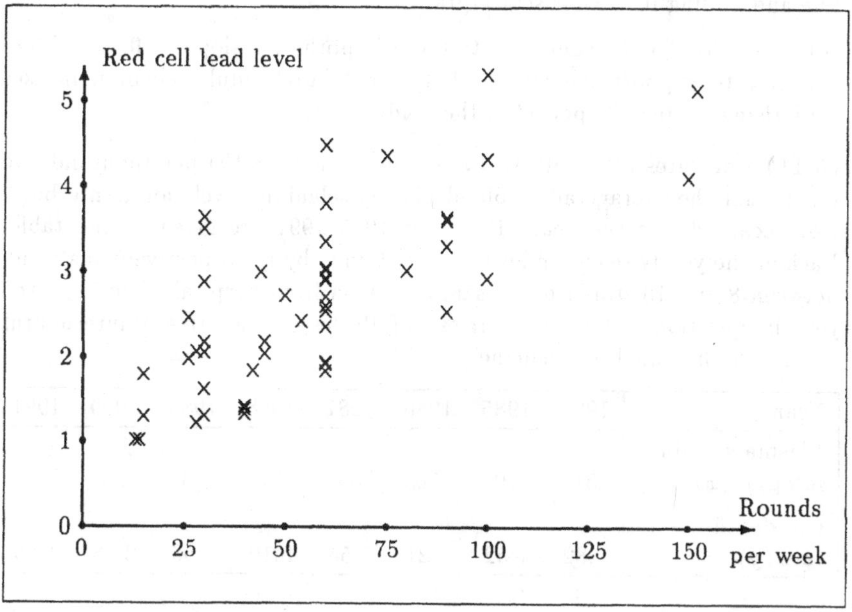

Figure 5.18: Red cell lead levels (in μmol/L) for 49 indoor small bore shooters at the end of a shooting season plotted against number of rounds fired per week. Trevor Walmsley is thanked for supplying the basic data used in George et al. (1993).

(c) Calculate the Pythagorean breakup

$$||\mathbf{y}||^2 = ||\bar{\mathbf{y}}||^2 + b^2||\mathbf{x} - \bar{\mathbf{x}}||^2 + ||\text{error vector}||^2$$

and use it to test the hypothesis $\beta = 0$. What are your conclusions?

(5.10)* In Table 4.1 each column of selenium levels is presented in the order of their sample numbers, which were assigned in chronological order. Unbeknown to the researchers at the time, the selenium intakes of Christchurch adults were rising during the period of the study (1989–1990) due to an increasing usage of imported, high-selenium wheat made possible by the deregulation of the New Zealand economy (Winterbourn et al., 1992). The percentage rise was estimated to be in the range 10–30% during the period 1989–1990 (using data in Winterbourn et al., 1992).

Note that this exercise could prove tedious for readers who do not have access to computers (see Appendix C for computing methods).

(a) Assign sample numbers 1–24 to the adult selenium levels in Table 4.1, numbering from top to bottom, and use graph paper to plot the selenium levels (y) against the sample numbers (x).

(b) Fit a regression line to these data, writing down the equation of the fitted line and adding it to your scattergram.

(c) Calculate the F value and test the hypothesis "slope = 0." Is there evidence to support the notion that Christchurch adult selenium intakes were rising during the period of the study?

(5.11) The rates of cot death (or sudden infant death) per thousand live births and the average adult blood plasma selenium levels for Canterbury, New Zealand, for the years 1981 and 1985–1991 are given in the table. Each of the yearly selenium levels was obtained by retrospectively analyzing between 8 and 16 stored blood samples taken from hospital patients in the year in question. Data by courtesy of Professor Christine Winterbourn, Christchurch School of Medicine.

Year	1981	1985	1986	1987	1988	1989	1990	1991
Plasma selenium in μg/L (x)	51	50	56	55	77	73	68	71
Cot death rate (y)	5.82	7.52	7.21	5.58	5.19	5.50	2.58	1.45

(a) On graph paper, draw a scattergram of y against x.

(b) Fit a regression line to the data. Write down the regression equation and add the fitted line to the scattergram.

(c) Write down the appropriate Pythagorean breakup and use it to test whether the regression slope is zero. What is your conclusion?

Note that a significant association between two variables does not prove cause and effect. The percentage of babies sleeping in a prone position (also seen as a risk factor) has decreased over the same time period, with the result that this variable also has a significant association with cot death rate.

The theory of the selenium/cot death researchers originated in work done 20 years ago which suggested a link between the local selenium deficiency and the high local rate of cot death (Money, 1970 and 1978). The theory is that the selenium deficiency lowers a baby's protection from stresses, including those related to poor sleeping position, infections, smoking of the mother during pregnancy, and so on. In other words, the deficiency predisposes the baby to death from other causes, and is the underlying cause which induces associations between cot death rate and other variables such as sleeping position, infections and smoking. The current debate in New Zealand is whether the alleviation of the selenium deficiency has *caused* the drop in cot death rate through alleviating vulnerability to stress, or whether publicity on subjects such as better sleeping position and the dangers of smoking have in themselves caused the drop in cot death rate without any involvement of selenium. The type of data presented in this exercise cannot distinguish between the two, so we may never know which is more true.

(5.12)* The Canterbury Regional Council, based in Christchurch, New Zealand, has recorded groundwater levels in monitoring bore number "M35/1080" for the last 40 years. In Table 5.4 we present the April and October groundwater levels for each year from 1953 to 1990; these are expressed as the distance in meters from ground level to the level of water in the bore. In the Southern Hemisphere summer straddles Christmas, and the annual summer drop in water level (known as the *drawdown*) is calculated as the drop in water level between October 1 and the following April 1. For example, in the summer which included the Christmas of 1953 the water level dropped from 12.2 m below ground level on October 1, 1953, to 13.2 m below ground level on April 1, 1954, resulting in a drawdown of 13.2−12.2 = 1.0 m (Table 5.4). Also presented in Table 5.4 is the rainfall for each October 1 to April 1 period from the meteorological station at Darfield, 20–30 km away.

In this exercise we shall use these data to shed some light on two hypotheses which have been debated between irrigators, university-based hydrologists, and Canterbury Regional Council hydrologists. These are:

(1) That summer drawdown is heavily influenced by summer rainfall, but not by the amount of water commonly extracted for irrigation purposes.

(2) That summer drawdown is related to the groundwater level at the start of summer (October 1), with the drawdown being low or zero if the October 1 level is low. The theory here is that in an average

Year	Water level (m) April	Water level (m) October	Drawdown (m) (Oct. – Apr.)	Rainfall (mm) (Oct. – Apr.)
1953	*	12.2	*	*
1954	13.2	12.6	1.0	406
1955	13.3	12.9	0.7	425
1956	13.2	14.0	0.3	281
1957	13.2	11.9	−0.8	536
1958	13.8	14.8	1.9	438
1959	12.9	13.9	−1.9	524
1960	14.1	13.0	0.2	314
1961	14.4	12.1	1.4	502
1962	13.6	12.6	1.5	284
1963	13.9	12.7	1.3	344
1964	13.8	14.2	1.1	270
1965	14.1	12.2	−0.1	447
1966	13.6	13.4	1.4	470
1967	14.5	13.9	1.1	388
1968	13.2	13.3	−0.7	457
1969	15.2	15.9	1.9	283
1970	15.6	13.9	−0.3	402
1971	15.1	14.6	1.2	257
1972	15.3	14.9	0.7	306
1973	15.4	12.9	0.5	312
1974	14.9	11.5	2.0	353
1975	12.7	10.8	1.2	546
1976	13.3	12.4	2.5	379
1977	14.2	11.5	1.8	348
1978	13.9	12.0	2.4	271
1979	12.6	11.9	0.6	488
1980	12.6	12.9	0.7	603
1981	14.1	12.5	1.2	336
1982	15.3	16.2	2.8	223
1983	16.2	13.9	0.0	355
1984	14.1	15.4	0.2	629
1985	15.7	16.2	0.3	289
1986	15.7	12.0	−0.5	656
1987	12.9	13.6	0.9	466
1988	15.2	15.4	1.6	275
1989	16.3	12.8	0.9	222
1990	15.0	*	2.2	383

Table 5.4: Groundwater levels in bore M35/1080 on April 1 and October 1 of each year, October to April drawdown, and rainfall for each October to April period. Groundwater data by courtesy of Mr Peter Callander, Canterbury Regional Council. Rainfall data is from the Darfield meteorological station.

year the groundwater level falls naturally from an annual high in October toward a minimum level set by the level of the nearby large Waimakariri river, regardless of how much water is extracted for irrigation purposes. This is based on the idea that the groundwater level will never fall below a certain minimum level, which depends on the distance from the river, since the "hydraulic gradient" (difference in water level) ensures that the groundwater is replenished by water from the river.

Unfortunately there is little precise data on the amount of water extracted by irrigators annually, so we cannot check whether there is a relationship between drawdown and amount of irrigation water extracted. However, we can examine the relationships between drawdown and both rainfall and the October 1 level. Note that this exercise could prove tedious for readers who do not have access to computers (see Appendix C for computing methods).

(a) To explore the first hypothesis, draw a scattergram of the summer drawdowns (y) versus the summer rainfalls (x) for the 37 summer periods summarized in Table 5.4.

(b) Fit a regression line to these data, write down its equation, and add the line to your scattergram.

(c) Test the hypothesis "slope = 0." What is your F value? How strong is the evidence that summer drawdown is associated with summer rainfall?

(d) To explore the second hypothesis, draw a scattergram of the summer drawdowns (y) versus the October 1 levels (x) for the 37 summer periods summarized in Table 5.4. Remember that each y value is associated with the x value on the *preceding* line of the table.

(e) Fit a regression line to these data, write down its equation, and add the line to your scattergram.

(f) Test the hypothesis "slope = 0." What is your F value? How strong is the evidence that summer drawdown is associated with the October 1 level?

(g) Use your regression line to calculate the x value at which the line crosses the x axis (corresponding to zero predicted drawdown).

(5.13) Show that the estimate of the slope b and the correlation coefficient r are related by the formula

$$r = \frac{b\,||\mathbf{x} - \bar{\mathbf{x}}||}{||\mathbf{y} - \bar{\mathbf{y}}||}$$

Chapter 6

Overview

In this brief summary chapter we reduce to essentials what the reader has learned in Chapters 1–5. In §6.1 we draw a "statistical triangle" for each of our introductory examples and relate our F and t tests to the lengths of the sides of this triangle. In §6.2 we draw the book to a conclusion. The chapter then ends with a few exercises.

6.1 Statistical Triangle

In Chapters 2–5 our F test statistic has always been calculated as the squared length of the projection onto a special direction divided by the average squared length of the projections onto the directions which correspond to the background variation. For our most recent regression example in §5.3 this is

$$F = \frac{(\mathbf{y}\cdot\mathbf{U}_2)^2}{\left[(\mathbf{y}\cdot\mathbf{U}_3)^2 + \cdots + (\mathbf{y}\cdot\mathbf{U}_{26})^2\right]/24} = \frac{b^2\|\mathbf{x}-\bar{\mathbf{x}}\|^2}{\|\mathbf{y}-\bar{\mathbf{y}}-b(\mathbf{x}-\bar{\mathbf{x}})\|^2/24} = \frac{A^2}{B^2/q}$$

where $A = |OM|$ and $B = |MS|$ in Figure 6.1, and q is the dimension of the error space, or the number of error degrees of freedom.

This calculation involves only the lengths of the sides of the triangle bounded by the vertices O, M and S in Figure 6.1 in addition to the number of error degrees of freedom. In recognition of the fundamental importance of this triangle we now give it the grand title "statistical triangle." The essential features of this triangle are illustrated in Figure 6.2.

Such a triangle can be drawn for all of the examples presented in this book, and for virtually all of the examples in the sister text (Saville and Wood, 1991). In Figure 6.3(a)–(d) we present the "statistical triangle" for

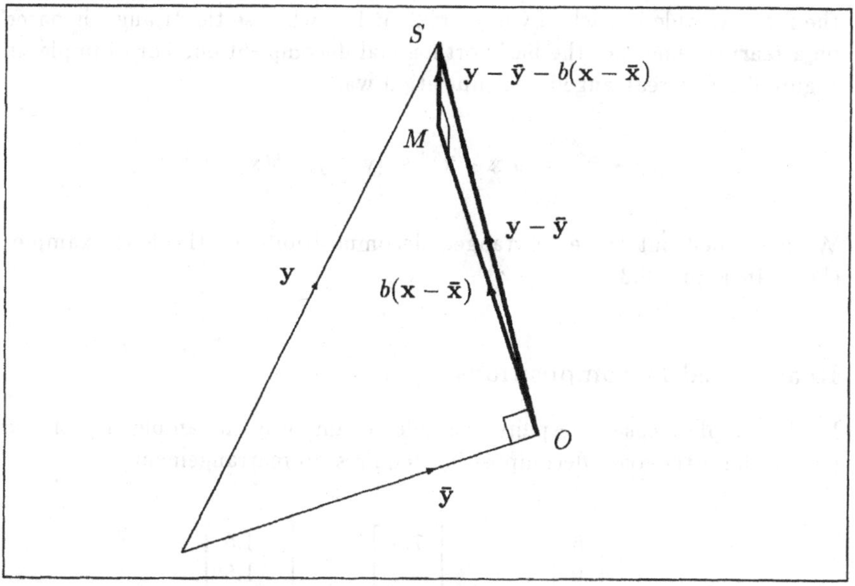

Figure 6.1: The orthogonal decomposition for the example in §5.3 with the "statistical triangle" OMS highlighted. This triangle is based on the rearranged decomposition $(\mathbf{y}-\bar{\mathbf{y}}) = b(\mathbf{x}-\bar{\mathbf{x}}) + [\mathbf{y}-\bar{\mathbf{y}}-b(\mathbf{x}-\bar{\mathbf{x}})]$. This figure is an embellished reproduction of Figure 5.13.

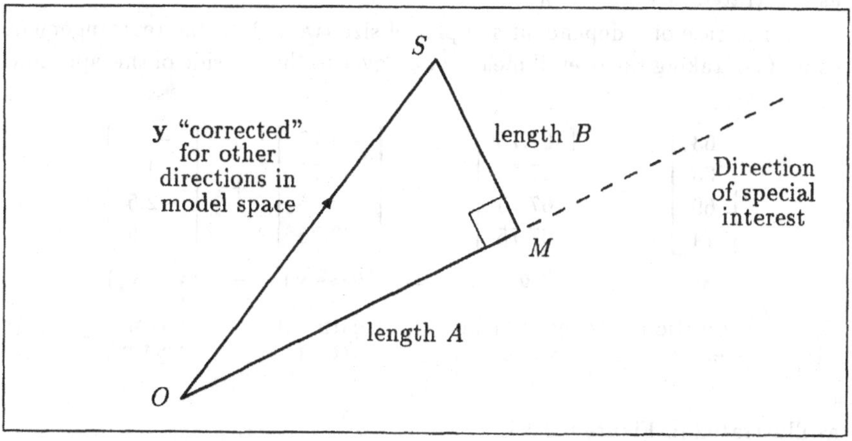

Figure 6.2: The "statistical triangle." This triangle is defined by two directions: (1) the direction of special interest; and (2) the direction of the "corrected" observation vector, $\mathbf{y}$ minus its projections onto all directions in the model space *except* the direction of special interest. The triangle lies in a $(q+1)$-dimensional subspace of N-space, where q is the number of error degrees of freedom. The test statistic F is A^2/B^2.

the first example in each of Chapters 2–5. In each case the triangle is based on a rearrangement of the basic orthogonal decomposition. For example, in Figure 6.1 this rearranged decomposition was

$$(\mathbf{y}-\bar{\mathbf{y}}) = b(\mathbf{x}-\bar{\mathbf{x}}) + [\mathbf{y}-\bar{\mathbf{y}}-b(\mathbf{x}-\bar{\mathbf{x}})]$$

We now spell out these rearranged decompositions for the four examples shown in Figure 6.3.

Rearranged Decompositions

In the simplest case, the paired samples example for a sample size of two (§2.2), the orthogonal decomposition requires no rearrangement:

$$\begin{bmatrix} 6 \\ 9 \end{bmatrix} = \begin{bmatrix} 7.5 \\ 7.5 \end{bmatrix} + \begin{bmatrix} -1.5 \\ 1.5 \end{bmatrix}$$

$$\underset{\text{observation vector}}{\mathbf{y}} = \underset{\text{model vector}}{\bar{\mathbf{y}}} + \underset{\text{error vector}}{(\mathbf{y}-\bar{\mathbf{y}})}$$

as illustrated in Figure 6.3(a).

In the case of independent samples of size two (§3.2), the rearrangement consists of taking the overall mean vector over to the left side of the equation:

$$\begin{bmatrix} 63 \\ 65 \\ 69 \\ 74 \end{bmatrix} - \begin{bmatrix} 67.75 \\ 67.75 \\ 67.75 \\ 67.75 \end{bmatrix} = \begin{bmatrix} -3.75 \\ -3.75 \\ 3.75 \\ 3.75 \end{bmatrix} + \begin{bmatrix} -1 \\ 1 \\ -2.5 \\ 2.5 \end{bmatrix}$$

$$\underset{\text{observation vector}}{\mathbf{y}} - \underset{\text{overall mean vector}}{\bar{\mathbf{y}}} = \underset{\text{treatment vector}}{(\bar{\mathbf{y}}_i-\bar{\mathbf{y}})} + \underset{\text{error vector}}{(\mathbf{y}-\bar{\mathbf{y}}_i)}$$

as illustrated in Figure 6.3(b).

In the case of three independent samples of size two (§4.2), there are two statistical triangles, one for testing whether the first contrast (adults minus babies) is zero, and another for testing whether the second contrast (full-term minus premature babies) is zero. To save space we shall discuss just the second of these two triangles. In this case the orthogonal decomposition is rearranged by taking the overall mean vector and the *first* contrast vector

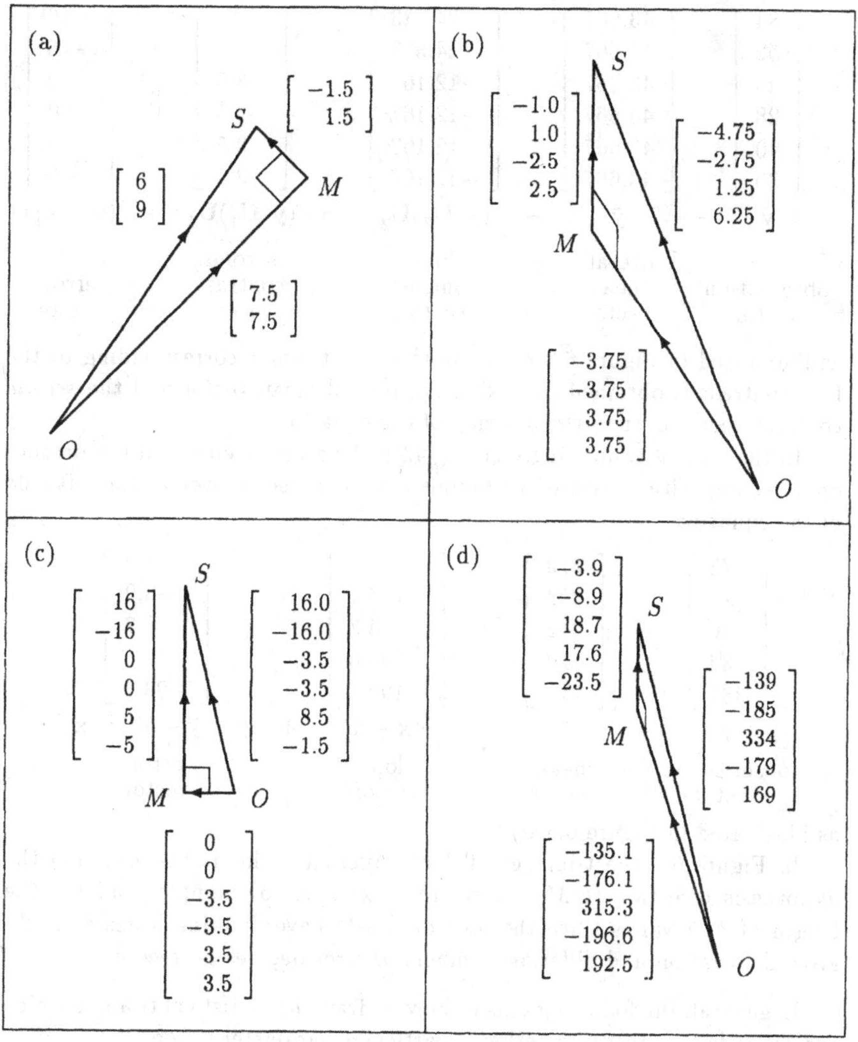

Figure 6.3: The "statistical triangles" for our introductory examples for: (a) paired samples; (b) two independent samples; (c) three independent samples; and (d) simple regression. Compare these triangles with the orthogonal decompositions shown in: (a) Figure 2.3; (b) Figure 3.3; (c) Figure 4.3; and (d) Figure 5.6. Note that we have subtracted from $\mathbf{y}$ its projections onto all directions in the model space *except* the direction corresponding to the current parameter of interest (there were 0, 1, 2 and 1 such directions in (a)–(d), respectively).

over to the left side of the equation:

$$\begin{bmatrix} 84 \\ 52 \\ 28 \\ 28 \\ 40 \\ 30 \end{bmatrix} - \begin{bmatrix} 43.667 \\ 43.667 \\ 43.667 \\ 43.667 \\ 43.667 \\ 43.667 \end{bmatrix} - \begin{bmatrix} 24.333 \\ 24.333 \\ -12.167 \\ -12.167 \\ -12.167 \\ -12.167 \end{bmatrix} = \begin{bmatrix} 0 \\ 0 \\ -3.5 \\ -3.5 \\ 3.5 \\ 3.5 \end{bmatrix} + \begin{bmatrix} 16 \\ -16 \\ 0 \\ 0 \\ 5 \\ -5 \end{bmatrix}$$

$$\mathbf{y} \quad - \quad \bar{\mathbf{y}} \quad - \quad (\mathbf{y}\cdot\mathbf{U}_2)\mathbf{U}_2 \quad = \quad (\mathbf{y}\cdot\mathbf{U}_3)\mathbf{U}_3 \quad + \quad (\mathbf{y}-\bar{\mathbf{y}}_i)$$

observation vector; overall mean vector; first contrast vector; second contrast vector; error vector

as illustrated in Figure 6.3(c). Note that the triangle corresponding to the first contrast is obtained by taking the overall mean vector and the *second* contrast vector over to the left side of the equation.

In the case of simple regression (§4.2), the rearrangement of the orthogonal decomposition consists of taking the mean vector over to the left side of the equation:

$$\begin{bmatrix} 73 \\ 27 \\ 546 \\ 33 \\ 381 \end{bmatrix} - \begin{bmatrix} 212 \\ 212 \\ 212 \\ 212 \\ 212 \end{bmatrix} = \begin{bmatrix} -135.1 \\ -176.1 \\ 315.3 \\ -196.6 \\ 192.5 \end{bmatrix} + \begin{bmatrix} -3.9 \\ -8.9 \\ 18.7 \\ 17.6 \\ -23.5 \end{bmatrix}$$

$$\mathbf{y} \quad - \quad \bar{\mathbf{y}} \quad = \quad b(\mathbf{x}-\bar{\mathbf{x}}) \quad + \quad [\mathbf{y}-\bar{\mathbf{y}}-b(\mathbf{x}-\bar{\mathbf{x}})]$$

observation vector; mean vector; slope vector; error vector

as illustrated in Figure 6.3(d).

In Figure 6.3 the triangles all look different. This is because: (1) the hypothesis direction (OM) varies from example to example; and (2) the length of MS varies since the four data sets have differing levels of background variation and differing numbers of error degrees of freedom.

In general, the following tells us how to draw the statistical triangle which corresponds to a test of whether a particular parameter is zero:

(1) The vector OM is the projection of the observation vector onto the direction corresponding to the parameter of interest.

(2) The vector OS is the observation vector minus its projections onto *all* directions in the model space *except* the direction corresponding to the parameter of interest. In the above examples there are 0, 1, 2 and 1 such directions.

(3) The vector MS is the error vector.

Clearly any two of these three vectors will suffice, since the third can be obtained by differencing.

Before calculating the F and t test statistics using the statistical triangle, we summarize the Pythagorean breakups corresponding to each of the above rearranged orthogonal decompositions.

Pythagorean Breakups

In the paired samples case the Pythagorean breakup is

$$\begin{aligned} 117 &= 112.5 + 4.5 \\ \|\mathbf{y}\|^2 &= \|\bar{\mathbf{y}}\|^2 + \|\mathbf{y} - \bar{\mathbf{y}}\|^2 \end{aligned}$$

as illustrated in Figure 6.4(a).

In the independent samples case the Pythagorean breakup is

$$\begin{aligned} 18431 - 18360.25 &= 56.25 + 14.5 \\ \|\mathbf{y}\|^2 - \|\bar{\mathbf{y}}\|^2 &= \|\bar{\mathbf{y}}_i - \bar{\mathbf{y}}\|^2 + \|\mathbf{y} - \bar{\mathbf{y}}_i\|^2 \end{aligned}$$

as illustrated in Figure 6.4(b).

In the case of three independent samples the Pythagorean breakup is

$$\begin{aligned} 13828 - 11440.7 - 1776.3 &= 49 + 562 \\ \|\mathbf{y}\|^2 - \|\bar{\mathbf{y}}\|^2 - (\mathbf{y}\cdot\mathbf{U}_2)^2 &= (\mathbf{y}\cdot\mathbf{U}_3)^2 + \|\mathbf{y} - \bar{\mathbf{y}}_i\|^2 \end{aligned}$$

as illustrated in Figure 6.4(c).

In the simple regression case the Pythagorean breakup is

$$\begin{aligned} 450424 - 224720 &= 224402 + 1302 \\ \|\mathbf{y}\|^2 - \|\bar{\mathbf{y}}\|^2 &= b^2\|\mathbf{x} - \bar{\mathbf{x}}\|^2 + \|\mathbf{y} - \bar{\mathbf{y}} - b(\mathbf{x} - \bar{\mathbf{x}})\|^2 \end{aligned}$$

as illustrated in Figure 6.4(d).

Test Statistics

In all four cases the F test statistic is given by the expression

$$F = \frac{A^2}{B^2/q}$$

where q is the number of error degrees of freedom. The calculated F values are summarized in the third column of Table 6.1.

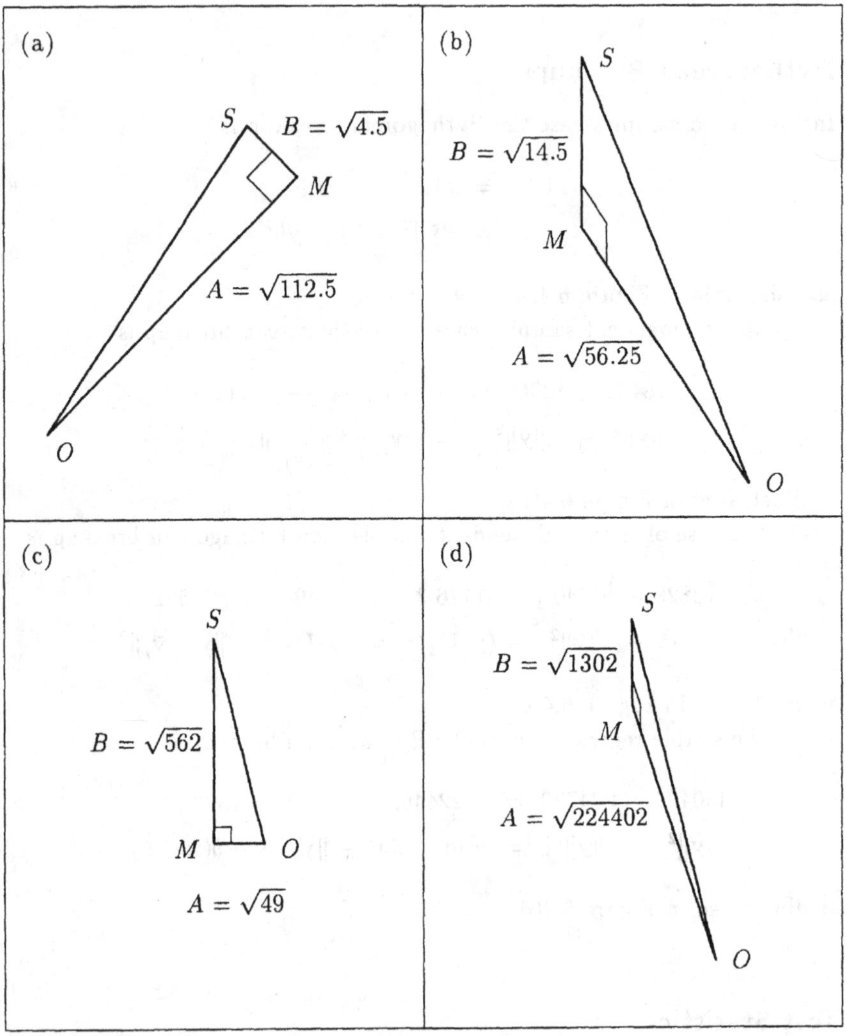

Figure 6.4: The lengths of two sides A and B of each statistical triangle. These are the basic quantities required to calculate the test statistics F and t in Table 6.1.

Similarly the t test statistic is given by the expression

$$t = \pm \frac{A}{B/\sqrt{q}}$$

where the resulting t value is positive if the observation vector has a component in the positive direction of the direction associated with the parameter of interest, or is negative if this is not the case. The calculated t values are summarized in the fourth column of Table 6.1.

	Error	Test statistics	
Section	d.f. (q)	F value	t value
2.2	1	25.00	5.000
3.2	2	7.76	2.785
4.2	3	0.26	0.511
5.2	3	517	22.74

Table 6.1: Two equivalent test statistics F and t, where $F = t^2$, for our four introductory examples.

The "critical values" of the distributions of $F_{1,q}$ and t_q are given in Table T.2. Comparison of the values in Table 6.1 with the appropriate critical values reveals that the two tests always yield the result.

In Appendix D we discuss a third equivalent test statistic, the angle θ at the vertex O of the statistical triangle. This test statistic leads directly to the p value via a simple geometric argument. Our current test statistics, F and t, are simply related to θ via the formulas $F = q\cot^2\theta$ and $t = \sqrt{q}\cot\theta$, obtained by substituting $\cot\theta = A/B$ in the formulas given earlier in this subsection. Toward the end of Appendix D we also introduce a fourth equivalent test statistic, $r = \cos\theta$; under the null hypothesis this follows the distribution of the correlation coefficient with q error degrees of freedom.

6.2 Conclusion

In conclusion, the geometric approach followed in this book has revealed the simplicity of the basic idea which underlies the statistical methods introduced by R.A. Fisher. The same basic approach is sufficient for all of the cases discussed in this book, as well as for those in its sister book (Saville and Wood, 1991). This reveals the fact that the underlying mathematics is the same for the following statistical methods: paired and independent samples t tests, analysis of variance, regression, and analysis of covariance.

Exercises

Note that solutions to exercises marked with an * are given in Appendix E.

(6.1)* (a) For the observation vector in 3-space used in §2.3, draw the statistical triangle following the examples shown in Figure 6.3, taking care to label the vertices (O, M and S).

(b) Redraw this figure as shown in Figure 6.4, giving the lengths of each side and again labeling the vertices.

(c) Use these lengths to calculate the appropriate F and t test statistics.

(d) Redo (a)–(c) using the observation vector in 9-space used in §2.4.

(6.2) (a) For the data in Exercise 2.6, draw the statistical triangle following the examples shown in Figure 6.3, taking care to label the vertices (O, M and S).

(b) Redraw this figure as shown in Figure 6.4, giving the lengths of each side and again labeling the vertices.

(c) Use these lengths to calculate the appropriate F and t test statistics.

(6.3) (a) For the observation vector in 8-space used in §3.3, draw the statistical triangle following the examples shown in Figure 6.3, taking care to label the vertices (O, M and S).

(b) Redraw this figure as shown in Figure 6.4, giving the lengths of each side and again labeling the vertices.

(c) Use these lengths to calculate the appropriate F and t test statistics.

(6.4) (a) For the data in Exercise 3.3, draw the statistical triangle following the examples shown in Figure 6.3, taking care to label the vertices (O, M and S).

(b) Redraw this figure as shown in Figure 6.4, giving the lengths of each side and again labeling the vertices.

(c) Use these lengths to calculate the appropriate F and t test statistics.

(6.5) (a) For the observation vector in 6-space used in §4.2, draw the statistical triangle corresponding to the *first* contrast (of adults versus babies). Follow the example of Figure 6.3, and label vertices. Note that some artistic licence is allowed!

(b) Redraw this figure as shown in Figure 6.4, giving the lengths of each side and again labeling the vertices.

(c) Use these lengths to calculate the appropriate F and t test statistics.

(6.6)* (a) For the data in Exercise 4.1, draw the statistical triangle corresponding to the first contrast, of sheep versus goats. Follow the example of Figure 6.3, and label vertices. Note that some artistic licence is allowed.

(b) Redraw this figure as shown in Figure 6.4, giving the lengths of each side and again labeling the vertices.

(c) Use these lengths to calculate the appropriate F and t test statistics.

(6.7)* (a) For the observation vector in 26-space used in §5.3, draw the appropriate statistical triangle. Follow the example of Figure 6.3, and label vertices.

(b) Redraw this figure as shown in Figure 6.4, giving the lengths of each side and again labeling the vertices.

(c) Use these lengths to calculate the appropriate F and t test statistics.

(6.8) (a) For the data in Exercise 5.4, draw the appropriate statistical triangle.

(b) Redraw this figure showing lengths as in Figure 6.4.

(c) Use these lengths to calculate the appropriate F and t test statistics.

Appendix A

Geometric Tool Kit

In this appendix we give a brief introduction to elementary vector geometry. Each of the most important geometric tools is introduced in relation to the data sets used in Chapters 1 and 3, namely, $\{1.3, 1.5\}$ and $\{63, 65, 69, 74\}$. These geometric tools are adequate for our purposes; for a more comprehensive discussion, however, the reader can refer to Chapter 2 of Saville and Wood (1991).

Definition of a Vector

The *vector* $\mathbf{y} = \begin{bmatrix} 1.3 \\ 1.5 \end{bmatrix}$ is an example of a vector in two-dimensional space (2-space). This is shown pictorially in Figure A.1. Note that vectors are written in nonitalic **bold** type to distinguish them from scalars.

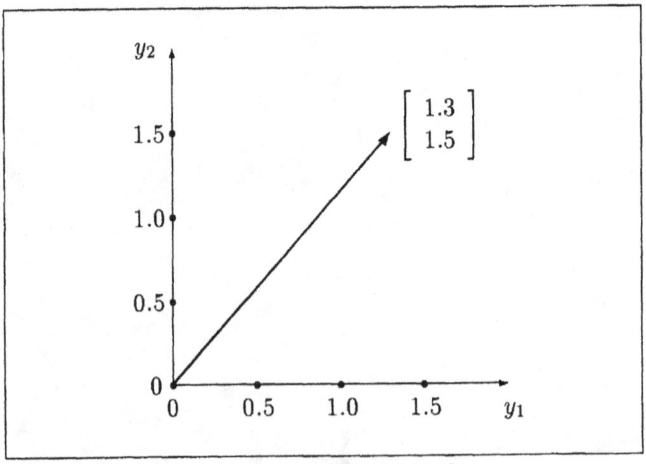

Figure A.1: The vector $\mathbf{y} = \begin{bmatrix} 1.3 \\ 1.5 \end{bmatrix}$ in 2-space.

The vector $\mathbf{y} = \begin{bmatrix} 63 \\ 65 \\ 69 \\ 74 \end{bmatrix}$ is an example of a vector in 4-space. In four and higher dimensions we can no longer represent vectors pictorially in an accurate manner; throughout this text, however, we indulge in a little artistic licence and draw diagrams involving vectors in four and higher dimensions.

In general, the vector $\mathbf{y} = \begin{bmatrix} y_1 \\ \vdots \\ y_N \end{bmatrix}$ in N-dimensional space is formally defined to be an ordered column of numbers or symbols.

A word on a space-saving notation. When a vector appears in a sentence we may decide to write it horizontally, so that for example the above vector would be written as $[y_1, \ldots, y_N]^{\mathrm{T}}$. The "T" denotes "transpose," indicating that the vector has been turned on its side.

Length of a Vector

The *length* of the vector $\mathbf{y} = [1.3, 1.5]^{\mathrm{T}}$ is defined to be

$$\|\mathbf{y}\| = \sqrt{1.3^2 + 1.5^2} = \sqrt{3.94} = 1.98$$

This is motivated by Pythagoras' Theorem, $a^2 = b^2 + c^2$, illustrated in Figure A.2.

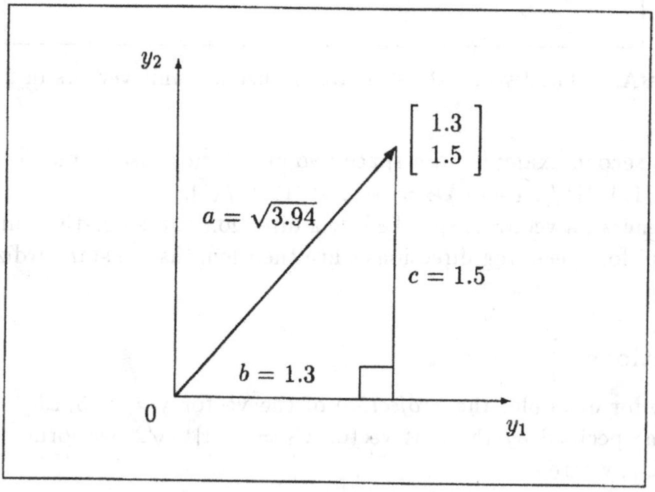

Figure A.2: The squared length of the vector $\mathbf{y} = [1.3, 1.5]^{\mathrm{T}}$ is given by $\|\mathbf{y}\|^2 = 1.3^2 + 1.5^2 = 3.94$.

As a second example, the length of the vector $\mathbf{y} = [63, 65, 69, 74]^T$ is given by $||\mathbf{y}|| = \sqrt{63^2 + 65^2 + 69^2 + 74^2} = \sqrt{18431} = 135.8$.

In general, the length of the vector $[y_1, \ldots, y_N]^T$ is defined to be

$$||\mathbf{y}|| = \sqrt{y_1^2 + \cdots + y_N^2}$$

Unit Vectors

A vector of length one is called a *unit vector.* For example, in 2-space the unit vectors which are most useful in statistical analysis are $\mathbf{U_1} = [1, 1]^T/\sqrt{2}$ and $\mathbf{U_2} = [-1, 1]^T/\sqrt{2}$; these are shown in Figure A.3.

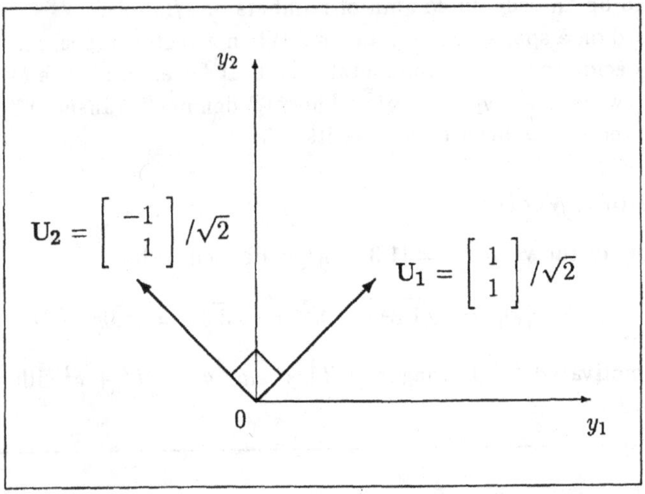

Figure A.3: The two most "statistically useful" unit vectors in 2-space.

As a second example, in 4-space two of the most useful unit vectors are $\mathbf{U_1} = [1, 1, 1, 1]^T/\sqrt{4}$ and $\mathbf{U_2} = [-1, -1, 1, 1]^T/\sqrt{4}$.

In general, a vector is specified by a direction and a length. Unit vectors are useful for specifying directions since their lengths are standardized to be one.

Projections

To find, for example, the *projection* of the vector $\mathbf{y} = [1.3, 1.5]^T$ onto the direction specified by the unit vector $\mathbf{U_1} = [1, 1]^T/\sqrt{2}$, we form the vector $(\mathbf{y} \cdot \mathbf{U_1})\mathbf{U_1}$, where

$$\mathbf{y} \cdot \mathbf{U_1} = \begin{bmatrix} 1.3 \\ 1.5 \end{bmatrix} \cdot \begin{bmatrix} 1/\sqrt{2} \\ 1/\sqrt{2} \end{bmatrix} = 1.3 \times (1/\sqrt{2}) + 1.5 \times (1/\sqrt{2}) = 2.8/\sqrt{2} = 1.4\sqrt{2}$$

Here the raised "·" (or "dot product") is shorthand for cross multiplication followed by addition. The resulting projection (vector) is

$$(\mathbf{y}\cdot\mathbf{U_1})\mathbf{U_1} = 1.4\sqrt{2}\begin{bmatrix}1/\sqrt{2}\\1/\sqrt{2}\end{bmatrix} = \begin{bmatrix}1.4\\1.4\end{bmatrix}$$

as illustrated in Figure A.4.

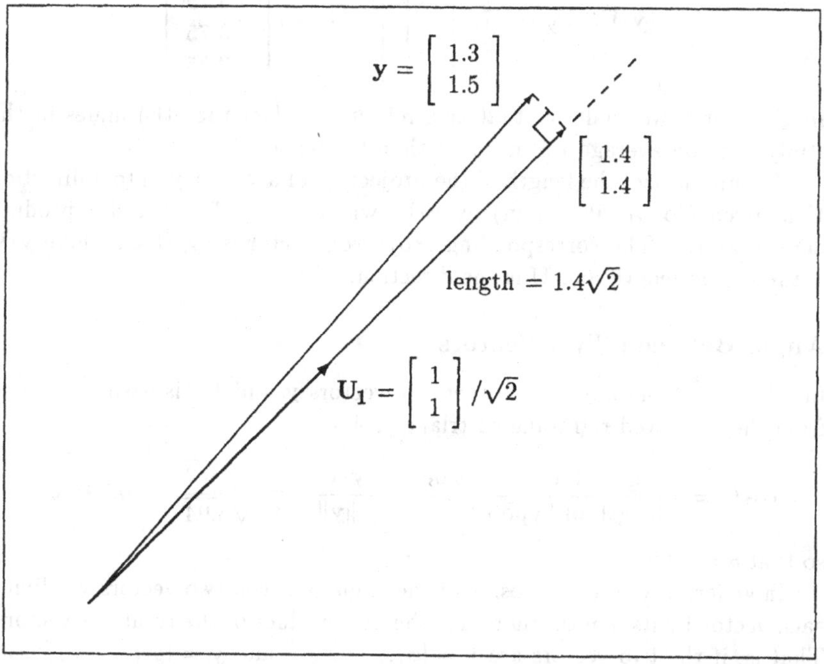

Figure A.4: The length of the projection of $\mathbf{y} = [1.3, 1.5]^T$ onto the direction $\mathbf{U_1} = [1,1]^T/\sqrt{2}$ is given by $\mathbf{y}\cdot\mathbf{U_1} = 1.3 \times (1/\sqrt{2}) + 1.5 \times (1/\sqrt{2}) = 2.8/\sqrt{2} = 1.4\sqrt{2}$. The corresponding projection vector is $(\mathbf{y}\cdot\mathbf{U_1})\mathbf{U_1} = 1.4\sqrt{2}[1,1]^T/\sqrt{2} = [1.4, 1.4]^T$.

The quantity $\mathbf{y}\cdot\mathbf{U_1}$ we loosely term the *projection length*; more correctly, its magnitude is the length of the projection, while its sign tells us whether the projection lies in the same, or the opposite direction to $\mathbf{U_1}$. Throughout this text we shall often be sloppy in the interests of brevity, referring to the projection length as $\mathbf{y}\cdot\mathbf{U_1}$ when it is more correctly $|\mathbf{y}\cdot\mathbf{U_1}|$.

As a second example we shall find the projection of $\mathbf{y} = [63, 65, 69, 74]^T$ onto the direction $\mathbf{U_2} = [-1,-1,1,1]^T/\sqrt{4}$. In this case the projection length is

$$\mathbf{y}\cdot\mathbf{U}_2 = \begin{bmatrix} 63 \\ 65 \\ 69 \\ 74 \end{bmatrix} \cdot \begin{bmatrix} -1 \\ -1 \\ 1 \\ 1 \end{bmatrix} / \sqrt{4} = \frac{-63-65+69+74}{2} = 7.5$$

The corresponding projection vector is

$$(\mathbf{y}\cdot\mathbf{U}_2)\mathbf{U}_2 = 7.5 \begin{bmatrix} -1 \\ -1 \\ 1 \\ 1 \end{bmatrix} / \sqrt{4} = \begin{bmatrix} -3.75 \\ -3.75 \\ 3.75 \\ 3.75 \end{bmatrix}$$

In Chapter 3 we shall see that this reflects the fact that the males in the study were on average 7.5 in. taller than the females.

To summarize, the length of the projection of a vector $\mathbf{y}$ onto a direction $\mathbf{U}$ is given (to within a sign) by $\mathbf{y}\cdot\mathbf{U}$, where the " · " is the dot product defined above. The corresponding projection vector is $(\mathbf{y}\cdot\mathbf{U})\mathbf{U}$, where $\mathbf{y}\cdot\mathbf{U}$ is the signed length and $\mathbf{U}$ is the direction.

Angle Between Two Vectors

In Figure A.4 the *angle* θ between the vectors $\mathbf{y}$ and $\mathbf{U}_1$ is easily obtained from the displayed right-angled triangle, since

$$\cos\theta = \frac{\text{length of adjacent side}}{\text{length of hypotenuse}} = \frac{\mathbf{y}\cdot\mathbf{U}_1}{\|\mathbf{y}\|} = \frac{1.4\sqrt{2}}{\sqrt{3.94}} = 0.997459$$

so that $\theta = 4.09°$.

In general, to find the cosine of the angle between two vectors we divide each vector by its length, then take the dot product of the resulting vectors. That is, if the two vectors are $\mathbf{x} = [x_1, \ldots, x_N]^{\mathrm{T}}$ and $\mathbf{y} = [y_1, \ldots, y_N]^{\mathrm{T}}$, we obtain the angle θ between them using the formula

$$\cos\theta = \frac{\mathbf{x}\cdot\mathbf{y}}{\|\mathbf{x}\| \times \|\mathbf{y}\|} = \frac{x_1y_1 + \cdots + x_Ny_N}{\sqrt{x_1^2 + \cdots + x_N^2} \times \sqrt{y_1^2 + \cdots + y_N^2}}$$

For example, in 4-space the angle between the vectors $[63, 65, 69, 74]^{\mathrm{T}}$ and $[1, 1, 1, 1]^{\mathrm{T}}$ is given by the formula

$$\cos\theta = \frac{\begin{bmatrix} 63 \\ 65 \\ 69 \\ 74 \end{bmatrix} \cdot \begin{bmatrix} 1 \\ 1 \\ 1 \\ 1 \end{bmatrix}}{\sqrt{18431} \times \sqrt{4}} = \frac{63+65+69+74}{\sqrt{18431} \times \sqrt{4}} = 0.99808$$

so that the required angle θ is 3.55°.

Orthogonality

Two vectors are said to be *orthogonal* if the angle θ between them is 90°. This occurs when $\cos\theta = 0$, or when the dot product of the two vectors is zero.

In 2-space, the vectors $\mathbf{U_1} = [1,1]^T/\sqrt{2}$ and $\mathbf{U_2} = [-1,1]^T/\sqrt{2}$ are orthogonal since $\mathbf{U_1}\cdot\mathbf{U_2} = (-1+1)/2 = 0$.

In 4-space, the vectors $\mathbf{U_1} = [1,1,1,1]^T/\sqrt{4}$ and $\mathbf{U_2} = [-1,-1,1,1]^T/\sqrt{4}$ are orthogonal since $\mathbf{U_1}\cdot\mathbf{U_2} = (-1-1+1+1)/4 = 0$.

Orthogonal Coordinate Systems

An *orthogonal coordinate system* for N-space is a set of N mutually orthogonal unit vectors $\mathbf{U_1},\ldots,\mathbf{U_N}$.

In 2-space the natural orthogonal coordinate system is the set $\mathbf{U_1} = \begin{bmatrix}1\\0\end{bmatrix}$, $\mathbf{U_2} = \begin{bmatrix}0\\1\end{bmatrix}$, as shown in Figure A.5(a). An alternative orthogonal coordinate system which is more useful for statistical analysis is the set $\mathbf{U_1} = \begin{bmatrix}1\\1\end{bmatrix}/\sqrt{2}$, $\mathbf{U_2} = \begin{bmatrix}-1\\1\end{bmatrix}/\sqrt{2}$, as shown in Figure A.5(b).

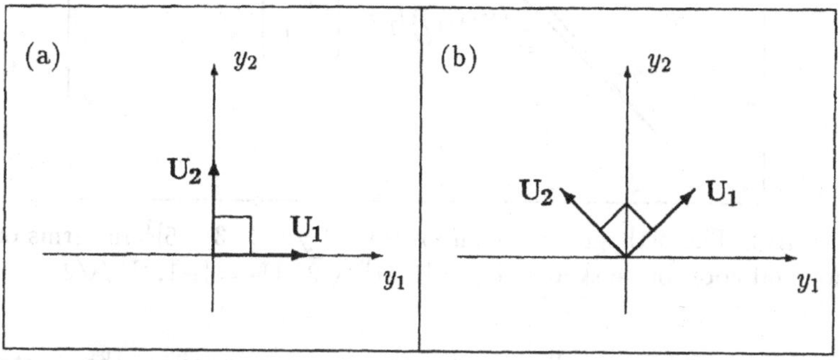

Figure A.5: Orthogonal coordinate systems for the plane.

In 4-space the orthogonal coordinate system which we use in Chapter 3 is

$$\mathbf{U_1} = \frac{1}{\sqrt{4}}\begin{bmatrix}1\\1\\1\\1\end{bmatrix},\ \mathbf{U_2} = \frac{1}{\sqrt{4}}\begin{bmatrix}-1\\-1\\1\\1\end{bmatrix},\ \mathbf{U_3} = \frac{1}{\sqrt{2}}\begin{bmatrix}-1\\1\\0\\0\end{bmatrix},\ \mathbf{U_4} = \frac{1}{\sqrt{2}}\begin{bmatrix}0\\0\\-1\\1\end{bmatrix}$$

Orthogonal Decomposition

A case of special interest to us is when a vector $\mathbf{y}$ is projected onto each of the directions specified by an orthogonal coordinate system $\mathbf{U_1},\ldots,\mathbf{U_N}$.

Here the vector $\mathbf{y}$ is the sum of these projections, i.e.,

$$\mathbf{y} = (\mathbf{y}\cdot\mathbf{U}_1)\mathbf{U}_1 + \cdots + (\mathbf{y}\cdot\mathbf{U}_N)\mathbf{U}_N$$

and the latter is called an *orthogonal decomposition* of $\mathbf{y}$.

For example, in 2-space we can project the vector $\mathbf{y} = [1.3, 1.5]^T$ onto the orthogonal coordinate system $\mathbf{U}_1 = \begin{bmatrix} 1 \\ 1 \end{bmatrix}/\sqrt{2}$, $\mathbf{U}_2 = \begin{bmatrix} -1 \\ 1 \end{bmatrix}/\sqrt{2}$ and obtain the orthogonal decomposition

$$\begin{array}{ccccc} \mathbf{y} & = & (\mathbf{y}\cdot\mathbf{U}_1)\mathbf{U}_1 & + & (\mathbf{y}\cdot\mathbf{U}_2)\mathbf{U}_2 \\ \begin{bmatrix} 1.3 \\ 1.5 \end{bmatrix} & = & \begin{bmatrix} 1.4 \\ 1.4 \end{bmatrix} & + & \begin{bmatrix} -0.1 \\ 0.1 \end{bmatrix} \end{array}$$

as illustrated in Figure A.6.

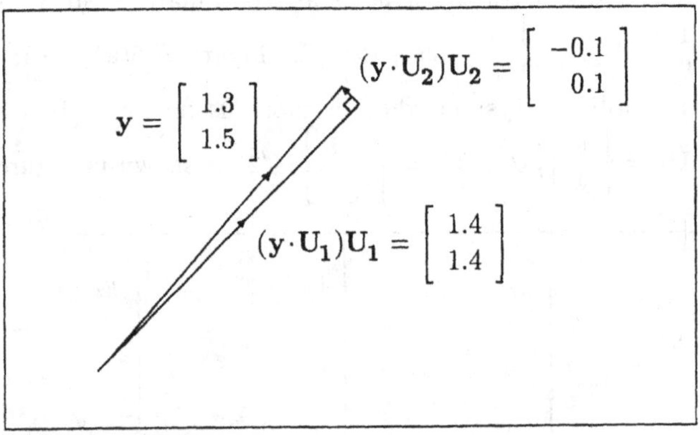

Figure A.6: The orthogonal decomposition of $\mathbf{y} = [1.3, 1.5]^T$ in terms of the orthogonal coordinate system $\mathbf{U}_1 = [1,1]^T/\sqrt{2}$, $\mathbf{U}_2 = [-1,1]^T/\sqrt{2}$.

As a second example, in 4-space we can project the vector $\mathbf{y} = [63, 65, 69, 74]^T$ onto the orthogonal coordinate system

$$\mathbf{U}_1 = \frac{1}{\sqrt{4}}\begin{bmatrix} 1 \\ 1 \\ 1 \\ 1 \end{bmatrix}, \quad \mathbf{U}_2 = \frac{1}{\sqrt{4}}\begin{bmatrix} -1 \\ -1 \\ 1 \\ 1 \end{bmatrix}, \quad \mathbf{U}_3 = \frac{1}{\sqrt{2}}\begin{bmatrix} -1 \\ 1 \\ 0 \\ 0 \end{bmatrix}, \quad \mathbf{U}_4 = \frac{1}{\sqrt{2}}\begin{bmatrix} 0 \\ 0 \\ -1 \\ 1 \end{bmatrix}$$

and obtain the orthogonal decomposition

$$\begin{array}{ccccccccc} \mathbf{y} & = & (\mathbf{y}\cdot\mathbf{U}_1)\mathbf{U}_1 & + & (\mathbf{y}\cdot\mathbf{U}_2)\mathbf{U}_2 & + & (\mathbf{y}\cdot\mathbf{U}_3)\mathbf{U}_3 & + & (\mathbf{y}\cdot\mathbf{U}_4)\mathbf{U}_4 \\ \begin{bmatrix} 63 \\ 65 \\ 69 \\ 74 \end{bmatrix} & = & \begin{bmatrix} 67.75 \\ 67.75 \\ 67.75 \\ 67.75 \end{bmatrix} & + & \begin{bmatrix} -3.75 \\ -3.75 \\ 3.75 \\ 3.75 \end{bmatrix} & + & \begin{bmatrix} -1 \\ 1 \\ 0 \\ 0 \end{bmatrix} & + & \begin{bmatrix} 0 \\ 0 \\ -2.5 \\ 2.5 \end{bmatrix} \end{array}$$

Pythagoras' Theorem

Pythagoras' Theorem in N-space says that the squared length of a vector can be expressed as the sum of the squared lengths of its projections onto the directions of an orthogonal coordinate system. That is,

$$\|\mathbf{y}\|^2 = (\mathbf{y}\cdot\mathbf{U_1})^2 + \cdots + (\mathbf{y}\cdot\mathbf{U_N})^2$$

In 2-space the orthogonal decomposition of the last subsection leads us to the Pythagorean breakup

$$\|\mathbf{y}\|^2 = (\mathbf{y}\cdot\mathbf{U_1})^2 + (\mathbf{y}\cdot\mathbf{U_2})^2$$

or,

$$1.3^2 + 1.5^2 = (1.4\sqrt{2})^2 + (0.1\sqrt{2})^2$$

or,

$$3.94 = 3.92 + 0.02$$

as illustrated in Figure A.7.

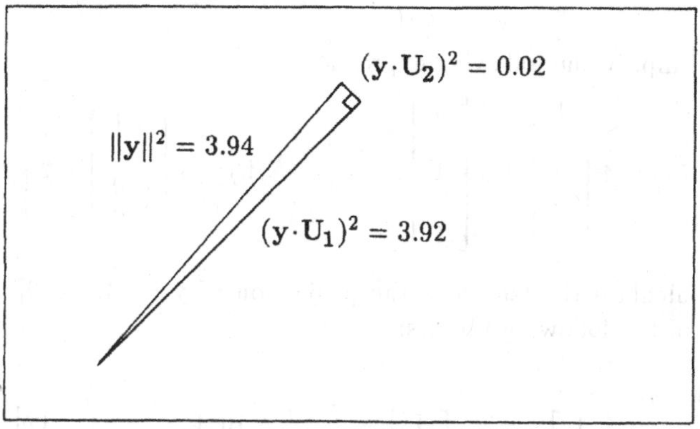

Figure A.7: The Pythagorean breakup of the squared length of $\mathbf{y} = [1.3, 1.5]^T$ in terms of the squared lengths of its projections onto $\mathbf{U_1} = [1,1]^T/\sqrt{2}$ and $\mathbf{U_2} = [-1,1]^T/\sqrt{2}$.

In 4-space the orthogonal decomposition of the last subsection leads us to the Pythagorean breakup

$$\|\mathbf{y}\|^2 = (\mathbf{y}\cdot\mathbf{U_1})^2 + (\mathbf{y}\cdot\mathbf{U_2})^2 + (\mathbf{y}\cdot\mathbf{U_3})^2 + (\mathbf{y}\cdot\mathbf{U_4})^2$$

or,

$$18431 = 18360.25 + 56.25 + 2 + 12.5$$

Exercises

Note that solutions to all of the exercises are given in Appendix E.

(A.1)* Find the lengths of the following vectors:

$$\begin{bmatrix} 4 \\ -1 \end{bmatrix} \quad \begin{bmatrix} 6 \\ 1 \\ 0 \\ -2 \\ 4 \end{bmatrix} \quad \begin{bmatrix} 1 \\ 1 \\ -2 \end{bmatrix} \quad \begin{bmatrix} 1 \\ 1 \\ 1 \\ 1 \\ 1 \end{bmatrix}.$$

(A.2)* Make the following vectors into unit vectors:

$$\begin{bmatrix} 1 \\ 1 \\ 1 \\ 1 \\ 1 \end{bmatrix} \quad \begin{bmatrix} -1 \\ 1 \\ 0 \\ 0 \end{bmatrix} \quad \begin{bmatrix} 3 \\ 3 \\ -2 \\ -2 \\ -2 \end{bmatrix} \quad \begin{bmatrix} -1 \\ 1 \\ -1 \\ 1 \end{bmatrix}$$

(A.3)* Simplify the following expressions:

$$\text{(a)} \quad 4\begin{bmatrix} 1 \\ 1 \\ 0 \\ 0 \\ 0 \end{bmatrix} + 5\begin{bmatrix} 0 \\ 0 \\ 1 \\ 1 \\ 1 \end{bmatrix}, \qquad \text{(b)} \quad 3\begin{bmatrix} -1 \\ 1 \\ 0 \\ 0 \end{bmatrix} - 2\begin{bmatrix} 0 \\ 0 \\ -1 \\ 1 \end{bmatrix}$$

(A.4)* Calculate the length of the projection of $\mathbf{y} = [4,5,8,9]^T$ onto the direction of the following vectors:

$$\text{(a)} \ \frac{1}{2}\begin{bmatrix} 1 \\ 1 \\ 1 \\ 1 \end{bmatrix}, \quad \text{(b)} \begin{bmatrix} 1 \\ 1 \\ 0 \\ 0 \end{bmatrix}, \quad \text{(c)} \begin{bmatrix} 10 \\ -10 \\ 0 \\ 0 \end{bmatrix}, \quad \text{(d)} \begin{bmatrix} 1 \\ -1 \\ -1 \\ 1 \end{bmatrix}$$

(A.5)* Find the angles between the following pairs of vectors:

(a) $[2,1]^T$ and $[1,3]^T$ in the plane.

(b) $[1,1]^T$ and $[-1,1]^T$ in the plane.

(c) $[1,1,1,1,1,1]^T$ and $[0,4,0,0,-7,0]^T$ in 6-space.

(d) $[5,2,1,3,7,0,2]^T$ and $[1,3,5,0,2,1,1]^T$ in 7-space.

(A.6)* Are the following sets of vectors orthogonal?

(a) $\begin{bmatrix}1\\1\\1\\1\\1\end{bmatrix}$ $\begin{bmatrix}2\\2\\2\\-3\\-3\end{bmatrix}$ $\begin{bmatrix}-1\\1\\0\\0\\0\end{bmatrix}$ $\begin{bmatrix}1\\1\\-2\\0\\0\end{bmatrix}$ $\begin{bmatrix}0\\0\\0\\-1\\1\end{bmatrix}$

(b) $\begin{bmatrix}-1\\1\\-1\\1\end{bmatrix}$ $\begin{bmatrix}-1\\-1\\1\\1\end{bmatrix}$ $\begin{bmatrix}-1\\1\\0\\0\end{bmatrix}$ $\begin{bmatrix}0\\0\\-1\\1\end{bmatrix}$

(c) $\begin{bmatrix}1\\1\\1\\1\end{bmatrix}$ $\begin{bmatrix}-1\\1\\-1\\1\end{bmatrix}$ $\begin{bmatrix}-1\\-1\\1\\1\end{bmatrix}$ $\begin{bmatrix}1\\-1\\-1\\1\end{bmatrix}$

(A.7)* Find a vector orthogonal to $[3,4]^T$.

(A.8)* Which of the following sets of vectors make up an orthogonal coordinate system? If not, why?

(a) $\frac{1}{\sqrt{3}}[1,1,1]^T$, $\frac{1}{\sqrt{6}}[-2,1,1]^T$, $\frac{1}{\sqrt{14}}[2,1,3]^T$

(b) $\frac{1}{\sqrt{4}}[1,1,1,1]^T$, $\frac{1}{\sqrt{20}}[-3,-1,1,3]^T$, $\frac{1}{\sqrt{4}}[1,-1,-1,1]^T$

(c) $\frac{1}{\sqrt{3}}[1,1,1]^T$, $\frac{1}{\sqrt{74}}[3,-7,4]^T$, $\frac{1}{\sqrt{222}}[11,-1,-10]^T$.

(A.9)* Given the vector $\mathbf{y} = [3,9,11,13]^{\mathrm{T}}$ and orthogonal coordinate system

$$\mathbf{U_1} = \frac{1}{2}\begin{bmatrix}1\\1\\1\\1\end{bmatrix},\quad \mathbf{U_2} = \frac{1}{2}\begin{bmatrix}-1\\-1\\1\\1\end{bmatrix},\quad \mathbf{U_3} = \frac{1}{2}\begin{bmatrix}-1\\1\\-1\\1\end{bmatrix},\quad \mathbf{U_4} = \frac{1}{2}\begin{bmatrix}1\\-1\\-1\\1\end{bmatrix}$$

write down the orthogonal decomposition of $\mathbf{y}$ in the form

$$\mathbf{y} = \text{constant} \times \mathbf{U_1} + \text{constant} \times \mathbf{U_2} + \text{constant} \times \mathbf{U_3} + \text{constant} \times \mathbf{U_4}$$

(A.10)* In 4-space consider $\mathbf{y} = [63,65,69,74]^{\mathrm{T}}$, the vector used in this appendix, in conjunction with an alternative to the coordinate system used so far:

$$\mathbf{U_1} = \frac{1}{\sqrt{2}}\begin{bmatrix}1\\1\\0\\0\end{bmatrix},\quad \mathbf{U_2} = \frac{1}{\sqrt{2}}\begin{bmatrix}0\\0\\1\\1\end{bmatrix},\quad \mathbf{U_3} = \frac{1}{\sqrt{2}}\begin{bmatrix}-1\\1\\0\\0\end{bmatrix},\quad \mathbf{U_4} = \frac{1}{\sqrt{2}}\begin{bmatrix}0\\0\\-1\\1\end{bmatrix}$$

For this alternative orthogonal coordinate system, write out in full:

(a) the orthogonal decomposition of $\mathbf{y}$ with respect to $\mathbf{U}_1, \ldots, \mathbf{U}_4$; and

(b) the breakup of $||\mathbf{y}||^2$ in terms of squared lengths of projections of $\mathbf{y}$ onto $\mathbf{U}_1, \ldots, \mathbf{U}_4$.

(A.11)* (a) Find the vectors $\mathbf{P}_1\mathbf{y}, \mathbf{P}_2\mathbf{y}, \mathbf{P}_3\mathbf{y}$ corresponding to the projection of $\mathbf{y} = [6, 5, 9]^T$ onto the orthogonal unit vectors

$$\mathbf{U}_1 = \frac{1}{\sqrt{3}}\begin{bmatrix} 1 \\ 1 \\ 1 \end{bmatrix}, \quad \mathbf{U}_2 = \frac{1}{\sqrt{42}}\begin{bmatrix} -4 \\ -1 \\ 5 \end{bmatrix}, \quad \mathbf{U}_3 = \frac{1}{\sqrt{14}}\begin{bmatrix} 2 \\ -3 \\ 1 \end{bmatrix}$$

(b) Write $\mathbf{y}$ as the sum of these projection vectors.

(c) Calculate the squared length of $\mathbf{y}$ and of each projection vector.

(d) Hence confirm Pythagoras' Theorem,

$$||\mathbf{y}||^2 = ||\mathbf{P}_1\mathbf{y}||^2 + ||\mathbf{P}_2\mathbf{y}||^2 + ||\mathbf{P}_3\mathbf{y}||^2$$

(e) Also confirm that

$$||\mathbf{P}_1\mathbf{y}||^2 = (\mathbf{y} \cdot \mathbf{U}_1)^2, \quad ||\mathbf{P}_2\mathbf{y}||^2 = (\mathbf{y} \cdot \mathbf{U}_2)^2 \text{ and } ||\mathbf{P}_3\mathbf{y}||^2 = (\mathbf{y} \cdot \mathbf{U}_3)^2$$

Appendix B

Statistical Tool Kit

In this appendix we give a brief introduction to the few statistical tools which are needed in our development. This small number of tools should be quite adequate for most readers; however, if the reader feels the need for additional tools he or she can refer to Chapter 3 of Saville and Wood (1991).

Normal Distributions

In this book we always deal with a particular type of "probability distribution" called the *normal* distribution. This is the distribution which occurs most commonly in biological and many other fields of work. As an example, Figure B.1 shows the relative frequency histogram for a population of refractometer readings of 10,000 onion bulbs, together with the fitted normal distribution. The distribution is bell-shaped, with medium values occurring more often than low or high values. The "probability density function" is

$$f(y) = \frac{1}{\sqrt{2\pi}\sigma} e^{-\frac{1}{2}\left(\frac{y-\mu}{\sigma}\right)^2}$$

where μ is the population *mean* and σ^2 is a measure of variability called the population *variance*. The mean is the average of the population values, and the variance is the average of the squared deviations of the population values about the population mean, i.e., the average of $(y-\mu)^2$, where y denotes an arbitrary population value.

Notation

The expression $N[\mu, \sigma^2]$ will be used to refer to the normal distribution with mean μ and variance σ^2.

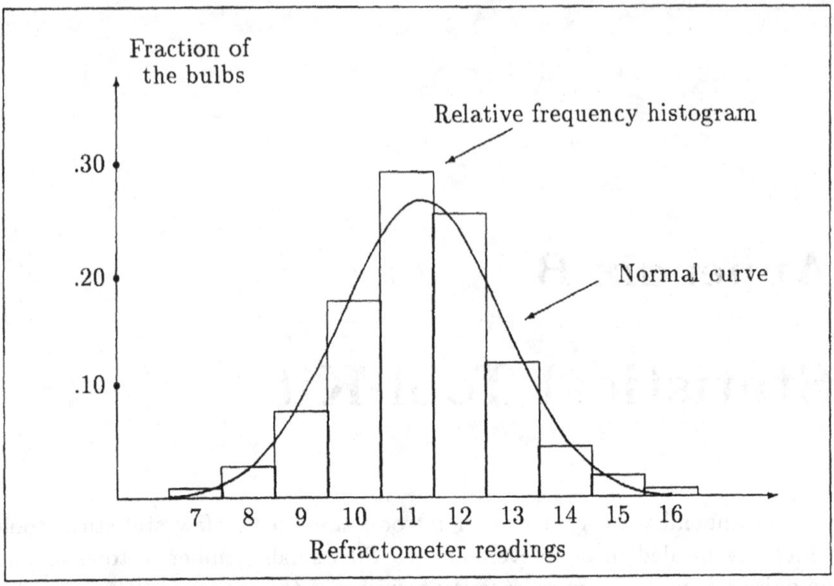

Figure B.1: The relative frequency histogram of refractometer readings of 10,000 onion bulbs with the theoretical normal distribution curve. Data from *Agricultural Experimentation: Design and Analysis*, by T.M. Little and F.J. Hills (1978), reproduced with kind permission from J. Wiley and Sons.

The F Distributions

Our standard method for testing an hypothesis such as $\mu = 0$ (as in Chapters 1 and 2) is to calculate a test statistic such as $(\mathbf{y}\cdot\mathbf{U}_1)^2/(\mathbf{y}\cdot\mathbf{U}_2)^2 = 3.92/0.02 = 196$. If the hypothesis is true, the test statistic comes from an $F_{1,q}$ distribution, where $q = 1$ in this simplest of examples. On the other hand, if the hypothesis is false, the test statistic is inflated, which causes us to reject the hypothesis provided the test statistic is large enough. To decide what is "large enough" we must have some knowledge of these fundamental $F_{1,q}$ distributions.

In this book we define the $F_{1,q}$ distribution to be the distribution of the ratio

$$\frac{u^2}{\left[\dfrac{v_1^2 + \cdots + v_q^2}{q}\right]}$$

where u and $v_1, \ldots, v_q$ are values independently drawn from a normal distribution with mean zero and variance σ^2. As examples, the $F_{1,2}$, $F_{1,4}$ and $F_{1,1000}$ distributions are shown in Figure B.2.

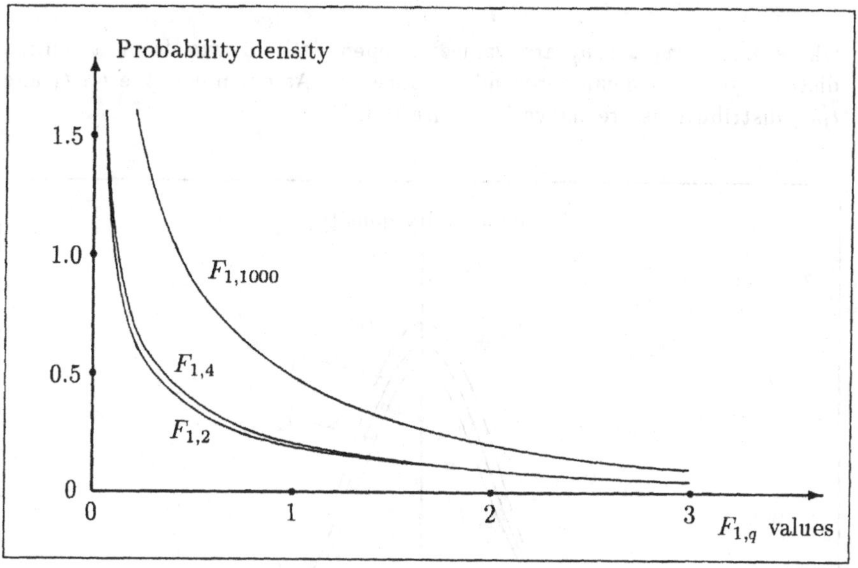

Figure B.2: A selection of $F_{1,q}$ distributions.

The 90, 95 and 99 percentiles of the $F_{1,q}$ distributions are given in Table T.2 for a wide range of values for q. When testing an hypothesis such as $\mu = 0$ versus a "two-sided" alternative hypothesis such as $\mu \neq 0$, we compare our test statistic, such as $(\mathbf{y} \cdot \mathbf{U}_1)^2/(\mathbf{y} \cdot \mathbf{U}_2)^2 = 196$ in §1.3, with the percentiles of the appropriate $F_{1,q}$ distribution ($F_{1,1}$ in our example) in order to decide whether our test statistic is unusually large. If our test statistic is not unusually large we accept that our hypothesis may be true, while if it is large in relation to the percentiles we reject the hypothesis. In our example the $F_{1,1}$ distribution has 90, 95 and 99 percentiles of 40, 161 and 4052, respectively, so that our test statistic of 196 is "large" in relation to the 95 percentile of 161. We therefore reject the hypothesis $\mu = 0$ at the "5% level of significance."

The t Distributions

An equivalent method of testing hypotheses is to take the signed square root of the above test statistic and compare the result with the distribution of the square root of the ratio given in the last subsection.

Formally, we define the t_q distribution to be the distribution of the ratio

$$\frac{u}{\sqrt{\frac{v_1^2 + \cdots + v_q^2}{q}}}$$

where u and $v_1, \ldots, v_q$ are values independently drawn from a normal distribution with mean zero and variance σ^2. As examples, the t_2, t_4 and t_{1000} distributions are shown in Figure B.3.

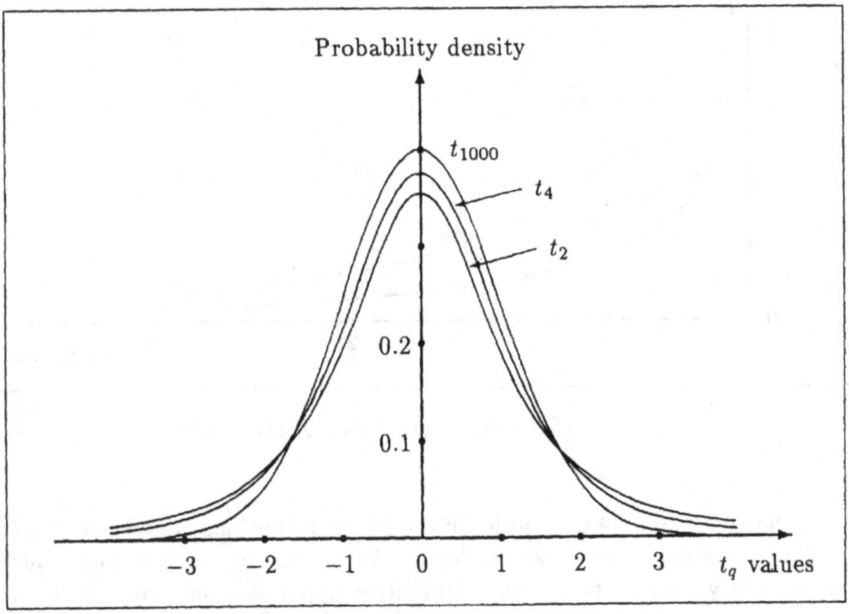

Figure B.3: A selection of t_q distributions.

The t_q distribution is symmetric about zero since its numerator, u, comes from a normal distribution with mean zero, which is symmetric about zero. Note that $t_q^2 = F_{1,q}$.

The 95, 97.5 and 99.5 percentiles of the t_q distributions are given in Table T.2 for a wide range of values for q. When testing a "two-sided" hypothesis such as $\mu = 0$ versus an alternative hypothesis such as $\mu \neq 0$, we compare the numerical value of our new test statistic, such as $\mathbf{y} \cdot \mathbf{U_1} / |\mathbf{y} \cdot \mathbf{U_2}| = 14$ in §1.3, with these upper percentiles of the appropriate t_q distribution (t_1 in our example) in order to decide whether our test statistic is unusually large. If our test statistic is not unusually large we accept that our hypothesis may be true, while if it is large in relation to the percentiles we reject the hypothesis. In our example the t_1 distribution has 95, 97.5 and 99.5 percentiles of 6.31,

12.71 and 63.66, respectively, so that our test statistic, $\mathbf{y}\cdot\mathbf{U}_1/|\mathbf{y}\cdot\mathbf{U}_2| = 14$, is "large" in relation to the 97.5 percentile of 12.71. We therefore reject the hypothesis $\mu = 0$ at the "5% level of significance."

Note that the alternative hypothesis for our two-sided test includes the possibilities $\mu < 0$ and $\mu > 0$. Taking the numerical value of our test statistic and comparing it with the 97.5 percentile is equivalent to checking whether the test statistic lies inside the range bounded by the 2.5 and 97.5 percentiles, that is, the range -12.71 to 12.71. Thus, if $\mu = 0$ the probability of false rejection is $2.5\% + 2.5\% = 5\%$.

Appendix C

Computing

Throughout this text readers are encouraged to do the bulk of the exercises without the aid of a computer. The exceptions are the larger exercises which would be tedious to do "by hand." The purpose of this appendix is to describe how to do the required calculations using a computer, to cater both for these larger exercises and for readers or teachers who prefer to use a computer throughout.

The computing is actually amazingly simple. A very unsophisticated spreadsheet will suffice, or if available, either a simple or multiple regression routine can be used. We now describe how to do the computing using either of these tools.

C.1 Computing Using a Simple Spreadsheet

To do the calculations in a spreadsheet, the reader needs to first input the vector of observations $\mathbf{y}$ into a column of the spreadsheet, then input the unit vectors $\mathbf{U}_i$ into further columns. The projection length $\mathbf{y} \cdot \mathbf{U}_1$ is then obtained by multiplying the "$\mathbf{y}$" column by the "$\mathbf{U}_1$" column, and summing. The squared length $(\mathbf{y} \cdot \mathbf{U}_1)^2$ is then obtained by squaring this sum. The same applies for $\mathbf{y} \cdot \mathbf{U}_2$, and so on. To make this perfectly clear, we now spell this out for each of the book's chapters.

Paired Samples

In the paired samples, or single population case, we need to calculate $\mathbf{y} \cdot \mathbf{U}_1 = \sqrt{n}\bar{y}$. We first put the observation vector $\mathbf{y}$ into column A of the spreadsheet, then a vector of ones into column B, then divide column B by $\sqrt{n}$ so that column B contains $\mathbf{U}_1$. We then multiply column A by column B, putting the result in column C. We then sum column C; this yields the projection length $\mathbf{y} \cdot \mathbf{U}_1$.

We are now in a position to compute the Pythagorean breakup

$$\|\mathbf{y}\|^2 = \|\bar{\mathbf{y}}\|^2 + \|\mathbf{y} - \bar{\mathbf{y}}\|^2$$

The first term, $\|\mathbf{y}\|^2$, is obtained by squaring column A and summing. The second term, $\|\bar{\mathbf{y}}\|^2 = (\mathbf{y}\cdot\mathbf{U}_1)^2$, is obtained by squaring $\mathbf{y}\cdot\mathbf{U}_1$ as obtained above. The third term, $\|\mathbf{y} - \bar{\mathbf{y}}\|^2$, is obtained by subtracting the second term from the first term. The Pythagorean breakup can then be used to calculate the test statistic $F = (\mathbf{y}\cdot\mathbf{U}_1)^2/[\|\mathbf{y} - \bar{\mathbf{y}}\|^2/(n-1)]$.

Independent Samples

In the independent samples case, we need to calculate $\mathbf{y}\cdot\mathbf{U}_1$ and $\mathbf{y}\cdot\mathbf{U}_2$. We first put the observation vector $\mathbf{y}$ into column A of the spreadsheet. To set up $\mathbf{U}_1$, we put a vector of $2n$ ones into column B and divide column B by $\sqrt{2n}$; column B then contains $\mathbf{U}_1$. To set up $\mathbf{U}_2$, we put a vector of n "-1"s and n "1"s into column C and divide column C by $\sqrt{2n}$; column C then contains $\mathbf{U}_2$. Next we multiply column A by column B, putting the result in column D. We then sum column D; this yields the projection length $\mathbf{y}\cdot\mathbf{U}_1$. Similarly, we multiply column A by column C, putting the result in column E and summing to obtain the projection length $\mathbf{y}\cdot\mathbf{U}_2$.

We are now in a position to compute the Pythagorean breakup

$$\|\mathbf{y}\|^2 = (\mathbf{y}\cdot\mathbf{U}_1)^2 + (\mathbf{y}\cdot\mathbf{U}_2)^2 + \|\mathbf{y} - \bar{\mathbf{y}}_i\|^2$$

The first term, $\|\mathbf{y}\|^2$, is obtained by squaring column A and summing. The second and third terms, $(\mathbf{y}\cdot\mathbf{U}_1)^2$ and $(\mathbf{y}\cdot\mathbf{U}_2)^2$, are obtained by squaring $\mathbf{y}\cdot\mathbf{U}_1$ and $\mathbf{y}\cdot\mathbf{U}_2$ as obtained above. The fourth term, $\|\mathbf{y} - \bar{\mathbf{y}}_i\|^2$, is obtained by subtracting the second and third terms from the first term. The Pythagorean breakup can then be used to calculate the test statistic $F = (\mathbf{y}\cdot\mathbf{U}_2)^2/\{\|\mathbf{y} - \bar{\mathbf{y}}_i\|^2/[2(n-1)]\}$.

Several Independent Samples

In the several independent samples, or analysis of variance case, we need to calculate $\mathbf{y}\cdot\mathbf{U}_1$ and $\mathbf{y}\cdot\mathbf{U}_2, \ldots, \mathbf{y}\cdot\mathbf{U}_k$, where k is the number of study populations. We first put the observation vector $\mathbf{y}$ into column A of the spreadsheet. We then put a vector of kn ones into column B and divide column B by $\sqrt{kn}$, so that column B contains $\mathbf{U}_1$. We then put the appropriate contrast unit vectors (as whole numbers) into columns C onward, and divide each column by an appropriate number (the square root of the sum of the squared column), so that columns C onward contain $\mathbf{U}_2, \ldots, \mathbf{U}_k$ (e.g., $\mathbf{U}_2$ and $\mathbf{U}_3$ on page 74). We then multiply column A by each of columns B onward, putting the results in spare columns. We then sum each of these last columns to obtain the projection lengths $\mathbf{y}\cdot\mathbf{U}_1, \ldots, \mathbf{y}\cdot\mathbf{U}_k$.

We are now in a position to compute the Pythagorean breakup

$$||\mathbf{y}||^2 = (\mathbf{y}\cdot\mathbf{U}_1)^2 + (\mathbf{y}\cdot\mathbf{U}_2)^2 + \cdots + (\mathbf{y}\cdot\mathbf{U}_k)^2 + ||\mathbf{y}-\bar{\mathbf{y}}_i||^2$$

The first term, $||\mathbf{y}||^2$, is obtained by squaring column A and summing. The first k terms on the right side of the equation, $(\mathbf{y}\cdot\mathbf{U}_1)^2,\ldots,(\mathbf{y}\cdot\mathbf{U}_k)^2$, are obtained by squaring $\mathbf{y}\cdot\mathbf{U}_1,\ldots,\mathbf{y}\cdot\mathbf{U}_k$, respectively. The last term, $||\mathbf{y}-\bar{\mathbf{y}}_i||^2$, is obtained by subtracting these k terms from the first term. The Pythagorean breakup can then be used to calculate the test statistics $F = (\mathbf{y}\cdot\mathbf{U}_i)^2/\{||\mathbf{y}-\bar{\mathbf{y}}_i||^2/[k(n-1)]\}$, where $\mathbf{y}\cdot\mathbf{U}_i$ ranges from $\mathbf{y}\cdot\mathbf{U}_2$ to $\mathbf{y}\cdot\mathbf{U}_k$.

Simple Regression

In the simple regression case, we need to calculate $\mathbf{y}\cdot\mathbf{U}_1$ and $\mathbf{y}\cdot\mathbf{U}_2$. We first put the observation vector $\mathbf{y}$ into column A of the spreadsheet. We then put a vector of ones into column B and divide column B by $\sqrt{n}$, so that column B contains $\mathbf{U}_1$. We then put the vector $\mathbf{x}$ into column C, calculate the mean of column C, and subtract this mean value from column C, putting the answer in column D, so that column D contains the vector $(\mathbf{x}-\bar{\mathbf{x}})$. We then square column D and put the answer in column E, then sum column E and take the square root of this sum; this yields the length of the vector $(\mathbf{x}-\bar{\mathbf{x}})$. Column D is then divided by this last quantity, with the answer going into column F, so that column F contains the vector $\mathbf{U}_2$. That is, we now have $\mathbf{y}$ in column A, $\mathbf{U}_1$ in column B and $\mathbf{U}_2$ in column F.

We then multiply column A by column B, putting the result in column G. We then sum column G; this yields the projection length $\mathbf{y}\cdot\mathbf{U}_1$. Similarly, we multiply column A by column F, putting the result in column H and summing to obtain the projection length $\mathbf{y}\cdot\mathbf{U}_2$.

We are now in a position to compute the Pythagorean breakup

$$||\mathbf{y}||^2 = (\mathbf{y}\cdot\mathbf{U}_1)^2 + (\mathbf{y}\cdot\mathbf{U}_2)^2 + ||\text{error vector}||^2$$

The first term, $||\mathbf{y}||^2$, is obtained by squaring column A and summing. The second and third terms, $(\mathbf{y}\cdot\mathbf{U}_1)^2$ and $(\mathbf{y}\cdot\mathbf{U}_2)^2$, are obtained by squaring $\mathbf{y}\cdot\mathbf{U}_1$ and $\mathbf{y}\cdot\mathbf{U}_2$ as obtained above. The fourth term, $||\text{error vector}||^2$, is obtained by subtracting the second and third terms from the first term. The Pythagorean breakup can then be used to calculate the test statistic $F = (\mathbf{y}\cdot\mathbf{U}_2)^2/[||\text{error vector}||^2/(n-2)]$.

If it is required, the equation of the fitted line $y = a + bx$ can be obtained using the formulas for the intercept, $a = \bar{y} - b\bar{x}$, and the slope, $b = \mathbf{y}\cdot(\mathbf{x}-\bar{\mathbf{x}})/||\mathbf{x}-\bar{\mathbf{x}}||^2$. Firstly, the slope b can be calculated by multiplying columns A and D and summing (this gives the numerator), and dividing by the sum of column E (the denominator). Secondly, the values $\bar{y}$ and $\bar{x}$ can be calculated as the means of columns A and C, respectively, and used in conjunction with b to calculate the intercept a.

C.2 Computing Using a Regression Routine

To do the calculations using a regression routine (which may be a standalone program or included as an option in a spreadsheet or a statistical computing package), the reader needs to either input the observation vector **y** and the unit vectors $\mathbf{U}_i$ into columns (in the case of a spreadsheet) or read them in as variables (in the other cases). Each projection length $\mathbf{y}\cdot\mathbf{U}_i$ is then obtained by "regressing" the **y** column or variable onto the appropriate $\mathbf{U}_i$ column or variable. Here the word "regress" is really just a synonym for the word "project." If a simple regression routine is used then **y** must be projected onto each $\mathbf{U}_i$ direction in turn, while if a multiple regression routine is used then **y** can be projected onto all of the $\mathbf{U}_i$ directions simultaneously. To make this perfectly clear, we again spell this out for each of the book's chapters.

Paired Samples

In the paired samples, or single population case, we first set up either spreadsheet columns or variables corresponding to **y** and $\mathbf{U}_1$. In the case of a spreadsheet we put the observation vector **y** into column A, then a vector of ones into column B, then divide column B by $\sqrt{n}$ so that column B contains $\mathbf{U}_1$. In the case of a standalone routine or a statistical package we set up variables corresponding to **y** and $\mathbf{U}_1$ in a similar manner. We then regress the column or variable corresponding to **y** onto the column or variable corresponding to $\mathbf{U}_1$ using the "regress" command. Here **y** is specified as the "dependent" variable and $\mathbf{U}_1$ as the "independent" variable, and it is also specified that the regression should be fitted "without any intercept or constant term" (sometimes called regression through the origin). In essence, we suppress the usual automatic fitting of the constant term so that we can then fit it explicitly. This may seem a little odd, and a hard way of doing some very simple arithmetic; however, we do it this way for consistency with later cases, so be patient! The result of this regression is the calculation of $\mathbf{y}\cdot\mathbf{U}_1$, which will be labeled as the regression coefficient corresponding to the independent variable $\mathbf{U}_1$.

Most regression routines will also automatically calculate the Pythagorean breakup

$$||\mathbf{y}||^2 = (\mathbf{y}\cdot\mathbf{U}_1)^2 + ||\mathbf{y} - \bar{\mathbf{y}}||^2$$

disguised as "sums of squares" in an "analysis of variance" table. If this is not done, then the first term, $||\mathbf{y}||^2$, can be obtained by squaring the **y** column or variable and summing. The second term, $(\mathbf{y}\cdot\mathbf{U}_1)^2$, can be obtained by squaring $\mathbf{y}\cdot\mathbf{U}_1$ as obtained above, and the third term can be obtained by subtraction.

Many regression routines will then automatically calculate the test statistic $F = (\mathbf{y}\cdot\mathbf{U}_1)^2/[||\mathbf{y} - \bar{\mathbf{y}}||^2/(n-1)]$. If this is not done, the Pythagorean breakup can be used as the basis for this calculation.

Independent Samples

In the independent samples case, we first set up either spreadsheet columns or variables corresponding to $\mathbf{y}$ and $\mathbf{U_2}$ (but not $\mathbf{U_1}$). In the case of a spreadsheet we first put the observation vector $\mathbf{y}$ into column A. For $\mathbf{U_2}$ we then put the appropriate vector of n "-1"s and n "1"s into column B, then divide column B by $\sqrt{2n}$ so that column B contains $\mathbf{U_2}$. In the case of a standalone routine or a statistical package we set up variables corresponding to $\mathbf{y}$ and $\mathbf{U_2}$ in a similar manner. We then regress the column or variable corresponding to $\mathbf{y}$ onto the column or variable corresponding to $\mathbf{U_2}$ using the "regress" command. Here $\mathbf{y}$ is specified as the "dependent" variable and $\mathbf{U_2}$ as the "independent" variable. It is not necessary to specify $\mathbf{U_1}$ since virtually all regression routines automatically include the intercept or constant term, which corresponds to automatically projecting onto the direction $\mathbf{U_1}$. The result of this regression is the calculation of $\mathbf{y}\cdot\mathbf{U_2}$, which will be labeled as the regression coefficient corresponding to the independent variable $\mathbf{U_2}$.

Most regression routines will automatically calculate the Pythagorean breakup as "sums of squares " in the form

$$\|\mathbf{y}-\bar{\mathbf{y}}\|^2 = (\mathbf{y}\cdot\mathbf{U_2})^2 + \|\mathbf{y}-\bar{\mathbf{y}}_i\|^2$$

where the term $(\mathbf{y}\cdot\mathbf{U_1})^2$ has been taken over to the left side of the equation. If this breakup is not printed, then the term on the left side, $\|\mathbf{y}-\bar{\mathbf{y}}\|^2$, can be obtained by squaring the $\mathbf{y}$ column or variable and summing, then subtracting $(\mathbf{y}\cdot\mathbf{U_1})^2 = 2n\bar{y}^2$, where $\bar{y}$ is the mean of y. The next term, $(\mathbf{y}\cdot\mathbf{U_2})^2$, can be obtained by squaring $\mathbf{y}\cdot\mathbf{U_2}$ as obtained above, and the third term can be obtained by subtraction.

Many regression routines will then automatically calculate the test statistic $F = (\mathbf{y}\cdot\mathbf{U_2})^2/\{\|\mathbf{y}-\bar{\mathbf{y}}_i\|^2/[2(n-1)]\}$. If this is not done, the Pythagorean breakup can be used as the basis for this calculation.

Several Independent Samples

In the several independent samples, or analysis of variance case, we first set up either spreadsheet columns or variables corresponding to $\mathbf{y}$ and $\mathbf{U_2},\ldots,\mathbf{U_k}$ (but not $\mathbf{U_1}$). In the case of a spreadsheet we first put the observation vector $\mathbf{y}$ into column A. We then put the appropriate contrast unit vectors (as whole numbers) into columns B onward, and divide each column by an appropriate number (the square root of the sum of the squared column), so that columns B onward contain $\mathbf{U_2},\ldots,\mathbf{U_k}$ (e.g., $\mathbf{U_2}$ and $\mathbf{U_3}$ on page 74). In the case of a standalone routine or a statistical package we set up variables corresponding to $\mathbf{y}$ and $\mathbf{U_2},\ldots,\mathbf{U_k}$ in a similar manner. If a multiple regression routine is available, we then use it to regress the column or variable corresponding to $\mathbf{y}$ onto the columns or variables corresponding to $\mathbf{U}_2, \ldots, \mathbf{U}_k$. Here $\mathbf{y}$ is specified as the "dependent" variable and $\mathbf{U}_2, \ldots, \mathbf{U}_k$ are specified as the "independent" variables. Again it is not usually necessary to specify $\mathbf{U}_1$. The result of this regression is the calculation of $\mathbf{y}\cdot\mathbf{U}_2, \ldots, \mathbf{y}\cdot\mathbf{U}_k$, which will be labeled as the regression coefficients corresponding to the independent variables $\mathbf{U}_2, \ldots, \mathbf{U}_k$. If only a simple regression routine is available, we must first regress $\mathbf{y}$ onto $\mathbf{U}_2$, then $\mathbf{y}$ onto $\mathbf{U}_3$, and so on, yielding $\mathbf{y}\cdot\mathbf{U}_2, \ldots, \mathbf{y}\cdot\mathbf{U}_k$.

Most multiple regression routines will automatically calculate the Pythagorean breakup as "sums of squares" in the form

$$||\mathbf{y} - \bar{\mathbf{y}}||^2 = \text{Regression sum of squares} + ||\mathbf{y} - \bar{\mathbf{y}}_i||^2$$

where the term $(\mathbf{y}\cdot\mathbf{U}_1)^2$ has been taken over to the left side of the equation. This can be converted to the more useful form

$$||\mathbf{y} - \bar{\mathbf{y}}||^2 = (\mathbf{y}\cdot\mathbf{U}_2)^2 + \cdots + (\mathbf{y}\cdot\mathbf{U}_k)^2 + ||\mathbf{y} - \bar{\mathbf{y}}_i||^2$$

using the projection lengths $\mathbf{y}\cdot\mathbf{U}_i$ obtained above (note here that rounding of these lengths may introduce minor inaccuracies). If a simple regression routine is used, the term on the left will be calculated by the routine and $\mathbf{y}\cdot\mathbf{U}_2, \ldots, \mathbf{y}\cdot\mathbf{U}_k$ will be available, so that the last term on the right can be obtained by subtraction. If no breakup is printed, then the term on the left side, $||\mathbf{y} - \bar{\mathbf{y}}||^2$, can be obtained by squaring the $\mathbf{y}$ column or variable and summing, then subtracting $(\mathbf{y}\cdot\mathbf{U}_1)^2 = kn\bar{y}^2$, where $\bar{y}$ is the mean of y. The terms $(\mathbf{y}\cdot\mathbf{U}_2)^2, \ldots, (\mathbf{y}\cdot\mathbf{U}_k)^2$ can be obtained by squaring $\mathbf{y}\cdot\mathbf{U}_2, \ldots, \mathbf{y}\cdot\mathbf{U}_k$ as obtained above, and the last term can be obtained by subtraction.

The Pythagorean breakup can then be used to calculate the test statistics $F = (\mathbf{y}\cdot\mathbf{U}_i)^2/\{||\mathbf{y} - \bar{\mathbf{y}}_i||^2/[k(n-1)]\}$, where $\mathbf{y}\cdot\mathbf{U}_i$ ranges from $\mathbf{y}\cdot\mathbf{U}_2$ to $\mathbf{y}\cdot\mathbf{U}_k$.

Simple Regression

In the simple regression case, we first set up either spreadsheet columns or variables corresponding to $\mathbf{y}$ and $\mathbf{U}_2$ (but not $\mathbf{U}_1$). In the case of a spreadsheet we first put the observation vector $\mathbf{y}$ into column A. We then put the vector $\mathbf{x}$ into column B, calculate the mean of column B, and subtract this mean value from column B, putting the answer in column C, so that column C contains the vector $(\mathbf{x} - \bar{\mathbf{x}})$. We then square column C and put the answer in column D, then sum column D and take the square root of this sum; this yields the length of the vector $(\mathbf{x} - \bar{\mathbf{x}})$. Column C is then divided by this last quantity, with the answer going into column E, so that column E contains the vector $\mathbf{U}_2$. That is, we now have $\mathbf{y}$ in column A and $\mathbf{U}_2$ in column E. In the case of a standalone routine or a statistical package we set up variables corresponding to $\mathbf{y}$ and $\mathbf{U}_2$ in a similar manner. We then regress the column or variable corresponding to $\mathbf{y}$ onto the column or variable corresponding to $\mathbf{U}_2$ using the "regress" command. Here $\mathbf{y}$ is specified as the "dependent" variable and $\mathbf{U}_2$ as the "independent" variable. Again it is not usually necessary to specify $\mathbf{U}_1$. The result of this regression is the calculation of $\mathbf{y}\cdot\mathbf{U}_2$, which will be labeled as the regression coefficient corresponding to the independent variable $\mathbf{U}_2$.

Most regression routines will automatically calculate the Pythagorean breakup as "sums of squares" in the form

$$||\mathbf{y} - \bar{\mathbf{y}}||^2 = (\mathbf{y}\cdot\mathbf{U}_2)^2 + ||\text{error vector}||^2$$

where the term $(\mathbf{y}\cdot\mathbf{U}_1)^2$ has been taken over to the left side of the equation. If this breakup is not printed, then the term on the left side, $||\mathbf{y} - \bar{\mathbf{y}}||^2$, can be obtained by squaring the $\mathbf{y}$ column or variable and summing, then subtracting $(\mathbf{y}\cdot\mathbf{U}_1)^2 = n\bar{y}^2$. The next term, $(\mathbf{y}\cdot\mathbf{U}_2)^2$, can be obtained by squaring $\mathbf{y}\cdot\mathbf{U}_2$ as obtained above, and the third term can be obtained by subtraction.

Many regression routines will then automatically calculate the test statistic $F = (\mathbf{y}\cdot\mathbf{U}_2)^2/[||\text{error vector}||^2/(n-2)]$. If this is not done, the Pythagorean breakup can be used as the basis for this calculation.

If it is required, the equation of the fitted line $y = a + bx$ can be obtained by regressing the column or variable corresponding to $\mathbf{y}$ onto the column or variable corresponding to $\mathbf{x}$. Note that this will often also produce most of the output referred to above, with the exception of the projection length $\mathbf{y}\cdot\mathbf{U}_2$ (though its square will be given). The reason we have not suggested this simpler "modus operandi" form the outset is that it lacks the same linkage to the geometric approach in the main body of this text. However, it is at least useful as a check on the calculation of $\mathbf{U}_2$.

C.3 Summary

In this appendix we have introduced computing methods which tie in with the explanations in the main body of this text. These are based either on an unsophisticated spreadsheet or on a regression routine. In the latter case only a simple regression routine is needed for Chapters 2, 3 and 5, while a multiple regression routine is an advantage for Chapter 4 (though a simple regression routine will suffice).

To tie in with the main chapters, each coordinate axis direction is specified by a *unit* vector $\mathbf{U}_i$. This is achieved by specifying the axis direction using whole numbers, then dividing the resulting vector by its length. In the case of regression routines, this last step is not essential in the sense that $(\mathbf{y}\cdot\mathbf{U}_i)^2$ will still be printed as the sum of squares. This is because the regression routine projects $\mathbf{y}$ onto the subspace spanned by the $\mathbf{U}_i$ directions, and this subspace is unaltered if the directions are not standardized to length one. However, if this standardization is not done, the calculated regression coefficients will no longer be $\mathbf{y}\cdot\mathbf{U}_i$ (this may or may not matter to the reader).

In our sister text (Saville and Wood, 1991), we have spelt out how to do the computing for each worked example using the statistical computing package Minitab. The key routine used is again the "regress" routine. These example programs can be referred to if there is any confusion over any aspect of this appendix, since the Minitab commands can usually be easily translated into spreadsheet commands.

Exercises

Note that the first six exercises are aimed at users of simple spreadsheets, while the remaining exercises are aimed at users of regression routines. Solutions to exercises marked with an * are given in Appendix E.

(C.1) (a) Take the nine-dimensional observation vector $\mathbf{y}$ used in §2.4 and put it into the first column of your spreadsheet. (Hint: It is a good idea to label each column and calculated quantity; for example, a "y" could be entered into the first cell of the first column.) Then put a column of nine ones into the second column, and divide this column by $\sqrt{9} = 3$; the unit vector $\mathbf{U}_1$ is now in this second column (which could be labeled "U1"). Then multiply columns 1 and 2, putting the result into column 3 (labeled "y x U1"). Then add up this third column and put the answer in the next plus one cell at the bottom of column 3; the answer is the projection length $\mathbf{y}\cdot\mathbf{U}_1$ (labeled "y.U1" in the empty cell above this quantity). As evidence of your work, print columns 1–3.

(b) To calculate the orthogonal decomposition of $\mathbf{y}$, multiply column 2 (containing $\mathbf{U}_1$) by the projection length $\mathbf{y}\cdot\mathbf{U}_1$ and put the answer into column 5 (leaving column 4 empty for now); column 5 now contains the projection vector $(\mathbf{y}\cdot\mathbf{U}_1)\mathbf{U}_1$, plus appropriate label. Then subtract column 5 from column 1 and put the answer into column 6; this now contains the error vector, $\mathbf{y} - (\mathbf{y}\cdot\mathbf{U}_1)\mathbf{U}_1$. Lastly, copy column 1 ($\mathbf{y}$) into column 4. Then print columns 4–6; these three columns represent the orthogonal decomposition $\mathbf{y} = (\mathbf{y}\cdot\mathbf{U}_1)\mathbf{U}_1 + [\mathbf{y} - (\mathbf{y}\cdot\mathbf{U}_1)\mathbf{U}_1]$.

(c) To calculate the corresponding Pythagorean breakup, first square each of columns 4–6, putting the answers into columns 7–9. Then sum each of the latter columns, putting the answers in the next plus one cell at the bottom of each of these columns. These three numbers represent the Pythagorean breakup $\|\mathbf{y}\|^2 = (\mathbf{y}\cdot\mathbf{U}_1)^2 + [\|\mathbf{y}\|^2 - (\mathbf{y}\cdot\mathbf{U}_1)^2]$. Print columns 7–9 and check that the last two numbers add up to the first.

(d) Lastly, calculate the appropriate F test statistic, and compare your answer with the percentiles of the appropriate F distribution. How strong is the evidence for a difference in height between male and female twins?

(C.2)* Mimic the procedure described in Exercise C.1 using the seven data values (changes in selenium level) given in Exercise 2.4(a).

(C.3) (a) Take the eight-dimensional observation vector $\mathbf{y}$ used in §3.3 and put it into the first column of your spreadsheet. Then put a column of eight ones into the second column, and divide this column by $\sqrt{8}$; the unit vector $\mathbf{U}_1$ is now in this second column. Then put a column of "−1"s and "1"s into column 3, and divide this column by $\sqrt{8}$; the unit vector $\mathbf{U}_2$ is now in column 3. Then multiply columns 1 and 2, putting the result into column 4. Then add up column 4 and put the answer in the next plus one cell at the bottom of column 4; the answer is the projection length $\mathbf{y}\cdot\mathbf{U}_1$. Label this quantity as described in Exercise C.1. Then multiply columns 1 and 3, putting the result into column 5. Then add up column 5 and put the answer in the next plus one cell at the bottom of column 5; the answer is the projection length $\mathbf{y}\cdot\mathbf{U}_2$ (label this also). As evidence of your work, print columns 1–5.

(b) To calculate the orthogonal decomposition of $\mathbf{y}$, first multiply column 2 (containing $\mathbf{U}_1$) by the projection length $\mathbf{y}\cdot\mathbf{U}_1$ and put the answer into column 7 (leaving column 6 empty for now); column 7 now contains the projection vector $(\mathbf{y}\cdot\mathbf{U}_1)\mathbf{U}_1$. Next multiply column 3 (containing $\mathbf{U}_2$) by the projection length $\mathbf{y}\cdot\mathbf{U}_2$ and put the answer into column 8; column 8 now contains the projection vector $(\mathbf{y}\cdot\mathbf{U}_2)\mathbf{U}_2$. Then subtract columns 7 and 8 from column 1 and put the answer into column 9; this column now contains the error vector $\mathbf{y}-(\mathbf{y}\cdot\mathbf{U}_1)\mathbf{U}_1-(\mathbf{y}\cdot\mathbf{U}_2)\mathbf{U}_2$. Lastly, copy column 1 ($\mathbf{y}$) into column 6. Then print columns 6–9; these columns represent the orthogonal decomposition $\mathbf{y}=(\mathbf{y}\cdot\mathbf{U}_1)\mathbf{U}_1+(\mathbf{y}\cdot\mathbf{U}_2)\mathbf{U}_2+[\mathbf{y}-(\mathbf{y}\cdot\mathbf{U}_1)\mathbf{U}_1-(\mathbf{y}\cdot\mathbf{U}_2)\mathbf{U}_2]$.

(c) To calculate the corresponding Pythagorean breakup, first square each of columns 6–9, putting the answers into columns 10–13. Then sum each of the latter columns, putting the answers in the next plus one cell at the bottom of each of these columns. These four numbers represent the Pythagorean breakup $||\mathbf{y}||^2=(\mathbf{y}\cdot\mathbf{U}_1)^2+(\mathbf{y}\cdot\mathbf{U}_2)^2+[||\mathbf{y}||^2-(\mathbf{y}\cdot\mathbf{U}_1)^2-(\mathbf{y}\cdot\mathbf{U}_2)^2]$. Print columns 10–13 and check that the last three numbers add up to the first.

(d) Lastly, calculate the appropriate F test statistic, and compare your answer with the percentiles of the appropriate F distribution. How strong is the evidence for a difference in height between males and females?

(C.4) Mimic the procedure described in Exercise C.3 using the ten load weights given in Exercise 3.5.

(C.5) Extend the procedure described in Exercise C.3 to the case of three independent samples, using for your observation vector the six selenium levels used in §4.2. What are the results of the two hypothesis tests?

(C.6) Adapt the procedure described in Exercise C.3 to the simple regression case, using for your observation vector the five air pollution levels used in §5.2. Note that the main change is that the vector $\mathbf{U}_2$ is now a unit vector derived from the x values.

(C.7)* (a) Set up the nine-dimensional observation vector $\mathbf{y}$ used in §2.4 as either a spreadsheet column or variable, and specify it to be the dependent variable for your regression routine. For your independent variable ($\mathbf{U}_1$), set up a column or variable of nine ones and divide it by $\sqrt{9} = 3$. Then regress the $\mathbf{y}$ variable onto the $\mathbf{U}_1$ variable using the regression routine, firstly specifying that there is "no constant (or intercept) term." The routine will then produce a regression coefficient which is the projection length $\mathbf{y} \cdot \mathbf{U}_1$. Print the regression output and circle this quantity.

(b) The Pythagorean breakup should be printed by the regression routine. If not, follow the procedure outlined in §C.2 in the "paired samples" subsection. Do your answers agree with those given in §2.4?

(c) The appropriate F test statistic may also be printed by the regression routine. If not, calculate it and compare your answer with the percentiles of the appropriate F distribution. How strong is the evidence for a difference in height between male and female twins?

(C.8) Mimic the procedure described in Exercise C.7 using the seven data values (changes in selenium level) given in Exercise 2.4(a).

(C.9) (a) Set up the eight-dimensional observation vector $\mathbf{y}$ used in §3.3 as either a spreadsheet column or variable, and specify it to be the dependent variable for your regression routine. For your independent variable ($\mathbf{U}_2$), set up a column or variable of "-1"s and "1"s, and divide it by $\sqrt{8}$. Then regress the $\mathbf{y}$ variable onto the $\mathbf{U}_2$ variable using the regression routine. The routine will automatically include a "constant (or intercept) term" which corresponds to automatic projection onto $\mathbf{U}_1$. The routine will then produce two regression coefficients, the second of which is the projection length $\mathbf{y} \cdot \mathbf{U}_2$. Print the regression output and circle this quantity.

(b) The Pythagorean breakup should be printed by the regression routine. The usual format, however, differs slightly from ours in that $||\bar{\mathbf{y}}||^2$ is subtracted from both sides of the equation (this does not affect the resulting F test statistic). If the breakup is not printed, follow the procedure outlined in §C.2 in the "independent samples" subsection. Do your answers agree with those given in §3.3?

(c) The appropriate F test statistic may also be printed by the regression routine. If not, calculate it. Compare your answer with the percentiles of the appropriate F distribution. How strong is the evidence for a difference in height between males and females?

(C.10) Mimic the procedure described in Exercise C.9 using the ten load weights given in Exercise 3.5.

(C.11) (a) Set up the six-dimensional observation vector $\mathbf{y}$ used in §4.2 as either a spreadsheet column or variable, and specify it to be the dependent variable for your regression routine.

If a *multiple regression* routine is available, as it usually is, proceed as follows. For your independent variables ($\mathbf{U}_2$ and $\mathbf{U}_3$), set up two columns or variables of contrast coefficients and divide each by its length. Then regress the $\mathbf{y}$ variable onto the $\mathbf{U}_2$ and $\mathbf{U}_3$ variables using the multiple regression routine (a $\mathbf{U}_1$ variable should not be specified since the routine automatically projects onto it, producing the constant, or intercept term). The routine will produce three regression coefficients, the second and third of which are the projection lengths $\mathbf{y}\cdot\mathbf{U}_2$ and $\mathbf{y}\cdot\mathbf{U}_3$. Print the regression output and circle these quantities.

If only a *simple regression* routine is available, proceed as in the last paragraph except that you will need to regress $\mathbf{y}$ onto $\mathbf{U}_2$, then $\mathbf{y}$ onto $\mathbf{U}_3$, in two separate regressions. The first projection length $\mathbf{y}\cdot\mathbf{U}_2$ will be the second regression coefficient output by the first regression, and the second projection length $\mathbf{y}\cdot\mathbf{U}_3$ will be the second regression coefficient output by the second regression. Print the two regression outputs and circle these two quantities.

(b) The Pythagorean breakup can now be calculated as described in §C.2 in the "several independent samples" subsection. Do your answers agree with those given in §4.2?

(c) The F test statistics can now be calculated, and compared with the percentiles of the appropriate F distribution. What are the results of the two hypothesis tests?

(C.12) Adapt the procedure described in Exercise C.9 to the simple regression case, using for your observation vector the five air pollution levels used in §5.2. Note that the main change is that the vector $\mathbf{U}_2$ is now a unit vector derived from the x values.

Appendix D

Alternative Test Statistic

In this appendix we introduce the idea of using an angle (θ) as the test statistic in place of the F or t test statistics which we have used in the main body of this textbook. This alternative test statistic always yields the same conclusions as the other two test statistics. We introduce it because it is conceptually much simpler than the traditional t or F, and has a very simple relationship with the probability value p.

D.1 Basic Idea

In order to understand the basic idea the reader is referred back to Figure 1.1, where we displayed our samples of size two as *points* in 2-space. Our new test statistic is motivated by redrawing this figure with our samples represented by *vectors* in 2-space (Figure D.1). In the case $\mu = 0$ the resulting vectors are uniformly distributed around the origin, as shown in Figure D.1(c). In the case $\mu = 1.2$, however, a typical vector of observations tends to be in a particular direction, the equiangular direction or the direction making an angle of 45° with both axes, as shown in Figure D.1(f). For our alternative approach we therefore phrase the geometric question as, "If the thermometer is unbiased ($\mu = 0$), how likely is it that the vector of observations would be at such a small angle with the equiangular direction?"

The answer is simple. Suppose our observation vector is $[y_1, y_2]^T = [1.3, 1.5]^T$, as in Chapter 1 and as shown in Figure D.2. In this case the angle θ can be shown to be 4.09°. Then if the mean μ is zero, all angles with the equiangular direction are equally likely, so it follows that the probability of observing as small an angle as $\theta = 4.09$, or smaller, is $p = \theta/90 = 4.09/90 = 0.045$, as illustrated in Figure D.2. This probability is less than the traditional cut-off point of 0.05, so we reject the idea that the mean is zero, and conclude that the thermometer is biased. (Note that we are carrying out a two-sided test, to allow for the possibility of either positive or negative biases.)

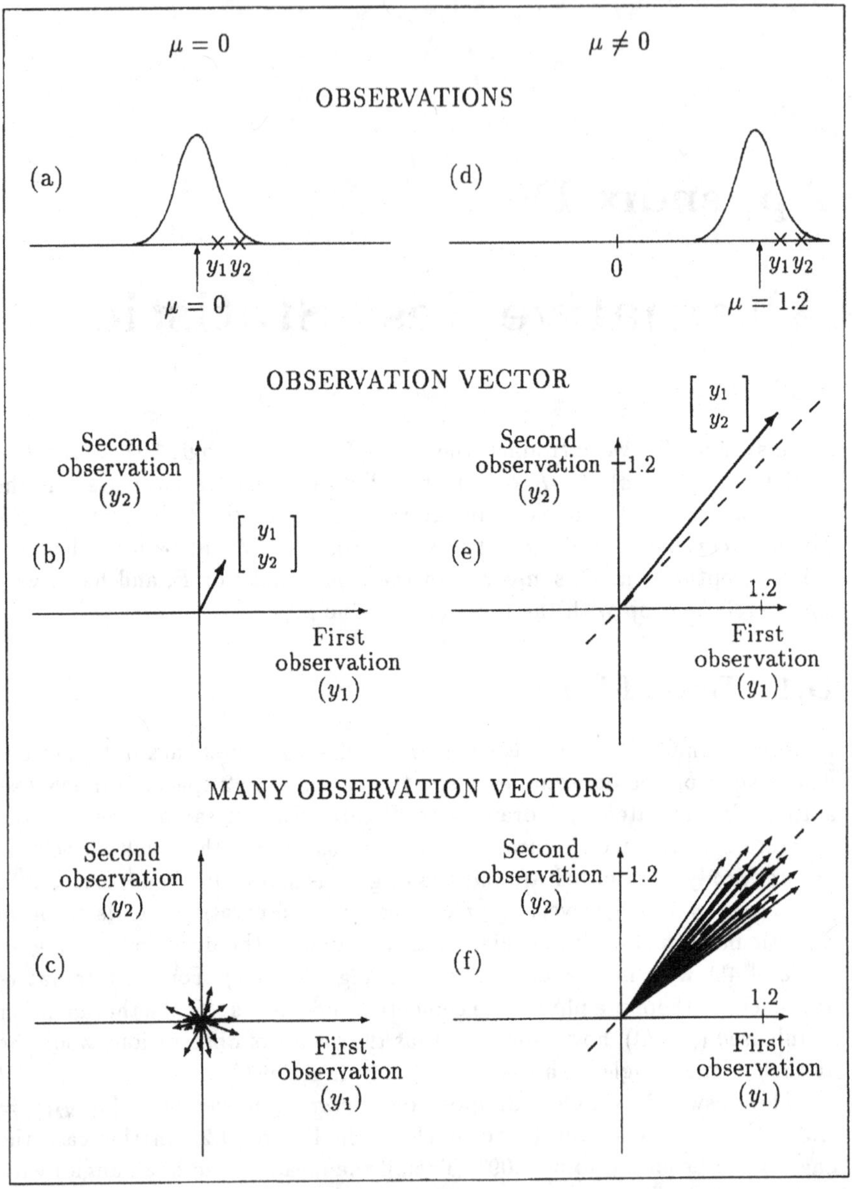

Figure D.1: The correspondence between samples of size two and vectors $[y_1, y_2]^T$ in 2-space. The top level shows a single sample, the middle level shows the single sample as a vector in space, and the bottom level shows many samples as many vectors in space. On the left the mean of the observations is zero, while on the right the mean is nonzero. Over many repetitions of the thermometer study the sample vectors will be uniformly distributed around the origin if $\mu = 0$ (as in (c)), and will make relatively small angles with the equiangular direction if $\mu \neq 0$ (as in (f)).

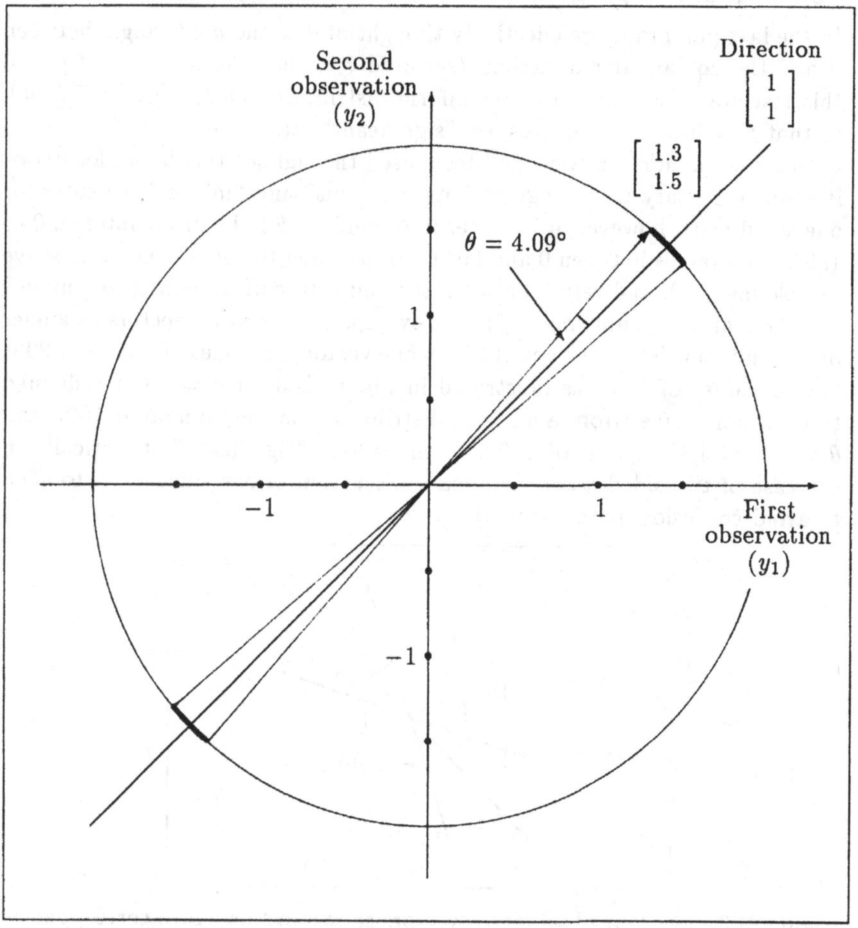

Figure D.2: Our vector of observations $[1.3, 1.5]^T$ makes an angle of 4.09° with the equiangular direction. If $\mu = 0$, what is the probability of observing an angle as small, or smaller than this? The answer is the ratio of the length of the highlighted arcs to the circumference of the circle ($p = 4 \times 4.09/360 = 4.09/90$). The highlighted arcs indicate angles with the equiangular direction less than or equal to that made by the vector of observations. The highlighting in the third quadrant is present since our alternative hypothesis does not rule out the possibility of a negative bias. Note that to be strictly correct p is a ratio of two integrals; each is an integral, from distance zero to infinity, of the appropriate arc length times the probability density for an arc at that distance. This ratio of integrals reduces to the above ratio of arc lengths, so we use the latter in its place, both here and from now on.

Some Housekeeping

In the last paragraph we effectively thought of θ as the *acute* angle between **y** and the equiangular direction, and used θ as our test statistic. If $\mu = 0$ this test statistic comes from a uniform distribution on the interval $[0, 90]$, so that θ values of 4.5° or less are "significant" statistically.

This is fine for the two-sided tests used throughout this book, for which it is not necessary to distinguish between "plus" and "minus." To cater for one-sided tests, however, it is better if we redefine θ to lie in the interval 0 to 180°, with values between 0 and 90° corresponding to "plus" (e.g., a positive sample mean $\bar{y}$), and values between 90° and 180° corresponding to "minus" (e.g., a negative sample mean $\bar{y}$). For example, observation vectors at angles of 25° and 335° have θ values of 25°, while vectors at angles of 155° and 205° have θ values of 155°, as illustrated in Figure D.3. If $\mu = 0$ our redefined test statistic comes from a uniform distribution on the interval $[0, 180]$, and θ values of 4.5° or less, or 175.5° or more, are "significant" statistically in the case of two-sided tests. (We leave discussion of one-sided tests to §D.8 to avoid confusion at this stage.)

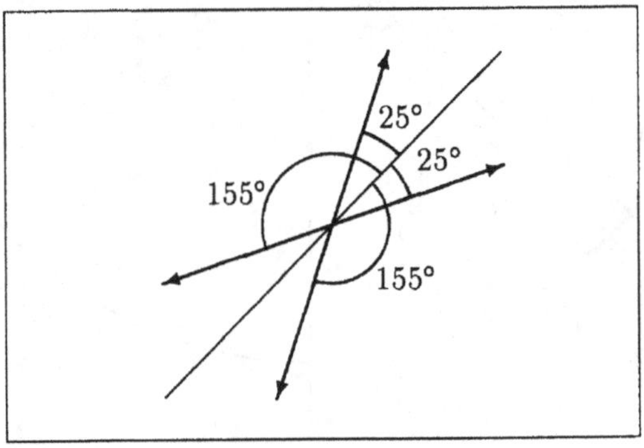

Figure D.3: The four observation vectors shown in this figure correspond to θ values of 25°, 25°, 155° and 155°, respectively.

A second housekeeping detail is that in more complicated cases we shall need to evaluate integrals which require the angle θ to be in *radians* rather than degrees. From now on we shall therefore work in radians.

Summary and Preview

In general terms our new basic idea is: Find the angle between the observation vector (appropriately "corrected" in all except the simplest cases) and the direction corresponding to the parameter of current interest, then calculate the probability of observing such a small angle under the hypothesis "parameter = 0."

By comparison, the basic idea we previously used was: isolate the direction associated with the parameter of interest, project onto it to obtain the projection length A, then divide by a quantity $B/\sqrt{q}$ which depends upon projection lengths for q "random error" directions; this yields the t test statistic.

In §D.7 we shall see that when the sample size is two the test statistics θ and t are linked via the formula $t = A/B = \cot\theta$ when $0 \leq \theta \leq \pi/2$ and $t = -A/B = \cot\theta$ when $\pi/2 < \theta \leq \pi$. This means that a small θ value corresponds to a large t value. In our example $t = 1.4\sqrt{2}/(0.1\sqrt{2}) = 14 = \cot 4.09°$. We also see in §D.7 that, in general, $t = A/(B/\sqrt{q}) = \sqrt{q}\cot\theta$ when $0 \leq \theta \leq \pi/2$ and $t = -A/(B/\sqrt{q}) = \sqrt{q}\cot\theta$ when $\pi/2 < \theta \leq \pi$, where q is the error degrees of freedom.

We now systematically apply our new basic idea to the worked examples presented in Chapters 2–5, demonstrating how the angle test statistic can be used in each of these examples.

D.2 Paired Samples

We first rework the examples in Chapter 2 using the angle test statistic.

Sample Size of Two

For our first analysis, for a sample size of two, we start by drawing Figure D.4, which displays our vector of observations $\mathbf{y} = [6, 9]^T$ in relation to the equiangular direction $[1, 1]^T$.

The angle θ between the observation vector and the equiangular direction is determined by

$$\cos\theta = \frac{\begin{bmatrix} 6 \\ 9 \end{bmatrix} \cdot \begin{bmatrix} 1 \\ 1 \end{bmatrix}}{\sqrt{6^2 + 9^2} \times \sqrt{2}} = \frac{15}{\sqrt{234}} = 0.9806$$

That is, the angle is $\theta = 0.197$ radians, or 11.3°.

Now if the population of differences in height is normally distributed with mean $\mu = 0$, it follows that the observation vector is equally likely to point in any direction. We shall not prove this here! Hence all angles with the equiangular direction (from 0 to π radians) are equally likely. What, then, is the probability of observing an angle as extreme, or more extreme, than $\theta = 0.197$ radians? The answer is

$$p = \frac{2\theta}{\pi} = \frac{0.197}{\pi/2} = 0.13$$

as illustrated in Figure D.4. Here we use the minimum of θ and $\pi - \theta$ in the formula since we are carrying out a two-sided test. This probability is not unusually small (in particular, it is greater than 0.05), so we conclude that we have found insufficient evidence to reject the idea that the population mean is zero. That is, we have not shown beyond reasonable doubt that there is, on average, a difference in height between the male and female twins in our study population.

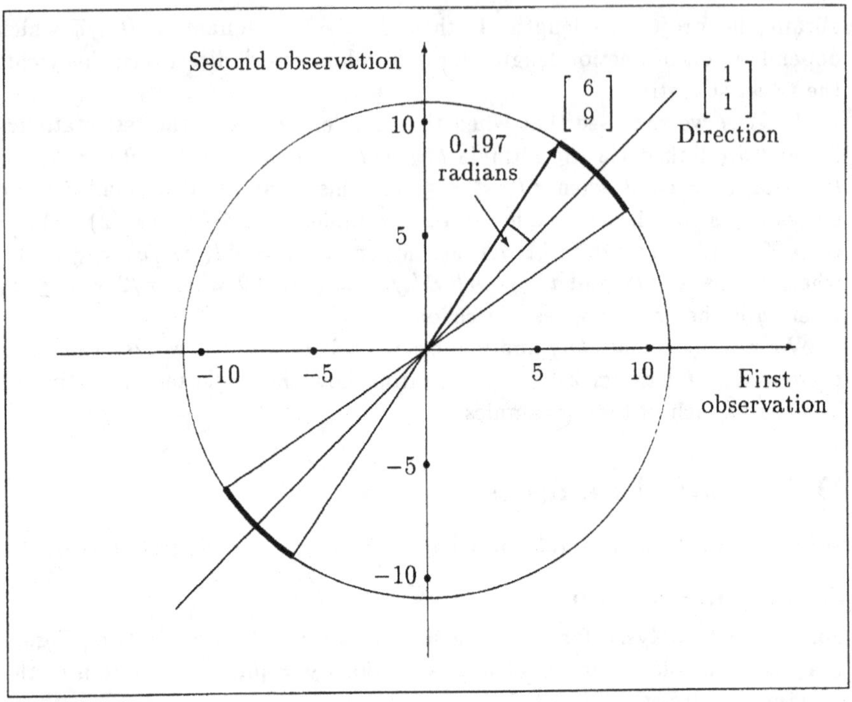

Figure D.4: Our vector of observations $[6, 9]^T$ makes an angle of 0.197 radians with the equiangular direction. If $\mu = 0$, what is the probability of observing an angle as small, or smaller than this? The answer is the ratio of the lengths of the highlighted arcs to the circumference of the circle, $p = 4 \times 0.197/(2\pi) = 0.197/(\pi/2) = 0.13$.

Sample Size of Three

For our second analysis we draw Figure D.5, which displays the vector of observations $\mathbf{y} = [7.5, 3, 9]^T$ in relation to the equiangular direction $[1, 1, 1]^T$. The angle θ between the observation vector and the equiangular direction is given by

$$\cos\theta = \frac{\begin{bmatrix}7.5\\3\\9\end{bmatrix} \cdot \begin{bmatrix}1\\1\\1\end{bmatrix}}{\sqrt{7.5^2 + 3^2 + 9^2} \times \sqrt{3}} = \frac{19.5}{\sqrt{438.75}} = 0.9309$$

That is, the angle is $\theta = 0.374$ radians, or 21.4°.

As in the previous example, if $\mu = 0$ all directions within a three-dimensional sphere are equally likely. However, this no longer equates to all angles between the observation vector and the $[1, 1, 1]^T$ direction being equally likely! To find the probability of observing an angle as small, or

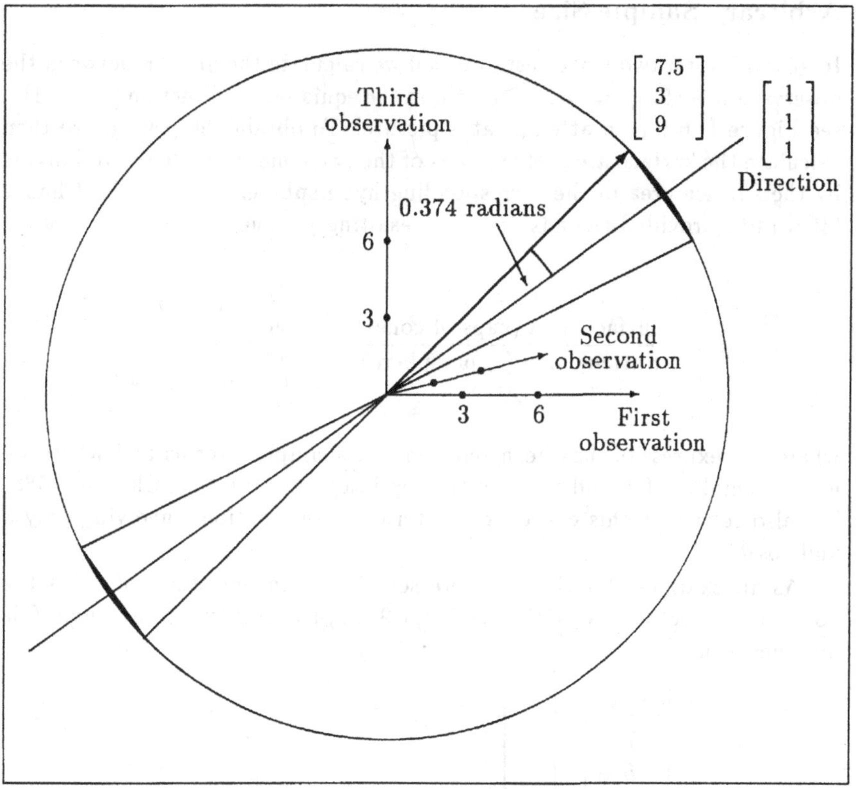

Figure D.5: Our vector of observations $[7.5, 3, 9]^T$ makes an angle of 0.374 radians with the equiangular direction. If $\mu = 0$, what is the probability of observing an angle as small, or smaller than this? The answer is the ratio of the surface area of the shaded caps of the two cones to the surface area of the sphere, $p = 1 - \cos 0.374 = 1 - 0.931 = 0.069$.

smaller, than $\theta = 0.374$ radians, we need to take the surface areas of the caps of the cones marked in Figure D.5 and divide by the surface area of the corresponding sphere. The answer has been given by Chance (1986), and is

$$p = \frac{\text{Surface area (caps of cones)}}{\text{Surface area (sphere)}} = 1 - \cos\theta = 1 - \cos 0.374 = 1 - 0.931$$

where we use the minimum of θ and $\pi - \theta$ in the formula. The resulting probability is $p = 0.069$. This probability is less than 0.10, so is somewhat unusual. We conclude that we have found some (weak) evidence to contradict the idea that the population mean is zero. That is, the data suggests that there *may* be, on average, a difference in height between the male and female twins in our study population, though the evidence is not strong.

Arbitrary Sample Size

In general, for a two-sided test of $\mu = 0$ we calculate the angle θ between the observation vector $\mathbf{y} = [y_1, \ldots, y_n]^T$ and the equiangular direction $[1, \ldots, 1]^T$; see Figure D.6 for an attempt at a picture! To obtain the p value we then calculate the surface areas of the caps of the two cones in n-space, and divide by the surface area of the corresponding hypersphere. Fortunately, Chance (1986) has provided the answer. The resulting p value is

$$p = \frac{\text{Surface area (caps of cones)}}{\text{Surface area (hypersphere)}} = \frac{\displaystyle\int_0^{\theta} \sin^{n-2} u \, du}{\displaystyle\int_0^{\pi/2} \sin^{n-2} u \, du}$$

where the expression has been reduced to its simplest terms and where we use the smaller of θ and $\pi - \theta$ in the top integral. Note that Chance (1986) has also rewritten this expression in terms of summations involving only θ and $\cos\theta$.

As an example, for the complete set of data in our study of twins the observation vector is $\mathbf{y} = [11.8, 7.5, 6, 7.9, 4.9, 4, 3, 10, 9]^T$, so the angle θ is obtained from

$$\cos\theta = \begin{bmatrix} 11.8 \\ 7.5 \\ 6 \\ 7.9 \\ 4.9 \\ 4 \\ 3 \\ 10 \\ 9 \end{bmatrix} \cdot \begin{bmatrix} 1 \\ 1 \\ 1 \\ 1 \\ 1 \\ 1 \\ 1 \\ 1 \\ 1 \end{bmatrix} / \sqrt{523.91 \times 9} = \frac{64.1}{\sqrt{4715.19}} = 0.9335$$

That is, the angle is $\theta = 0.367$ radians, or $21.0°$. The resulting p value is

$$p = \frac{\displaystyle\int_0^{0.367} \sin^7 u \, du}{\displaystyle\int_0^{\pi/2} \sin^7 u \, du} = 0.00008$$

where a numerical algorithm was used to evaluate the two integrals.

Our p value of 0.00008 is very low, so we conclude that there is very strong evidence to suggest that the population mean is not zero. That is, the data strongly suggests that there is, on average, a difference in height between the male and female twins in our study population.

Note that the formulas $p = \theta/[\pi/2]$ and $p = 1 - \cos\theta$ for sample sizes of $n = 2$ and 3, respectively, can be obtained by substituting into the above general formula.

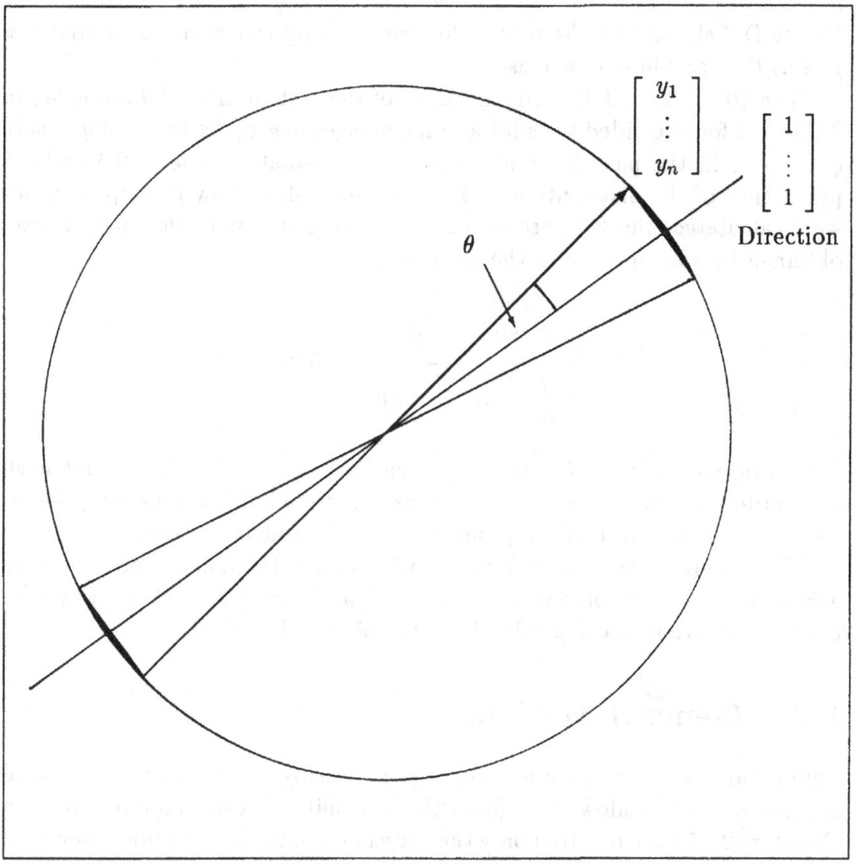

Figure D.6: Our vector of observations $[y_1, \ldots, y_n]^{\mathrm{T}}$ makes an angle of θ with the equiangular direction. If $\mu = 0$, what is the probability of observing an angle as small, or smaller than this? The answer is the ratio of the surface area of the shaded caps of the cones to the surface area of the entire hypersphere.

Distribution of θ

The distribution of the angle θ under the null hypothesis $\mu = 0$ is shown in Figure D.7 for sample sizes of $n = 2$, 3 and 4. The probability density which we have graphed is

$$f(\theta) = \frac{\sin^{n-2}\theta}{2\int_0^{\pi/2} \sin^{n-2} u \, du}$$

Figure D.7 shows how the probability increasingly concentrates around $\theta = \pi/2$ as the sample size increases.

The 10%, 5% and 1% critical values of the distribution of θ are given in Table T.2 for two-sided tests for a range of error degrees of freedom q, where $q = n - 1$ in the paired samples case. These equate to the 5, 2.5 and 0.5 percentiles of the distribution of θ. As an example of how these percentiles were calculated, the 2.5 percentiles of the distributions of the angle θ were obtained by solving for θ in the equation

$$\frac{\int_0^{\theta} \sin^{n-2} u \, du}{2\int_0^{\pi/2} \sin^{n-2} u \, du} = 0.025$$

for a range of values of n (or equivalently, $q = n - 1$). Note that each percentile tends to $\pi/2$ as the sample size tends to infinity, reflecting the increasing concentration of the probability density function around this point.

These percentiles will only need to be used if the reader does not have access to a suitable computer package such as Maple (Char et al., 1991) for direct calculation of the p value from the observed angle θ.

D.3 General Method

Before moving on to the independent samples case, we spell out the general method which we follow throughout this appendix. Recall that our F tests in Chapters 2–5 have involved only the projection onto the direction associated with the parameter of interest, plus projections onto q error space directions. In terms of this appendix the consequence is that we consider only the $(q+1)$-dimensional picture associated with these directions. This picture is the "statistical triangle" discussed in §6.1.

In the paired samples case the picture has been of the entire N-space since our model space has been of dimension one. In general, however, the model space will be of dimension p; in this case the procedure will be to subtract from the observation vector its projections onto all directions in the model space *except* the direction associated with the parameter of interest. We then treat this $(q+1)$-dimensional "corrected" observation vector in the same way that we have treated the observation vector in §D.2.

The above may be hard to comprehend in abstract terms. We trust it will become clear as we work through more examples.

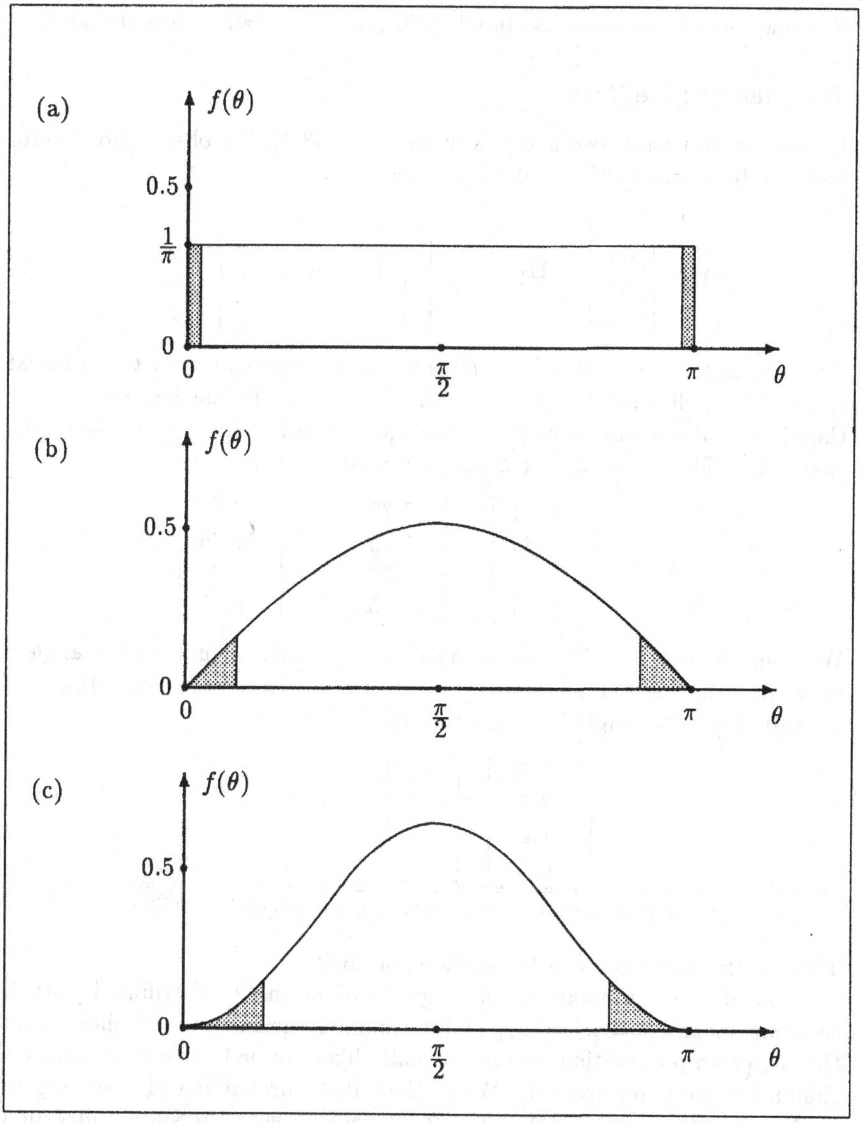

Figure D.7: The probability density $f(\theta)$ of the distribution of the angle θ for sample sizes of: (a) $n = 2$; (b) $n = 3$; and (c) $n = 4$, corresponding to $q = 1$, 2 and 3 error degrees of freedom. The left-hand shaded area is 2.5% of the area under the curve; its right-hand boundary is the 2.5 percentile of the distribution. Similarly for the right-hand shaded area.

D.4 Independent Samples

We now rework the examples in Chapter 3 using the angle test statistic.

Samples of Size Two

In the simplest case, two samples of size two (§3.2), the observation vector and the directions in the model space are

$$\mathbf{y} = \begin{bmatrix} 63 \\ 65 \\ 69 \\ 74 \end{bmatrix}, \quad \mathbf{U_1} = \frac{1}{\sqrt{4}}\begin{bmatrix} 1 \\ 1 \\ 1 \\ 1 \end{bmatrix}, \quad \mathbf{U_2} = \frac{1}{\sqrt{4}}\begin{bmatrix} -1 \\ -1 \\ 1 \\ 1 \end{bmatrix}$$

Here $\mathbf{U_2}$ is the direction of special interest, corresponding to the contrast $(\mu_2 - \mu_1)$. Following the general method outlined in the last section, we therefore subtract from $\mathbf{y}$ its projection onto the other direction in the model space, $\mathbf{U_1}$. This yields a "corrected" observation vector

$$\mathbf{y} - \bar{\mathbf{y}} = \begin{bmatrix} 63 \\ 65 \\ 69 \\ 74 \end{bmatrix} - \begin{bmatrix} 67.75 \\ 67.75 \\ 67.75 \\ 67.75 \end{bmatrix} = \begin{bmatrix} -4.75 \\ -2.75 \\ 1.25 \\ 6.25 \end{bmatrix}$$

We then proceed as in §D.2, drawing Figure D.8 and calculating the angle θ between the corrected observation vector and the direction of interest, $\mathbf{U_2}$. This angle is determined by

$$\cos\theta = \frac{\begin{bmatrix} -4.75 \\ -2.75 \\ 1.25 \\ 6.25 \end{bmatrix} \cdot \begin{bmatrix} -1 \\ -1 \\ 1 \\ 1 \end{bmatrix}}{\sqrt{(-4.75)^2 + (-2.75)^2 + 1.25^2 + 6.25^2} \times \sqrt{4}} = \frac{15}{\sqrt{283}} = 0.8917$$

That is, the angle is $\theta = 0.470$ radians, or 26.9°.

Now if both populations of heights are normally distributed with a common mean ($\mu = \mu_1 = \mu_2$) and a common variance σ^2, it follows that the corrected observation vector is equally likely to lie in any direction. We shall not digress to prove this! What, then, is the probability of observing an angle as small, or smaller than $\theta = 0.470$ radians with the special direction $[-1, -1, 1, 1]^T$? The answer is

$$p = \frac{\int_0^{\theta} \sin^{q-1} u \, du}{\int_0^{\pi/2} \sin^{q-1} u \, du} = \frac{\int_0^{0.470} \sin u \, du}{\int_0^{\pi/2} \sin u \, du} = 1 - \cos 0.470 = 0.108$$

where q is the number of error degrees of freedom (2 here). This probability is not unusually small, so we have not shown beyond reasonable doubt that there is, on average, any difference in height between the female and male students in our study populations. Note that this is the same as the result obtained in §3.2.

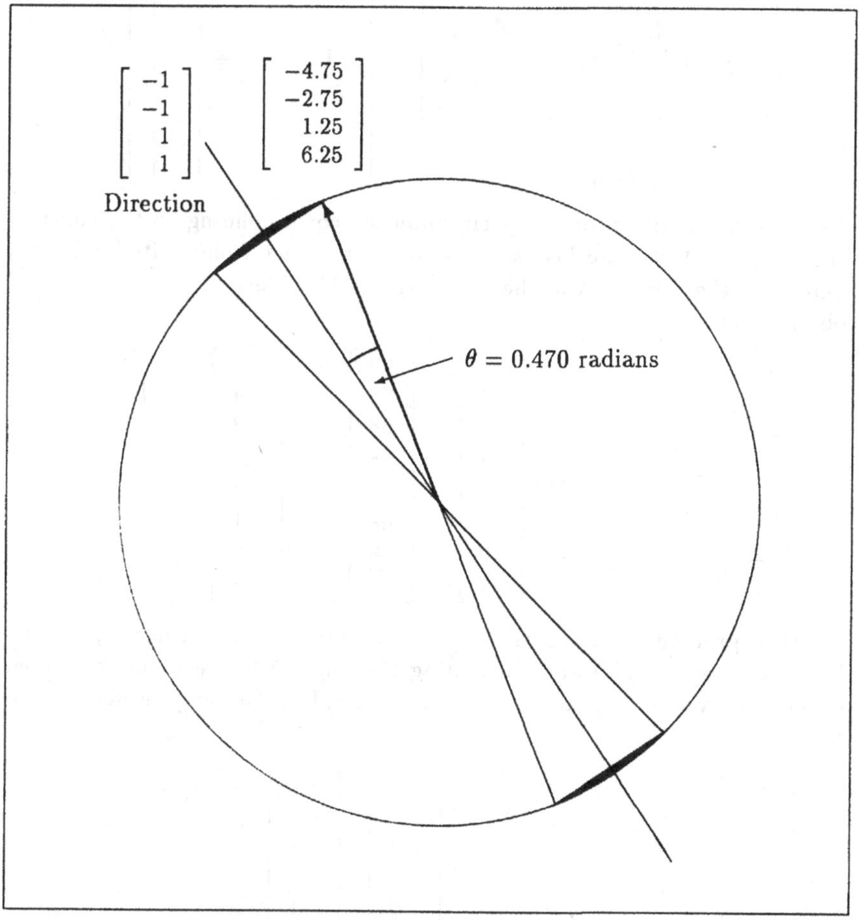

Figure D.8: Our corrected vector of observations $[-4.75, -2.75, 1.25, 6.25]^T$ makes an angle of 0.470 radians with the direction of special interest $[-1, -1, 1, 1]^T$. If $\mu_1 = \mu_2$, what is the probability of observing an angle as small, or smaller than this? The answer is the ratio of the surface area of the two shaded cone caps to the surface area of the sphere, $p = 1 - \cos 0.470 = 1 - 0.892 = 0.108$.

General Case

In the general case of two samples of size n and in particular our example with $n = 4$ (§3.3), the observation vector and the directions in the model space are

$$\mathbf{y} = \begin{bmatrix} 60 \\ 64 \\ 65 \\ 68 \\ 70 \\ 69 \\ 77 \\ 71 \end{bmatrix}, \quad \mathbf{U_1} = \frac{1}{\sqrt{8}} \begin{bmatrix} 1 \\ 1 \\ 1 \\ 1 \\ 1 \\ 1 \\ 1 \\ 1 \end{bmatrix}, \quad \mathbf{U_2} = \frac{1}{\sqrt{8}} \begin{bmatrix} -1 \\ -1 \\ -1 \\ -1 \\ 1 \\ 1 \\ 1 \\ 1 \end{bmatrix}$$

Here $\mathbf{U_2}$ is the direction of special interest, corresponding to the contrast $(\mu_2 - \mu_1)$. As in the last subsection we subtract from $\mathbf{y}$ its projection onto the other direction in the model space, $\mathbf{U_1}$. This yields a "corrected" observation vector

$$\mathbf{y} - \bar{\mathbf{y}} = \begin{bmatrix} 60 \\ 64 \\ 65 \\ 68 \\ 70 \\ 69 \\ 77 \\ 71 \end{bmatrix} - \begin{bmatrix} 68 \\ 68 \\ 68 \\ 68 \\ 68 \\ 68 \\ 68 \\ 68 \end{bmatrix} = \begin{bmatrix} -8 \\ -4 \\ -3 \\ 0 \\ 2 \\ 1 \\ 9 \\ 3 \end{bmatrix}$$

We then proceed as previously, drawing a figure very much like Figure D.8 (we shall not do this) and calculating the angle θ between this corrected observation vector and the direction of interest, $\mathbf{U_2}$. This angle is determined by

$$\begin{aligned} \cos\theta &= \frac{\begin{bmatrix} -8 \\ -4 \\ -3 \\ 0 \\ 2 \\ 1 \\ 9 \\ 3 \end{bmatrix} \cdot \begin{bmatrix} -1 \\ -1 \\ -1 \\ -1 \\ 1 \\ 1 \\ 1 \\ 1 \end{bmatrix}}{\sqrt{(-8)^2 + (-4)^2 + (-3)^2 + 0^2 + 2^2 + 1^2 + 9^2 + 3^2} \times \sqrt{8}} \\ &= \frac{30}{\sqrt{1472}} = 0.7819 \end{aligned}$$

That is, the angle is $\theta = 0.673$ radians, or 38.6°.

As previously, if $\mu_1 = \mu_2$ the corrected observation vector is equally likely to lie in any direction. What, then, is the probability of observing an angle as small, or smaller than $\theta = 0.673$ radians between it and the special direction $[-1, -1, -1, -1, 1, 1, 1, 1]^T$? The answer is

$$p = \frac{\int_0^{\theta} \sin^{q-1} u\, du}{\int_0^{\pi/2} \sin^{q-1} u\, du} = \frac{\int_0^{0.673} \sin^5 u\, du}{\int_0^{\pi/2} \sin^5 u\, du} = 0.022$$

where $q = 6$ is the number of error degrees of freedom and the integrals were evaluated using Maple. This probability is unusually small, so we conclude that there is, on average, a difference in height between the female and male students in our study populations. Note that this is the same as the result obtained in §3.3.

D.5 Several Independent Samples

We now rework the examples in Chapter 4 using the angle test statistic.

Samples of Size Two

In the simplest case, three samples of size two (§4.2), the observation vector and the directions in the model space are

$$\mathbf{y} = \begin{bmatrix} 84 \\ 52 \\ 28 \\ 28 \\ 40 \\ 30 \end{bmatrix}, \quad \mathbf{U_1} = \frac{1}{\sqrt{6}} \begin{bmatrix} 1 \\ 1 \\ 1 \\ 1 \\ 1 \\ 1 \end{bmatrix}, \quad \mathbf{U_2} = \frac{1}{\sqrt{12}} \begin{bmatrix} 2 \\ 2 \\ -1 \\ -1 \\ -1 \\ -1 \end{bmatrix}, \quad \mathbf{U_3} = \frac{1}{\sqrt{4}} \begin{bmatrix} 0 \\ 0 \\ -1 \\ -1 \\ 1 \\ 1 \end{bmatrix}$$

Here $\mathbf{U_2}$ and $\mathbf{U_3}$ are the directions of special interest, corresponding to the contrasts $[\mu_1 - (\mu_2 + \mu_3)/2]$ and $(\mu_3 - \mu_2)$. Corresponding to these are two questions of interest which we must now investigate separately.

We begin with the first question of interest: Is $\mu_1 = (\mu_2 + \mu_3)/2$? The corresponding direction of special interest is $\mathbf{U_2} = [2, 2, -1, -1, -1, -1]^T/\sqrt{12}$. Following the general method outlined in §D.3, we subtract from $\mathbf{y}$ its projections onto the other directions in the model space, $\mathbf{U_1}$ and $\mathbf{U_3}$. This yields a "corrected" observation vector

$$\mathbf{y} - (\mathbf{y} \cdot \mathbf{U_1})\mathbf{U_1} - (\mathbf{y} \cdot \mathbf{U_3})\mathbf{U_3} = \begin{bmatrix} 84 \\ 52 \\ 28 \\ 28 \\ 40 \\ 30 \end{bmatrix} - \begin{bmatrix} 43.7 \\ 43.7 \\ 43.7 \\ 43.7 \\ 43.7 \\ 43.7 \end{bmatrix} - \begin{bmatrix} 0 \\ 0 \\ -3.5 \\ -3.5 \\ 3.5 \\ 3.5 \end{bmatrix} = \begin{bmatrix} 40.3 \\ 8.3 \\ -12.2 \\ -12.2 \\ -7.2 \\ -17.2 \end{bmatrix}$$

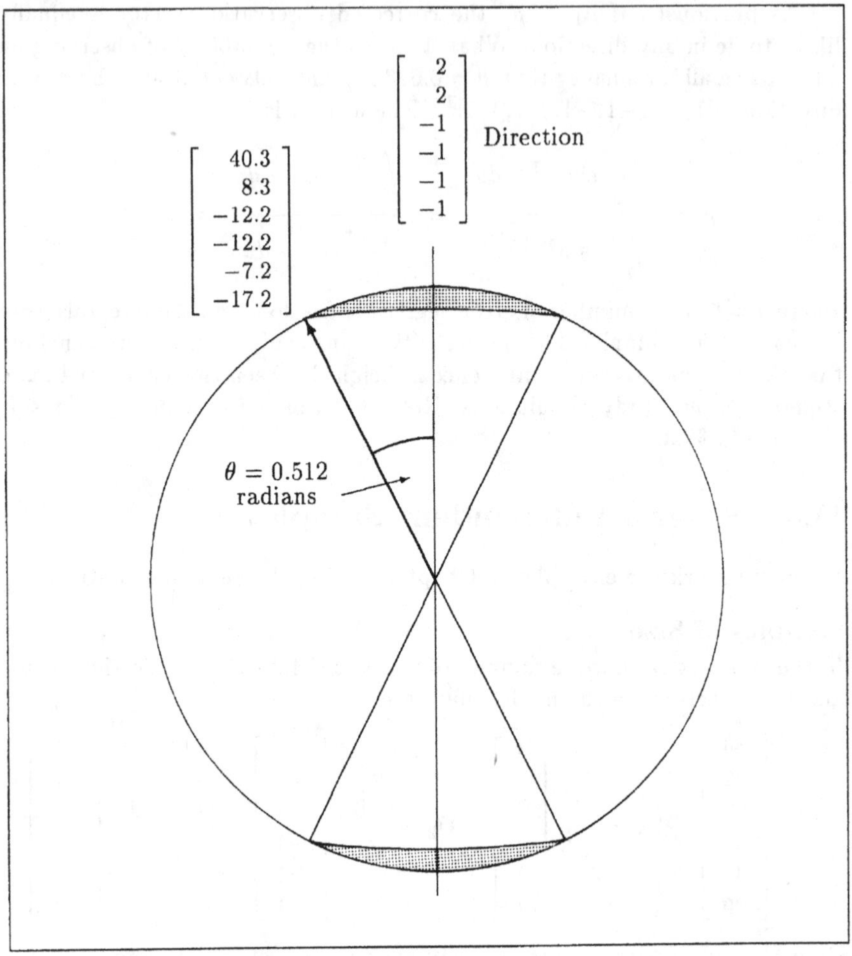

Figure D.9: The vector $[40.3, 8.3, -12.2, -12.2, -7.2, -17.2]^T$, our corrected observation vector, makes an angle of 0.512 radians with the direction of special interest $[2, 2, -1, -1, -1, -1]^T$. If $\mu_1 = (\mu_2 + \mu_3)/2$, what is the probability of observing an angle as small, or smaller than this? The answer is the ratio of the surface area of the two shaded cone caps to the surface area of the entire hypersphere, $p = 0.054$.

We then proceed as in §D.4, drawing Figure D.9 and calculating the angle θ between the corrected observation vector and the direction of interest, $\mathbf{U_2}$. This angle is determined by

$$\cos\theta = \frac{\begin{bmatrix} 40.3 \\ 8.3 \\ -12.2 \\ -12.2 \\ -7.2 \\ -17.2 \end{bmatrix} \cdot \begin{bmatrix} 2 \\ 2 \\ -1 \\ -1 \\ -1 \\ -1 \end{bmatrix}}{\sqrt{40.3^2 + 8.3^2 + (-12.2)^2 + (-12.2)^2 + (-7.2)^2 + (-17.2)^2} \times \sqrt{12}} = \frac{146}{\sqrt{28060}} = 0.8716$$

That is, the angle is $\theta = 0.512$ radians, or 29.4°.

Now if the three populations of selenium levels are all normally distributed with a common variance σ^2, and if the contrast $(2\mu_1 - \mu_2 - \mu_3)$ is zero (i.e., $\mu_1 = (\mu_2 + \mu_3)/2$), it follows that the corrected observation vector is equally likely to lie in any direction. What, then, is the probability of observing an angle as small, or smaller than $\theta = 0.512$ radians between it and the special direction $[2, 2, -1, -1, -1, -1]^{\mathrm{T}}$? The answer is

$$p = \frac{\int_0^{\theta} \sin^{q-1} u \, du}{\int_0^{\pi/2} \sin^{q-1} u \, du} = \frac{\int_0^{0.512} \sin^2 u \, du}{\int_0^{\pi/2} \sin^2 u \, du} = 0.054$$

where q is the number of error degrees of freedom (3 here). This probability is somewhat small ($p < 0.10$), so we conclude that there is some (weak) evidence of a difference in mean selenium level between our populations of adults and babies. Note that this is the same as the result obtained in §4.2.

We now investigate the second question of interest: Is $\mu_2 = \mu_3$? The corresponding direction of special interest is $\mathbf{U_3} = [0, 0, -1, -1, 1, 1]^{\mathrm{T}}/\sqrt{4}$. In this case we subtract from $\mathbf{y}$ its projections onto $\mathbf{U_1}$ and $\mathbf{U_2}$. This yields a "corrected" observation vector

$$\mathbf{y} - (\mathbf{y}\cdot\mathbf{U_1})\mathbf{U_1} - (\mathbf{y}\cdot\mathbf{U_2})\mathbf{U_2} = \begin{bmatrix} 84 \\ 52 \\ 28 \\ 28 \\ 40 \\ 30 \end{bmatrix} - \begin{bmatrix} 43.7 \\ 43.7 \\ 43.7 \\ 43.7 \\ 43.7 \\ 43.7 \end{bmatrix} - \begin{bmatrix} 24.3 \\ 24.3 \\ -12.2 \\ -12.2 \\ -12.2 \\ -12.2 \end{bmatrix} = \begin{bmatrix} 16.0 \\ -16.0 \\ -3.5 \\ -3.5 \\ 8.5 \\ -1.5 \end{bmatrix}$$

We then draw Figure D.10 and calculate the angle θ between the corrected observation vector and the direction of interest, $\mathbf{U_3}$. This angle is determined by

$$\cos\theta = \frac{\begin{bmatrix} 16.0 \\ -16.0 \\ -3.5 \\ -3.5 \\ 8.5 \\ -1.5 \end{bmatrix} \cdot \begin{bmatrix} 0 \\ 0 \\ -1 \\ -1 \\ 1 \\ 1 \end{bmatrix}}{\sqrt{16^2 + (-16)^2 + (-3.5)^2 + (-3.5)^2 + 8.5^2 + (-1.5)^2} \times \sqrt{4}} = \frac{14}{\sqrt{2444}} = 0.2832$$

That is, the angle is $\theta = 1.284$ radians, or 73.5°.

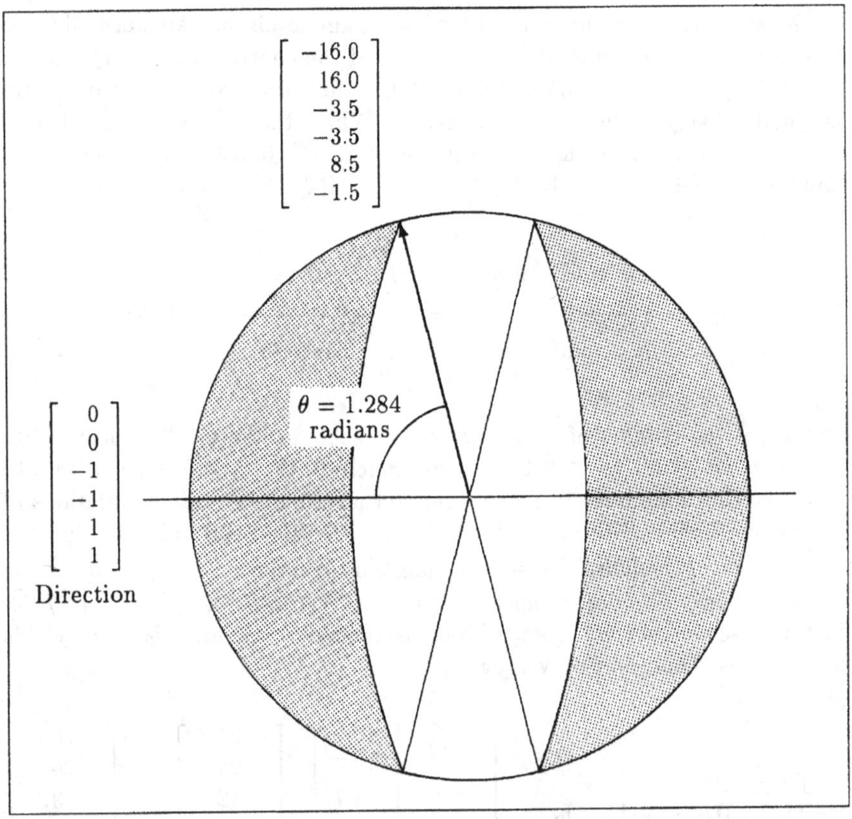

Figure D.10: The vector $[-16, 16, -3.5, -3.5, 8.5, -1.5]^{\mathrm{T}}$, our corrected observation vector, makes an angle of 1.284 radians with the direction of special interest $[0, 0, -1, -1, 1, 1]^{\mathrm{T}}$. If $\mu_2 = \mu_3$, what is the probability of observing an angle as small, or smaller than this? The answer is the ratio of the surface area of the two shaded cone caps to the surface area of the entire hypersphere, $p = 0.64$.

Now, if the three populations of selenium levels are all normally distributed with a common variance σ^2, and if the contrast $\mu_3 - \mu_2$ is zero (i.e., $\mu_2 = \mu_3$), it follows that the corrected observation vector is equally likely to lie in any direction. What, then, is the probability of observing an angle as small, or smaller than $\theta = 1.284$ radians with the special direction $[0, 0, -1, -1, 1, 1]^T$? The answer is

$$p = \frac{\int_0^\theta \sin^{q-1} u \, du}{\int_0^{\pi/2} \sin^{q-1} u \, du} = \frac{\int_0^{1.284} \sin^2 u \, du}{\int_0^{\pi/2} \sin^2 u \, du} = 0.64$$

This probability is not unusually small, so we have not shown that there is, on average, any difference in mean selenium level between our populations of premature and full-term babies. Note that this is the same as the result obtained in §4.2.

Full Data Set

For the full data set, with three samples of size 24 (§4.3), the observation vector and the directions in the model space are

$$\mathbf{y} = \begin{bmatrix} 60 \\ \vdots \\ 24 \\ \vdots \\ 31 \\ \vdots \end{bmatrix}, \quad \mathbf{U_1} = \frac{1}{\sqrt{72}} \begin{bmatrix} 1 \\ \vdots \\ 1 \\ \vdots \\ 1 \\ \vdots \end{bmatrix}, \quad \mathbf{U_2} = \frac{1}{\sqrt{144}} \begin{bmatrix} 2 \\ \vdots \\ -1 \\ \vdots \\ -1 \\ \vdots \end{bmatrix}, \quad \mathbf{U_3} = \frac{1}{\sqrt{48}} \begin{bmatrix} 0 \\ \vdots \\ -1 \\ \vdots \\ 1 \\ \vdots \end{bmatrix}$$

Here $\mathbf{U_2}$ and $\mathbf{U_3}$ are the directions of special interest, corresponding to the contrasts $[\mu_1 - (\mu_2 + \mu_3)/2]$ and $(\mu_3 - \mu_2)$ and to the two questions of interest.

We first investigate the question: Is $\mu_1 = (\mu_2 + \mu_3)/2$? The corresponding direction of special interest is $\mathbf{U_2} = [2, \ldots, -1, \ldots, -1, \ldots]^T / \sqrt{144}$. Following the usual method we subtract from $\mathbf{y}$ its projections onto the other directions in the model space, $\mathbf{U_1}$ and $\mathbf{U_3}$. This yields a "corrected" observation vector

$$\mathbf{y} - (\mathbf{y} \cdot \mathbf{U_1})\mathbf{U_1} - (\mathbf{y} \cdot \mathbf{U_3})\mathbf{U_3} = \begin{bmatrix} 60 \\ \vdots \\ 24 \\ \vdots \\ 31 \\ \vdots \end{bmatrix} - \begin{bmatrix} 44.6 \\ \vdots \\ 44.6 \\ \vdots \\ 44.6 \\ \vdots \end{bmatrix} - \begin{bmatrix} 0 \\ \vdots \\ 4.1 \\ \vdots \\ -4.1 \\ \vdots \end{bmatrix} = \begin{bmatrix} 15.4 \\ \vdots \\ -24.6 \\ \vdots \\ -9.5 \\ \vdots \end{bmatrix}$$

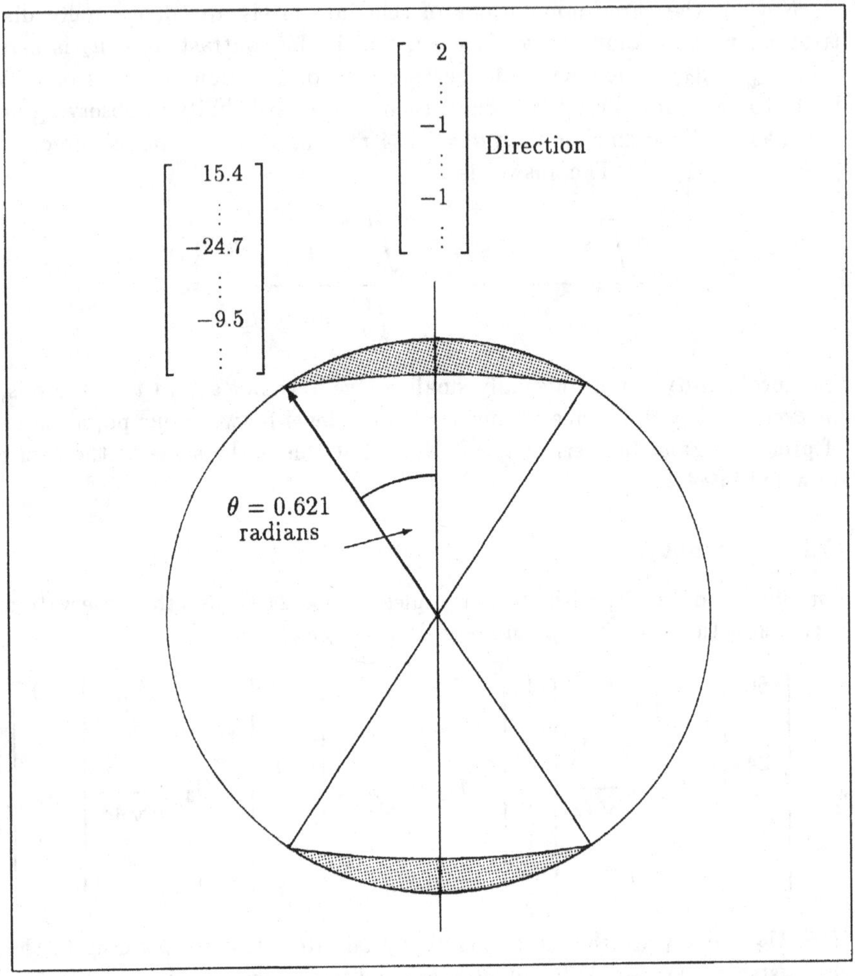

Figure D.11: The vector $[15.4,\ldots,-24.6,\ldots,-9.5,\ldots]^T$, our corrected observation vector, makes an angle of 0.621 radians with the direction of special interest $[2,\ldots,-1,\ldots,-1,\ldots]^T$. If $\mu_1 = (\mu_2 + \mu_3)/2$, what is the probability of observing an angle as small, or smaller than this? The answer is the ratio of the surface area of the two shaded cone caps to the surface area of the entire hypersphere, $p = 6.9 \times 10^{-18}$.

We then proceed as previously, drawing Figure D.11 and calculating the angle θ between the corrected observation vector and the direction of interest, $\mathbf{U_2}$. This angle is determined by

$$\cos\theta = \frac{\begin{bmatrix} 15.4 \\ \vdots \\ -24.6 \\ \vdots \\ -9.5 \\ \vdots \end{bmatrix} \cdot \begin{bmatrix} 2 \\ \vdots \\ -1 \\ \vdots \\ -1 \\ \vdots \end{bmatrix}}{\sqrt{15.4^2 + \cdots + (-24.6)^2 + \cdots + (-9.5)^2 + \cdots} \times \sqrt{144}}$$

$$= \frac{1934}{\sqrt{39259.5 \times 144}} = 0.8134$$

That is, the angle is $\theta = 0.621$ radians, or $35.6°$.

Now if the three populations of selenium levels are all normally distributed with a common variance σ^2, and if the contrast $(2\mu_1 - \mu_2 - \mu_3)$ is zero (i.e., $\mu_1 = (\mu_2 + \mu_3)/2$), it follows that the corrected observation vector is equally likely to lie in any direction. What, then, is the probability of observing as small, or smaller, an angle as $\theta = 0.621$ radians with the special direction $[2, \ldots, -1, \ldots, -1, \ldots]^T$? The answer is

$$p = \frac{\int_0^{\theta} \sin^{q-1} u \, du}{\int_0^{\pi/2} \sin^{q-1} u \, du} = \frac{\int_0^{0.621} \sin^{68} u \, du}{\int_0^{\pi/2} \sin^{68} u \, du} = 6.9 \times 10^{-18}$$

where $q = 69$ is the number of error degrees of freedom. This probability is extremely small, so we conclude that that there is very strong evidence of a difference in mean selenium level between our populations of adults and babies. Note that this is the same as the result obtained in §4.3.

We now investigate the second question of interest: Is $\mu_2 = \mu_3$? The corresponding direction of special interest is $\mathbf{U_3} = [0, \ldots, -1, \ldots, 1, \ldots]^T/\sqrt{48}$. In this case we subtract from $\mathbf{y}$ its projections onto $\mathbf{U_1}$ and $\mathbf{U_2}$. This yields a "corrected" observation vector

$$\mathbf{y} - (\mathbf{y} \cdot \mathbf{U_1})\mathbf{U_1} - (\mathbf{y} \cdot \mathbf{U_2})\mathbf{U_2} = \begin{bmatrix} 60 \\ \vdots \\ 24 \\ \vdots \\ 31 \\ \vdots \end{bmatrix} - \begin{bmatrix} 44.6 \\ \vdots \\ 44.6 \\ \vdots \\ 44.6 \\ \vdots \end{bmatrix} - \begin{bmatrix} 26.9 \\ \vdots \\ -13.4 \\ \vdots \\ -13.4 \\ \vdots \end{bmatrix} = \begin{bmatrix} -11.4 \\ \vdots \\ -7.1 \\ \vdots \\ -0.1 \\ \vdots \end{bmatrix}$$

We then draw Figure D.12 and calculate the angle θ between the corrected observation vector and the direction of interest, $\mathbf{U_3}$. This angle is determined by

$$\cos\theta = \frac{\begin{bmatrix} -11.4 \\ \vdots \\ -7.1 \\ \vdots \\ -0.1 \\ \vdots \end{bmatrix} \cdot \begin{bmatrix} 0 \\ \vdots \\ -1 \\ \vdots \\ 1 \\ \vdots \end{bmatrix}}{\sqrt{(-11.4)^2 + \cdots + (-7.1)^2 + \cdots + (-0.1)^2 + \cdots} \times \sqrt{48}}$$

$$= \frac{-196}{\sqrt{14085 \times 48}} = -0.2384$$

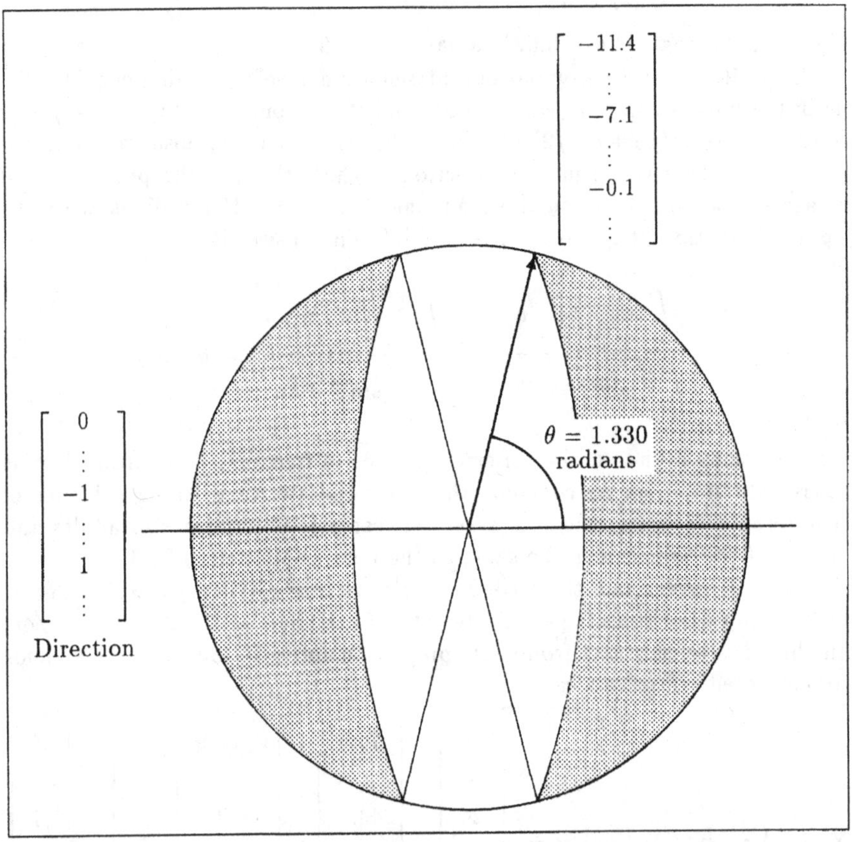

Figure D.12: The vector $[-11.4, \ldots, -7.1, \ldots, -0.1, \ldots]^{\mathrm{T}}$, our corrected observation vector, makes an angle of $1.812 = \pi - 1.330$ radians with the direction of special interest $[0, \ldots, -1, \ldots, 1, \ldots]^{\mathrm{T}}$. If $\mu_2 = \mu_3$, what is the probability of observing an angle as small, or smaller than 1.330 radians? The answer is the ratio of the surface area of the two shaded cone caps to the surface area of the entire hypersphere, $p = 0.045$.

That is, the angle is $\theta = 1.812$ radians, or $103.8°$. Since we are carrying out a two-sided test we treat this in the same manner as an angle of $\pi - \theta = 1.330$ radians, or $(180 - 103.8)° = 76.2°$.

Now if the three populations of selenium levels are all normally distributed with a common variance σ^2, and if the contrast $\mu_3 - \mu_2$ is zero (i.e., $\mu_2 = \mu_3$), it follows that the corrected observation vector is equally likely to lie in any direction. What, then, is the probability of observing an angle as small, or smaller than $\theta = 1.330$ radians with the special direction $[0, \ldots, -1, \ldots, 1, \ldots]^T$? The answer is

$$p = \frac{\int_0^{\theta} \sin^{q-1} u \, du}{\int_0^{\pi/2} \sin^{q-1} u \, du} = \frac{\int_0^{1.330} \sin^{68} u \, du}{\int_0^{\pi/2} \sin^{68} u \, du} = 0.045$$

This probability is unusually small ($p < 0.05$), so we conclude that there is evidence of a difference in mean selenium level between our populations of premature and full-term babies. Note that this is the same as the result obtained in §4.3, and recall that we must view this result with some suspicion as discussed in §4.3.

D.6 Simple Regression

We now rework the examples in Chapter 5 using the angle test statistic.

Sample of Size Five

In the case of a sample of size five (§5.2) the observation vector and the directions in the model space are

$$\mathbf{y} = \begin{bmatrix} 73 \\ 27 \\ 546 \\ 33 \\ 381 \end{bmatrix}, \quad \mathbf{U_1} = \frac{1}{\sqrt{5}} \begin{bmatrix} 1 \\ 1 \\ 1 \\ 1 \\ 1 \end{bmatrix}, \quad \mathbf{U_2} = \frac{1}{\sqrt{5.352}} \begin{bmatrix} -0.66 \\ -0.86 \\ 1.54 \\ -0.96 \\ 0.94 \end{bmatrix}$$

Here $\mathbf{U_2}$ is the direction of special interest, associated with the slope β.

Following the general method outlined in §D.3, we subtract from $\mathbf{y}$ its projection onto the other direction in the model space, $\mathbf{U_1}$. This yields a "corrected" observation vector

$$\mathbf{y} - (\mathbf{y} \cdot \mathbf{U_1})\mathbf{U_1} = \begin{bmatrix} 73 \\ 27 \\ 546 \\ 33 \\ 381 \end{bmatrix} - \begin{bmatrix} 212 \\ 212 \\ 212 \\ 212 \\ 212 \end{bmatrix} = \begin{bmatrix} -139 \\ -185 \\ 334 \\ -179 \\ 169 \end{bmatrix}$$

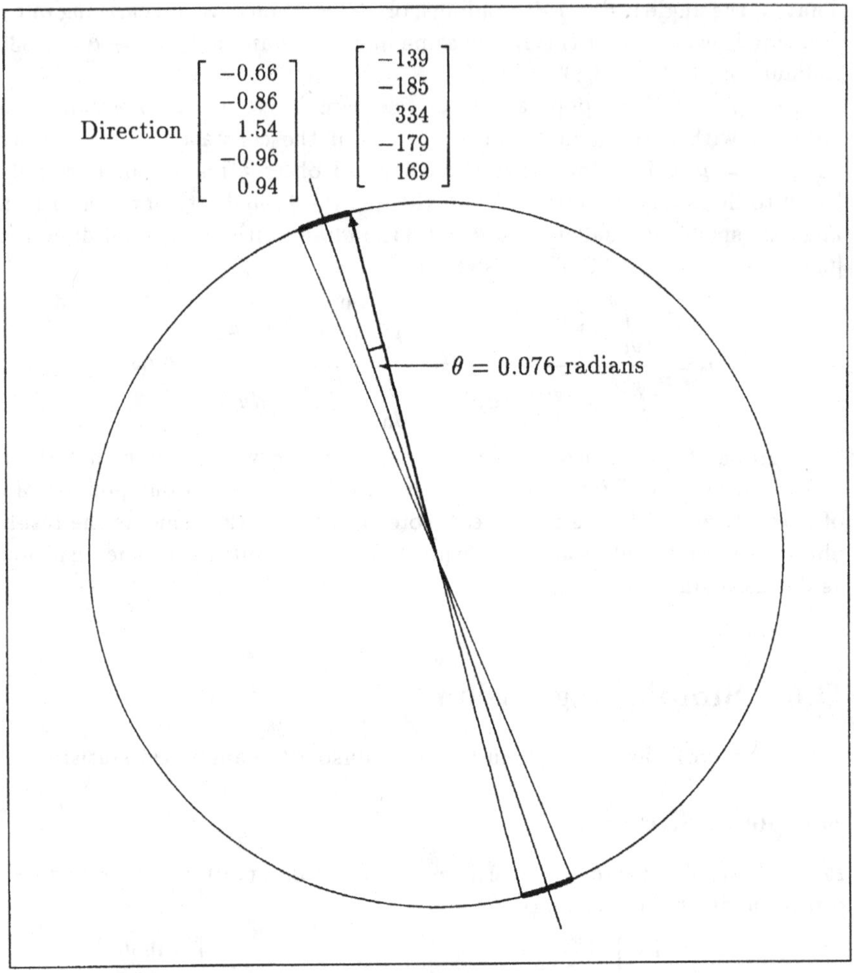

Figure D.13: Our corrected observation vector $[-139, -185, 334, -179, 169]^{\mathrm{T}}$ makes an angle of 0.076 radians with the direction of special interest $(\mathbf{x}-\bar{\mathbf{x}}) = [-0.66, -0.86, 1.54, -0.96, 0.94]^{\mathrm{T}}$. If the slope $\beta = 0$, what is the probability of observing an angle as small, or smaller than this? The answer is the ratio of the surface area of the two shaded cone caps to the surface area of the entire hypersphere, $p = 0.00019$.

We then proceed as previously, drawing Figure D.13 and calculating the angle θ between the corrected observation vector and the direction of interest, $\mathbf{U_2}$. This angle is determined by

$$\cos\theta = \frac{\begin{bmatrix} -139 \\ -185 \\ 334 \\ -179 \\ 169 \end{bmatrix} \cdot \begin{bmatrix} -0.66 \\ -0.86 \\ 1.54 \\ -0.96 \\ 0.94 \end{bmatrix}}{\sqrt{(-139)^2 + (-185)^2 + 334^2 + (-179)^2 + 169^2} \times \sqrt{5.352}} = \frac{1095.9}{\sqrt{225704 \times 5.352}} = 0.9971$$

That is, the angle is $\theta = 0.076$ radians, or 4.4°.

We have already assumed that the pollution levels are normally distributed about true values lying on the line $y = \alpha + \beta x$ with a common variance σ^2. If we also assume that the slope β is zero, then we can show that the corrected observation vector is equally likely to lie in any direction. In this case, what is the probability of observing an angle as small, or smaller than $\theta = 0.076$ radians with the special direction $(\mathbf{x} - \bar{\mathbf{x}}) = [-0.66, -0.86, 1.54, -0.96, 0.94]^T$? The answer is

$$p = \frac{\int_0^{\theta} \sin^{q-1} u \, du}{\int_0^{\pi/2} \sin^{q-1} u \, du} = \frac{\int_0^{0.076} \sin^2 u \, du}{\int_0^{\pi/2} \sin^2 u \, du} = 0.00019$$

where q is the number of error degrees of freedom (3 here). This probability is very small ($p < 0.001$), so we conclude that the true slope β is not zero, and that pollution level is related to the inversion effect. Note that this is the same as the result obtained in §5.2.

General Case

In the general case of a sample of size n, and, in particular, our example with $n = 26$ (§5.3), the observation vector and the directions in the model space are

$$\mathbf{y} = \begin{bmatrix} 290 \\ \vdots \\ 207 \end{bmatrix}, \quad \mathbf{U}_1 = \frac{1}{\sqrt{26}} \begin{bmatrix} 1 \\ \vdots \\ 1 \end{bmatrix}, \quad \mathbf{U}_2 = \frac{1}{\sqrt{25.9646}} \begin{bmatrix} 0.454 \\ \vdots \\ 1.954 \end{bmatrix}$$

Here $\mathbf{U}_2$ is the direction of special interest, associated with the slope β.

As previously, we subtract from $\mathbf{y}$ its projection onto the other direction in the model space, $\mathbf{U}_1$. This yields a "corrected" observation vector

$$\mathbf{y} - (\mathbf{y} \cdot \mathbf{U}_1)\mathbf{U}_1 = \begin{bmatrix} 290 \\ \vdots \\ 207 \end{bmatrix} - \begin{bmatrix} 148.3 \\ \vdots \\ 148.3 \end{bmatrix} = \begin{bmatrix} 141.7 \\ \vdots \\ 58.7 \end{bmatrix}$$

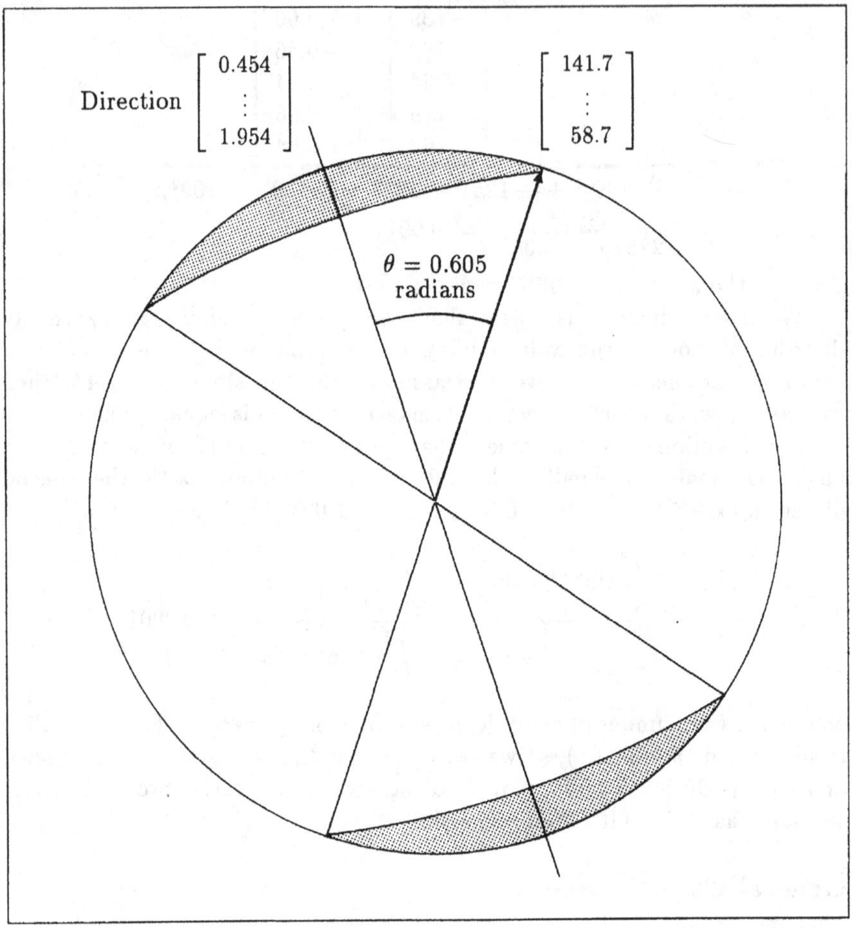

Figure D.14: Our corrected observation vector $[141.7, \ldots, 58.7]^T$ makes an angle of 0.605 radians with the direction of special interest $(\mathbf{x} - \bar{\mathbf{x}}) = [0.454, \ldots, 1.954]^T$. If the slope $\beta = 0$, what is the probability of observing an angle as small, or smaller than this? The answer is the ratio of the surface area of the two shaded cone caps to the surface area of the entire hypersphere, $p = 2.5 \times 10^{-7}$.

We then draw Figure D.14 and calculate the angle θ between the corrected observation vector and the direction of interest, $\mathbf{U_2}$. This angle is determined by

$$\cos\theta = \frac{\begin{bmatrix} 141.7 \\ \vdots \\ 58.7 \end{bmatrix} \cdot \begin{bmatrix} 0.454 \\ \vdots \\ 1.954 \end{bmatrix}}{\sqrt{141.7^2 + \cdots + 58.7^2} \times \sqrt{25.9646}}$$

$$= \frac{3035.1}{\sqrt{524720 \times 25.9646}} = 0.822$$

That is, the angle is $\theta = 0.605$ radians, or 34.7°.

As in the previous subsection, if the slope β is zero the corrected observation vector is equally likely to lie in any direction. In this case, what is the probability of observing an angle as small, or smaller than $\theta = 0.605$ radians with the special direction $(\mathbf{x} - \bar{\mathbf{x}}) = [0.454, \ldots, 1.954]^T$? The answer is

$$p = \frac{\int_0^{\theta} \sin^{q-1} u \, du}{\int_0^{\pi/2} \sin^{q-1} u \, du} = \frac{\int_0^{0.605} \sin^{23} u \, du}{\int_0^{\pi/2} \sin^{23} u \, du} = 2.5 \times 10^{-7}$$

where q is the number of error degrees of freedom (24 here). This probability is extremely small, so we conclude that there is very strong evidence that the true slope β is not zero, and that pollution level is related to the inversion effect. Note that this is the same as the result obtained in §5.3.

Notice also that $\cos\theta = 0.822$ is the correlation coefficient defined at the end of §5.3.

This completes the reanalysis of the examples in Chapters 2–5 using the angle test statistic.

D.7 Equivalence of Test Statistics

In this section we revisit the "statistical triangles" displayed in Figure 6.4, and explain the relationships between our new angle test statistic θ and our previous test statistics F and t. We also briefly introduce a fourth test statistic $r = \cos\theta$, and explain its relationship to the other three.

Statistical Triangle

In Figure D.15 we redisplay the general statistical triangle shown in Figure 6.2 with the addition of the angle θ and the length of the third side C. Then in Figure D.16 we redisplay the four statistical triangles in Figure 6.4 with these same additions.

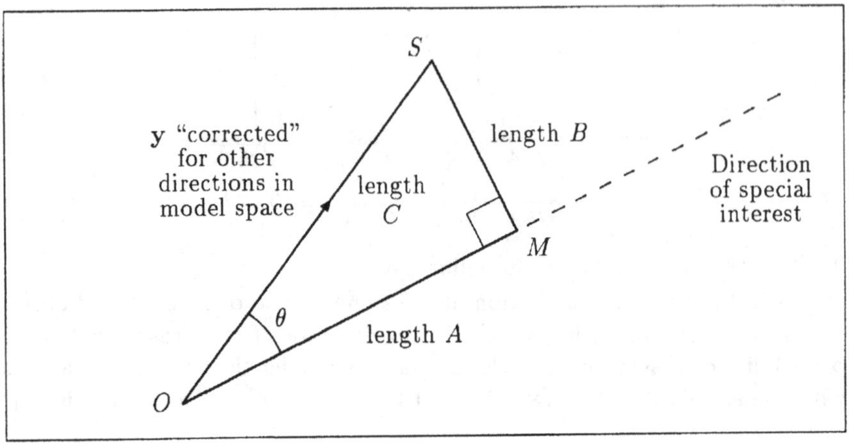

Figure D.15: The "statistical triangle." This triangle is defined by: (1) the direction of special interest; and (2) the observation vector **y** minus its projections onto all directions in the model space *except* the direction of special interest. It lies in a $(q+1)$-dimensional subspace of N-space, where q is the number of error degrees of freedom.

Test Statistics

In all four cases shown in Figure D.16 the F test statistic is given by the expression

$$F = \frac{A^2}{B^2/q} = q\cot^2\theta$$

where q is the number of error degrees of freedom. The calculated F values are summarized in the third column of Table D.1.

Similarly the t test statistic is given by the expression

$$t = \pm\frac{A}{B/\sqrt{q}} = \sqrt{q}\cot\theta$$

where the resulting t value is positive if the observation vector has a component in the positive direction of the direction associated with the parameter of interest, or is negative if this is not the case. Here θ is restricted to the range 0 to π radians, with values less than $\pi/2$ corresponding to positive t values and values between $\pi/2$ and π corresponding to negative t values. The calculated t values are summarized in the fourth column of Table D.1.

The θ test statistic is simply the angle θ itself. The calculated θ values are summarized in the fifth column of Table D.1. In our examples these are all acute angles since all of our corrected observation vectors had components in the positive direction of the parameter of interest.

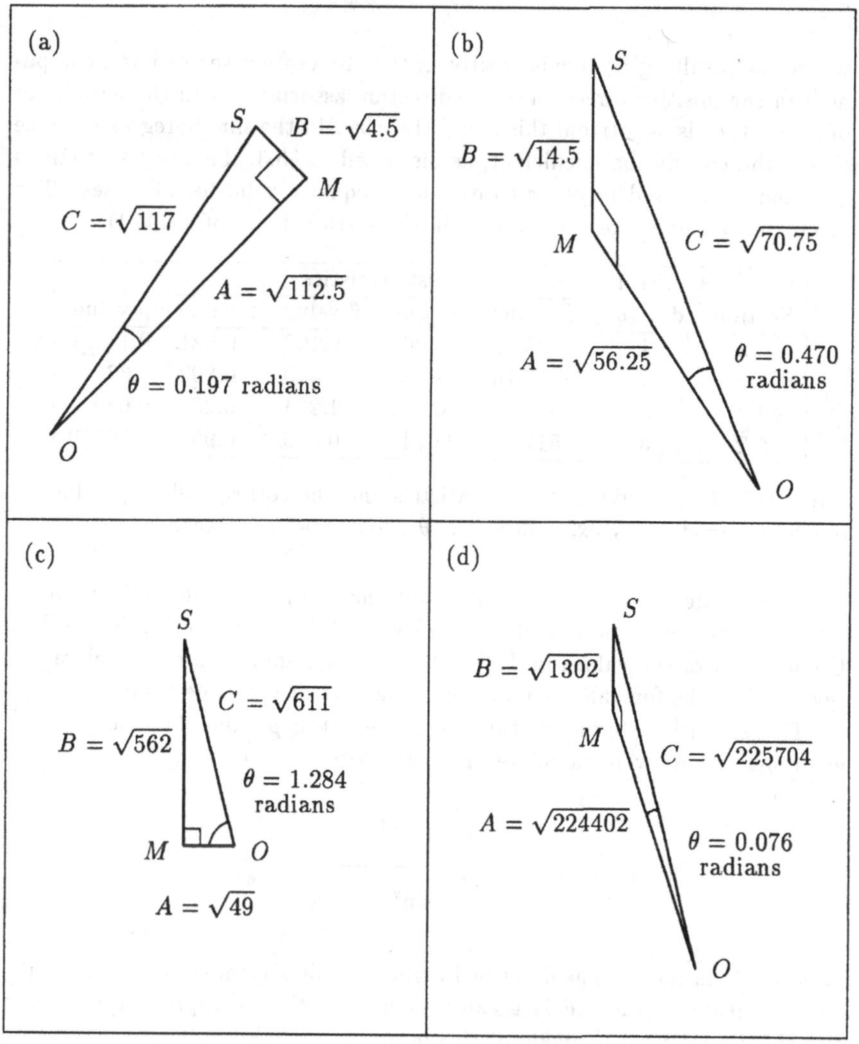

Figure D.16: The lengths of the sides of each statistical triangle together with the angle between the "corrected" observation vector and the direction corresponding to the parameter of interest. These are the basic quantities required to calculate the test statistics in Table D.1. Note that this is Figure 6.4 with the addition of the lengths C and the angles θ.

The r test statistic is given by the expression

$$r = \pm\frac{A}{C} = \cos\theta$$

where the resulting r value is positive if the observation vector has a component in the positive direction of the direction associated with the parameter of interest, or is negative if this is not the case. In the simple regression case this is the correlation coefficient, as discussed in §5.3. This test statistic is not commonly used in other cases, but is equally valid for all cases. The calculated r values are summarized in the sixth column of Table D.1.

	Error	Test statistics				
Section	d.f. (q)	F value	t value	θ value	r value	p value
2.2	1	25.00	5.000	0.197	0.981	0.13
3.2	2	7.76	2.785	0.470	0.892	0.37
4.2	3	0.26	0.511	1.284	0.283	0.64
5.2	3	517	22.74	0.076	0.997	0.00019

Table D.1: Four equivalent test statistics and the corresponding p value for our four introductory examples. The θ values are in radians.

The "critical values" of the distributions of $F_{1,q}$, t_q, the angle θ with q d.f., and the correlation coefficient (r) with q d.f. are given in Table T.2. Comparison of the values in Table D.1 with the appropriate critical values reveals that the four alternative tests always yield the same result.

The seventh column of Table D.1 gives the p value for each of our examples. These were calculated from the formula

$$p = \frac{\int_0^{\theta} \sin^{q-1} u \, du}{\int_0^{\pi/2} \sin^{q-1} u \, du}$$

(where θ is in radians) as described earlier in this appendix. In this formula we substitute $\pi - \theta$ if θ is greater than $\pi/2$; this is appropriate for the two-sided tests used throughout this book.

Note that the skinnier the triangle in Figure D.16 in terms of the angle at the point O, the lower the p value in Table D.1. For example, the angle of 0.076 radians in (d) corresponds to a p value of 0.00019, while the angle of 1.284 radians in (c) corresponds to a p value of 0.64. In general, this skinnyness comparison must be moderated by taking into account the number of error degrees of freedom (3 in both (c) and (d)).

The relationships between the four test statistics and the p value are summarized in Figure D.17.

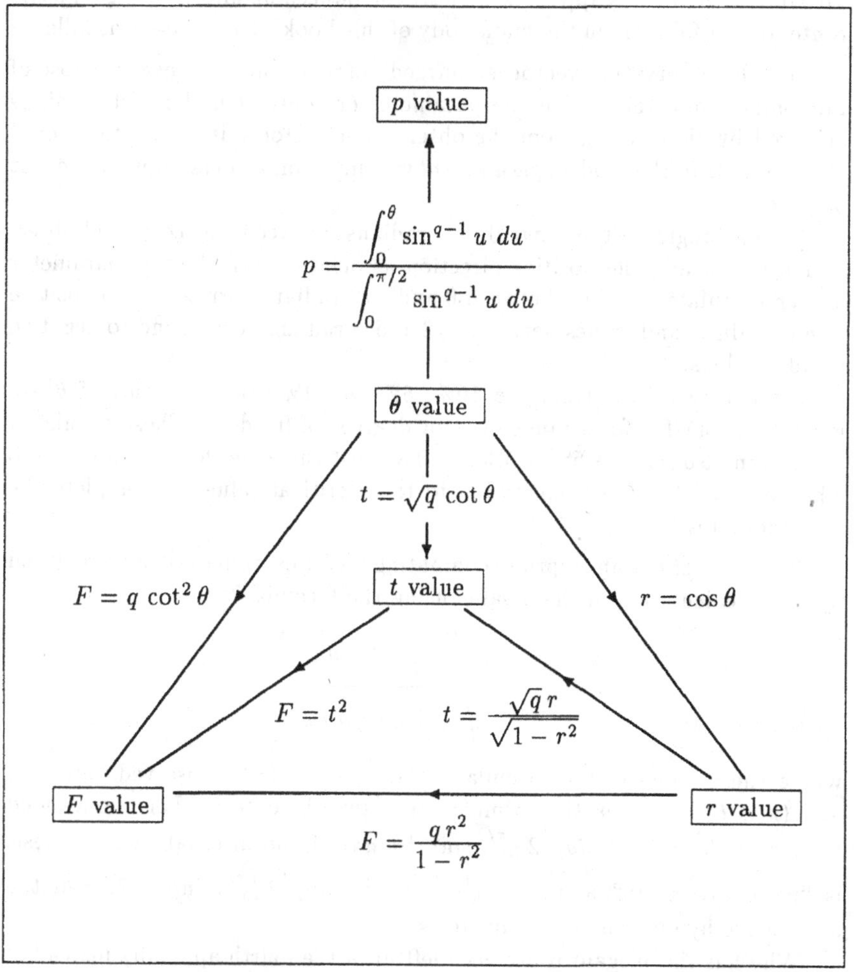

Figure D.17: The relationship between the F value, the t value, the r value, the θ value and the p value. Any one of these quantities determines the other four, with the proviso that p and F do not determine the sign of t and r and do not distinguish between θ and $\pi - \theta$. Note that θ is restricted to the range 0 to π radians, with values less than $\pi/2$ corresponding to positive t and r values, and values between $\pi/2$ and π corresponding to negative t and r values. Note also that the p value formula applies to two-sided hypothesis tests, with $\pi - \theta$ being substituted for θ if the latter is greater than $\pi/2$.

D.8 Summary

To summarize, in this appendix we have introduced an alternative geometric route to that followed in the main body of this book. This route is as follows:

(1) The observation vector is "purged" of any influences except those of randomness and the size of the parameter or contrast under test. This is achieved by subtracting from the observation vector $\mathbf{y}$ its projections onto all directions in the model space *except* the direction associated with the test parameter.

(2) The angle, in the range 0 to π radians, between this corrected observation vector and the positive direction associated with the test parameter is then calculated. Values between 0 and $\pi/2$ radians correspond to positive t and r values, and values between $\pi/2$ and π radians correspond to negative t and r values.

(3) For two-sided tests the 10%, 5% and 1% critical values of θ are given in Table T.2 for a range of error degrees of freedom. These tabulated values can also serve as 5%, 2.5% and 0.5% critical values for one-sided tests. The observed θ value is compared to these critical values to complete the hypothesis test.

If the reader has appropriate computing facilities, step (3) can be replaced by direct calculation of the p value using the formula

$$p = \frac{\displaystyle\int_0^{\theta} \sin^{q-1} u \; du}{\displaystyle\int_0^{\pi/2} \sin^{q-1} u \; du}$$

where the θ value in the formula is the smaller of the observed value (θ) and ($\pi - \theta$). This is the formula for a two-sided test. For a one-sided test, $p = \int_0^{\theta} \sin^{q-1} u \; du \Big/ \left[2 \int_0^{\pi/2} \sin^{q-1} u \; du\right]$ if the alternative hypothesis is "parameter > 0," and $p = \int_0^{\pi-\theta} \sin^{q-1} u \; du \Big/ \left[2 \int_0^{\pi/2} \sin^{q-1} u \; du\right]$ if the alternative hypothesis is "parameter < 0."

Why has this elegant route been left until the fourth appendix instead of being given pride of place in this book? The answer is that step (1) requires the knowledge of the underlying geometry which is taught in Chapters 1–5. Without this knowledge the reader would have difficulty "getting to first base." Conversely, with this knowledge the route described in this appendix is immensely appealing, and should be seriously considered by anyone teaching these statistical methods to students with a liking for mathematics.

For further reading, the reader is referred first to the key paper by Chance (1986), where the link between the θ value and the p value for the case of the correlation coefficient is derived. Box, Hunter and Hunter (1978) discuss the single population case on pages 197–203. Box (1978) discusses the regression through the origin case on pages 124–128.

Class Exercise

In this class exercise we use the data provided in the class exercise in Chapter 3 to illustrate the basic idea of this appendix. We first simulate the case $\mu_1 = \mu_2$, then the case $\mu_1 \neq \mu_2$. To keep things simple we take samples of size one.

(a) To simulate the case $\mu_1 = \mu_2$ (a "dummy run"), each class member is asked to use the random numbers in Table T.1 to draw two samples of size one (y_{11} and y_{21}) from Group 1 in the table of heartbeat rates in the class exercise in Chapter 3.

(b) Each class member is then asked to set up axes $x = 0$ to 60, $y = 0$ to 60 on a piece of graph paper, and plot their second value y_{21} against their first value y_{11} using a small dot as the plotting symbol.

(c) The class member is then asked to mark the population mean [we will pretend it is $(\mu, \mu) = (33, 33)$] as a larger dot on the graph paper, and draw a vector from this larger dot to the small dot.

(d) The class member is then asked to repeat the above until they have ten vectors on their graph paper. Are these vectors approximately evenly scattered from 0 to 360°?

At this point the teacher can collate the class results to achieve a better long-term average. To do so, the teacher can count how many of the vectors are at an angle of 0–90° with the positive x axis, how many are at an angle of 90–180°, how many are at an angle of 180–270°, and how many are at an angle of 270–360°. Are these numbers approximately equal?

(e) To simulate the case $\mu_1 \neq \mu_2$ (the "real thing"), each class member is now asked to use the random numbers in Table T.1 to draw two samples of size one (y_{11} and y_{21}), the first from Group 1 and the second from Group 2 in the table of heartbeat rates in the class exercise in Chapter 3.

(f) Each class member is again asked to set up identical axes $x = 0$ to 60, $y = 0$ to 60 on a fresh piece of graph paper, and plot their second value y_{21} against their first value y_{11} using a small dot as the plotting symbol.

(g) The class member is then asked to mark the mean of the two population means [we will pretend it is $(\mu, \mu) = (38, 38)$] as a larger dot on the graph paper, and draw a vector from this larger dot to the small dot.

(h) The class member is then asked to repeat (e)–(g) until they have ten vectors on their graph paper. Are these vectors approximately evenly scattered from 0° to 360°?

(i) Each class member is now asked to draw three dashed lines onto each of their two pieces of graph paper. The first is the equiangular line in the direction $[1, 1]^T$ (at 45° to the x axis), through the origin. The second is the line perpendicular to this, in the direction $[-1, 1]^T$ (at 45° to the negative x axis), again through the origin. The third is parallel to the latter line, but translated so it goes through the population mean (μ, μ).

(j) Each class member is then asked to compare their two graphs. What is the difference between the cases $\mu_1 = \mu_2$ and $\mu_1 \neq \mu_2$ in terms of the directions in which the vectors lie?

At this point the teacher can again collate the class results to achieve a better long-term average. To do so, the teacher can count how many of the vectors are at an angle of 0–90° with the positive x axis, how many are at an angle of 90–180°, how many are at an angle of 180–270°, and how many are at an angle of 270–360°. Are these numbers approximately equal? Which directions are most common?

Comment: This exercise is intended to show that the $[-1, 1]^T$ direction is the direction of special interest when the test hypothesis is $\mu_1 = \mu_2$. With samples of size one we are unable to progress any further in terms of constructing a test statistic since we have no directions which can be used for assessing the error term; however, the exercise is useful since we can easily draw vectors in 2-space.

Exercises

Note that solutions to exercises marked with an * are given in Appendix E.

(D.1)* (a) Using the data in Exercise 2.2, write down the observation vector **y** and the direction of special interest **U**.

(b) Calculate the angle θ between **y** and **U**.

(c) Carry out a two-sided test of the hypothesis $\mu = 0$, where μ is the mean percentage decrease in production due to nitrogen fertilizer. What are the 10%, 5% and 1% critical values of the distribution of the test statistic? What are your conclusions?

(d) In this exercise the error degrees of freedom are low enough for the p value integrals to be evaluated without recourse to a numerical integration routine. Chance (1986) provided formulas which lead to the following p value (for four error degrees of freedom):

$$p = 1 - r - r(1 - r^2)/2$$

where $r = \cos\theta$. Use your θ value from (b) to calculate p. Is your calculated p value in accord with your conclusions in (c)?

(e) (Optional) Readers with access to a suitable numerical integration routine can calculate the p value by integration. Is the answer the same as in (d)? [If not, try using more decimal places for θ and r.]

(D.2) (a) Using the data in Exercise 2.5, draw up the "statistical triangle" corresponding to a test of the hypothesis $\mu = 0$, where μ is the mean weight change of the herd during the month. Include calculated lengths and the calculated angle.

(b) Calculate the F value, t value, θ value and r value ($= \cos\theta$) from the triangle in (a). What are the corresponding 10%, 5% and 1% critical values for a two-sided test of the hypothesis $\mu = 0$? Do your four test statistics all yield the same result? Is there any evidence of a weight change during the month?

(c) Confirm the relationships between F, t, θ and r shown in Figure D.17 using the values you calculated in (b).

(d) (Optional) Readers with access to a suitable numerical integration routine can calculate the p value. Does your result agree with your findings in (b)?

(D.3)* (a) For a paired samples data set of size n with observations denoted by $y_1, y_2, \ldots, y_n$ calculate a general expression for $\cos\theta$ and hence show that the r value ($= \cos\theta$) is given by the algebraic expression

$$r = \frac{\sqrt{n}\bar{y}}{\sqrt{(n-1)s^2 + n\bar{y}^2}}$$

where $s^2 = \sum_{i=1}^{n}(y_i - \bar{y})^2/(n-1)$ is the sample variance.

(b) Confirm this formula for the $n = 2$ example worked through in §2.2 and §D.2.

(D.4) Using the data in Exercise 2.8, test the hypothesis $\mu_S = 10$ by calculating the θ value and comparing it with the appropriate critical values for θ. What is your conclusion?

(D.5) (a) Using the data in Exercise 3.2, write down the observation vector **y** and the direction of special interest **U**.

(b) "Correct" **y** as described in §D.3.

(c) Calculate the angle θ between the corrected observation vector and the direction of special interest **U**.

(d) Carry out a two-sided test of the hypothesis $\mu_1 = \mu_2$. What are the appropriate critical values of the distribution of the test statistic θ? What are your conclusions?

(e) In this exercise there are four error degrees of freedom, so that the formula given in Exercise D.1 can again be used. Use your θ value from (c) to calculate p in this manner. Is your calculated p value in accord with your conclusions in (d)?

(f) (Optional) Readers with access to a suitable numerical integration routine can calculate the p value by integration. Is the answer the same as in (e)? [If not, try using more decimal places for θ and r.]

(D.6)* (a) Using the data in Exercise 3.5, draw up the "statistical triangle" corresponding to a test of the hypothesis $\mu_1 = \mu_2$. Include calculated lengths and angle.

(b) Calculate the F value, t value, θ value and r value ($= \cos\theta$) from the triangle in (a). What are the corresponding critical values for a two-sided test of the hypothesis $\mu_1 = \mu_2$? Do your four test statistics all yield the same result? What is this result?

(c) Confirm the relationships between F, t, θ and r shown in Figure D.17 using the values you calculated in (b).

(d) (Optional) Readers with access to a suitable numerical integration routine can calculate the p value. Does your result agree with your findings in (b)?

(D.7) (a) For a general independent samples data set with two samples of size n, with observations denoted by $y_{11}, y_{12}, \ldots, y_{1n}$ and $y_{21}, y_{22}, \ldots, y_{2n}$, respectively, calculate a general expression for $\cos\theta$ and therefore show that

$$r = \frac{\sqrt{n}(\bar{y}_2 - \bar{y}_1)}{\sqrt{n(\bar{y}_2 - \bar{y}_1)^2 + 4(n-1)s^2}}$$

where $s^2 = \sum_{i=1}^{2} \sum_{j=1}^{n} (y_{ij} - \bar{y}_i)^2 / [2(n-1)]$ is the pooled sample variance.

(b) Confirm this formula for the $n = 2$ example worked through in §3.2 and §D.4.

(D.8) (a) Using the data in Exercise 3.8(a), write down the observation vector **y** and the direction of special interest **U**.

(b) "Correct" **y** as described in §D.3.

(c) Calculate the angle θ between the corrected observation vector and the direction of special interest **U**.

(d) Carry out a two-sided test of the hypothesis $\mu_1 = \mu_2$. What are the appropriate critical values of the distribution of the test statistic θ? What are your conclusions?

(e) (Optional) Readers with access to a suitable numerical integration routine can calculate the p value. Does your result agree with your findings in (d)?

(D.9) In §D.5 with three samples of size two, the error degrees of freedom are low enough for the p value integrals to be evaluated without recourse to a numerical integration routine. Chance (1986) provided formulas which lead to the following p value (for three error degrees of freedom):

$$p = \frac{2}{\pi}\left[\theta - \cos\theta\sqrt{1-\cos^2\theta}\right]$$

Use the values for θ and $\cos\theta$ given in §D.5 to calculate p. Is the answer the same as that given in §D.5?

(D.10) (a) Using the data in Exercise 4.1, write down the observation vector **y** and the directions corresponding to the three questions of interest.

(b) In this exercise we shall test just the first of the three hypotheses of interest: (A) Do sheep differ from goats? With this in mind, "correct" **y** as described in §D.3.

(c) Calculate the angle θ between the corrected observation vector and the direction corresponding to our test hypothesis.

(d) Carry out a two-sided test of the hypothesis $\mu_1 + \mu_2 = \mu_3 + \mu_4$. What are the appropriate critical values of the distribution of the test statistic θ? What are your conclusions?

(e) In this exercise there are four error degrees of freedom, so that the formula given in Exercise D.1 can again be used. Use your θ value from (c) to calculate p in this manner. Is your calculated p value in accord with your conclusions in (d)?

(f) (Optional) Readers with access to a suitable numerical integration routine can calculate the p value by integration. Is the answer the same as in (e)? [If not, try using more decimal places for θ and r.]

(D.11)* (a) Using the data in Exercise 4.3, draw up the "statistical triangle" corresponding to a test of the hypothesis $\mu_1 + \mu_3 = \mu_2 + \mu_4$ (Siamese versus ordinary cats). Include calculated lengths and angle.

(b) Calculate the F value, t value, θ value and r value ($= \cos\theta$) from the triangle in (a). What are the corresponding critical values for a two-sided test of the hypothesis $\mu_1 = \mu_2$? Do your four test statistics all yield the same result? What is this result?

(c) Confirm the relationships between F, t, θ and r shown in Figure D.17 using the values you calculated in (b).

(d) (Optional) Readers with access to a suitable numerical integration routine can calculate the p value. Does your result agree with your findings in (b)?

(D.12) In §D.6 with a sample size of five there are three error degrees of freedom, so the formula for the p value given in Exercise D.9 again applies. Use the values for θ and $\cos\theta$ given in §D.6 to calculate the p value, except that you will need more accuracy so use $\cos\theta = 0.99711$ and the corresponding θ value. Is the answer the same as that given in §D.6?

(D.13) (a) Using the data in Exercise 5.4, write down the observation vector $\mathbf{y}$ and the direction of special interest $\mathbf{U}$.

(b) "Correct" $\mathbf{y}$ as described in §D.3.

(c) Calculate the angle θ between the corrected observation vector and the direction of special interest $\mathbf{U}$.

(d) Carry out a two-sided test of the hypothesis $\beta = 0$. What are the appropriate critical values of the distribution of the test statistic θ? What are your conclusions?

(e) In this exercise there are four error degrees of freedom, so that the formula given in Exercise D.1 can again be used. Use your θ value from (c) to calculate p in this manner. Is your calculated p value in accord with your conclusions in (d)?

(f) (Optional) Readers with access to a suitable numerical integration routine can calculate the p value by integration. Is the answer the same as in (e)? [If not, try using more decimal places for θ and r.]

(D.14) (a) Using the data in Exercise 5.5, draw up the "statistical triangle" corresponding to a test of the hypothesis $\beta = 0$. Include calculated lengths and angle.

(b) Calculate the F value, t value, θ value and r value ($= \cos\theta$) from the triangle in (a). What are the corresponding critical values for a two-sided test of the hypothesis $\beta = 0$? Do your four test statistics all yield the same result? What is this result?

(c) Confirm the relationships between F, t, θ and r shown in Figure D.17 using the values you calculated in (b).

(d) In this exercise there are four error degrees of freedom, so that the formula given in Exercise D.1 can again be used. Use your θ value from (b) to calculate the p value in this manner. Is your calculated p value in accord with your results in (b)?

(e) (Optional) Readers with access to a suitable numerical integration routine can calculate the p value by integration. Is the answer the same as in (d)? [If not, try using more decimal places for θ and r.]

(D.15) Using the data in Exercise 5.8, test the hypothesis $\beta = 0$ by calculating the θ value and comparing it with the appropriate critical values for θ. What is your conclusion?

(D.16)* Using the data in Exercise 5.9, test the hypothesis $\beta = 0$ by calculating the test statistic $r = \cos\theta$ and comparing it with the appropriate critical values for the correlation coefficient. What is your conclusion?

(D.17) (a) For a general simple regression data set of size n, with y values denoted by $y_1, y_2, \ldots, y_n$ and x values denoted by $x_1, x_2, \ldots, x_n$, calculate a general expression for $\cos\theta$ and hence show that the r value is given by

$$r = \frac{\sum_{i=1}^{n}(x_i - \bar{x})(y_i - \bar{y})}{\sqrt{\sum_{i=1}^{n}(x_i - \bar{x})^2 \sum_{i=1}^{n}(y_i - \bar{y})^2}}$$

(b) Confirm this formula for the $n = 5$ example worked through in §5.2 and §D.6.

(D.18) In §D.7 we explained how the distribution of the correlation coefficient (r) can be used to test hypotheses in cases other than simple regression. We now guide the reader through one such example.

In §3.2, the hypothesis $\mu_1 = \mu_2$ was tested using an independent samples t (or F) test. The equivalent test using the r value involves working out how highly the variable of four "corrected" observations ($y_{ij} - \bar{y}$ values) correlates with the variable defined by the values -1, -1, 1, 1 (a high correlation is evidence that $\mu_1 \neq \mu_2$). Do the appropriate calculation using the formula given on page 124, and compare your calculated r value with the appropriate percentiles in Table T.2. What is your conclusion? Does it agree with that at the end of §3.2?

Appendix E

Solutions to Exercises

We now provide solutions for the exercises marked with an asterisk.

E.1 Solutions for Chapter 2 Exercises

(2.1) (a) The length of the projection of $\mathbf{y}$ onto $\mathbf{U}_2$ is

$$\mathbf{y}\cdot\mathbf{U}_2 = \begin{bmatrix} 7.5 \\ 3 \\ 9 \end{bmatrix} \cdot \frac{1}{\sqrt{42}} \begin{bmatrix} -4 \\ -1 \\ 5 \end{bmatrix} = \frac{7.5\times(-4) + 3\times(-1) + 9\times 5}{\sqrt{42}} = \frac{12}{\sqrt{42}}$$

Hence the corresponding projection vector is

$$(\mathbf{y}\cdot\mathbf{U}_2)\mathbf{U}_2 = \frac{12}{\sqrt{42}}\frac{1}{\sqrt{42}} \begin{bmatrix} -4 \\ -1 \\ 5 \end{bmatrix} = \frac{12}{42} \begin{bmatrix} -4 \\ -1 \\ 5 \end{bmatrix} = \begin{bmatrix} -1.14 \\ -0.29 \\ 1.43 \end{bmatrix}$$

The length of the projection of $\mathbf{y}$ onto $\mathbf{U}_3$ is

$$\mathbf{y}\cdot\mathbf{U}_3 = \begin{bmatrix} 7.5 \\ 3 \\ 9 \end{bmatrix} \cdot \frac{1}{\sqrt{14}} \begin{bmatrix} 2 \\ -3 \\ 1 \end{bmatrix} = \frac{7.5\times 2 + 3\times(-3) + 9\times 1}{\sqrt{14}} = \frac{15}{\sqrt{14}}$$

Hence the corresponding projection vector is

$$(\mathbf{y}\cdot\mathbf{U}_3)\mathbf{U}_3 = \frac{15}{\sqrt{14}}\frac{1}{\sqrt{14}} \begin{bmatrix} 2 \\ -3 \\ 1 \end{bmatrix} = \frac{15}{14} \begin{bmatrix} 2 \\ -3 \\ 1 \end{bmatrix} = \begin{bmatrix} 2.14 \\ -3.21 \\ 1.07 \end{bmatrix}$$

Hence

$$(\mathbf{y}\cdot\mathbf{U}_2)\mathbf{U}_2 + (\mathbf{y}\cdot\mathbf{U}_3)\mathbf{U}_3 = \begin{bmatrix} -1.14 \\ -0.29 \\ 1.43 \end{bmatrix} + \begin{bmatrix} 2.14 \\ -3.21 \\ 1.07 \end{bmatrix} = \begin{bmatrix} 1.00 \\ -3.50 \\ 2.50 \end{bmatrix}$$

which equals the vector $(\mathbf{y} - \bar{\mathbf{y}})$ given in §2.3.

(b) The first projection length was found in §2.3 to be $\mathbf{y}\cdot\mathbf{U}_1 = 19.5/\sqrt{3}$. The other two projection lengths were found in (a). Hence

$$F = \frac{\left(\frac{19.5}{\sqrt{3}}\right)^2}{\left[\left(\frac{12}{\sqrt{42}}\right)^2 + \left(\frac{15}{\sqrt{14}}\right)^2\right]/2} = \frac{\frac{19.5^2}{3}}{\left[\frac{12^2}{42} + \frac{15^2}{14}\right]/2} = \frac{126.75}{9.75} = 13.0$$

(2.3) (a) There are infinitely many correct answers to this question. One orthogonal coordinate system for 8-space can be obtained by following the pattern outlined at the start of §2.4. This is:

$$\begin{array}{cccccccc}
\mathbf{U}_1 & \mathbf{U}_2 & \mathbf{U}_3 & \mathbf{U}_4 & \mathbf{U}_5 & \mathbf{U}_6 & \mathbf{U}_7 & \mathbf{U}_8 \\
\dfrac{\begin{bmatrix}1\\1\\1\\1\\1\\1\\1\\1\end{bmatrix}}{\sqrt{8}} &
\dfrac{\begin{bmatrix}-1\\1\\0\\0\\0\\0\\0\\0\end{bmatrix}}{\sqrt{2}} &
\dfrac{\begin{bmatrix}-1\\-1\\2\\0\\0\\0\\0\\0\end{bmatrix}}{\sqrt{6}} &
\dfrac{\begin{bmatrix}-1\\-1\\-1\\3\\0\\0\\0\\0\end{bmatrix}}{\sqrt{12}} &
\dfrac{\begin{bmatrix}-1\\-1\\-1\\-1\\4\\0\\0\\0\end{bmatrix}}{\sqrt{20}} &
\dfrac{\begin{bmatrix}-1\\-1\\-1\\-1\\-1\\5\\0\\0\end{bmatrix}}{\sqrt{30}} &
\dfrac{\begin{bmatrix}-1\\-1\\-1\\-1\\-1\\-1\\6\\0\end{bmatrix}}{\sqrt{42}} &
\dfrac{\begin{bmatrix}-1\\-1\\-1\\-1\\-1\\-1\\-1\\7\end{bmatrix}}{\sqrt{56}}
\end{array}$$

In general, the easiest way of generating an orthogonal coordinate system is to specify an overall average direction ($\mathbf{U}_1$), then divide the observations into two groups, comparing observations in the first group with those in the second group ($\mathbf{U}_2$), before going on to subdivide each group into smaller groups until all groups are of size one. The following orthogonal coordinate system was obtained by splitting into two groups of size four, then splitting each of these into two subgroups of size two, then splitting each of these into two subgroups of size one.

$$\begin{array}{cccccccc}
\mathbf{U}_1 & \mathbf{U}_2 & \mathbf{U}_3 & \mathbf{U}_4 & \mathbf{U}_5 & \mathbf{U}_6 & \mathbf{U}_7 & \mathbf{U}_8 \\
\dfrac{\begin{bmatrix}1\\1\\1\\1\\1\\1\\1\\1\end{bmatrix}}{\sqrt{8}} &
\dfrac{\begin{bmatrix}-1\\-1\\-1\\-1\\1\\1\\1\\1\end{bmatrix}}{\sqrt{8}} &
\dfrac{\begin{bmatrix}-1\\-1\\1\\1\\0\\0\\0\\0\end{bmatrix}}{\sqrt{4}} &
\dfrac{\begin{bmatrix}0\\0\\0\\0\\-1\\-1\\1\\1\end{bmatrix}}{\sqrt{4}} &
\dfrac{\begin{bmatrix}-1\\1\\0\\0\\0\\0\\0\\0\end{bmatrix}}{\sqrt{2}} &
\dfrac{\begin{bmatrix}0\\0\\-1\\1\\0\\0\\0\\0\end{bmatrix}}{\sqrt{2}} &
\dfrac{\begin{bmatrix}0\\0\\0\\0\\-1\\1\\0\\0\end{bmatrix}}{\sqrt{2}} &
\dfrac{\begin{bmatrix}0\\0\\0\\0\\0\\0\\-1\\1\end{bmatrix}}{\sqrt{2}}
\end{array}$$

Note that the first of our two orthogonal coordinate systems can also be thought of in this way if we reorder the axes in the order $\mathbf{U}_1$, $\mathbf{U}_8$, $\mathbf{U}_7$, $\mathbf{U}_6$,

$\mathbf{U}_5$, $\mathbf{U}_4$, $\mathbf{U}_3$ and $\mathbf{U}_2$. With this order, $\mathbf{U}_8$ can be thought of as a comparison of two groups, one of size seven and one of size one. Then $\mathbf{U}_7$ can be thought of as a comparison of two subgroups of the group of size seven, with the subgroups being of size six and one. The pattern then continues until both subgroups are of size one.

(b) Again there are an infinite number of correct answers. For example, we can again follow the pattern of §2.4 to obtain

$$\mathbf{U}_1 = \frac{\begin{bmatrix} 1 \\ 1 \\ 1 \\ 1 \\ 1 \end{bmatrix}}{\sqrt{5}} \quad \mathbf{U}_2 = \frac{\begin{bmatrix} -1 \\ 1 \\ 0 \\ 0 \\ 0 \end{bmatrix}}{\sqrt{2}} \quad \mathbf{U}_3 = \frac{\begin{bmatrix} -1 \\ -1 \\ 2 \\ 0 \\ 0 \end{bmatrix}}{\sqrt{6}} \quad \mathbf{U}_4 = \frac{\begin{bmatrix} -1 \\ -1 \\ -1 \\ 3 \\ 0 \end{bmatrix}}{\sqrt{12}} \quad \mathbf{U}_5 = \frac{\begin{bmatrix} -1 \\ -1 \\ -1 \\ -1 \\ 4 \end{bmatrix}}{\sqrt{20}}$$

Note that this corresponds to an initial split into groups of sizes four and one, respectively ($\mathbf{U}_5$). The only other choice is to split into groups of sizes two and three, as in Exercise 2.2. To make our system different to that system, we make the initial split differently, for example, by splitting the first three away from the last two. The result is as follows:

$$\mathbf{U}_1 = \frac{\begin{bmatrix} 1 \\ 1 \\ 1 \\ 1 \\ 1 \end{bmatrix}}{\sqrt{5}} \quad \mathbf{U}_2 = \frac{\begin{bmatrix} -2 \\ -2 \\ -2 \\ 3 \\ 3 \end{bmatrix}}{\sqrt{30}} \quad \mathbf{U}_3 = \frac{\begin{bmatrix} -1 \\ -1 \\ 2 \\ 0 \\ 0 \end{bmatrix}}{\sqrt{6}} \quad \mathbf{U}_4 = \frac{\begin{bmatrix} -1 \\ 1 \\ 0 \\ 0 \\ 0 \end{bmatrix}}{\sqrt{2}} \quad \mathbf{U}_5 = \frac{\begin{bmatrix} 0 \\ 0 \\ 0 \\ -1 \\ 1 \end{bmatrix}}{\sqrt{2}}$$

(2.6) (a) The resulting observation vector, to one decimal place, is $\mathbf{y} = [12.7, 12.0, 6.3, 27.0, 7.3]^{\mathrm{T}}$.

(b) The appropriate orthogonal decomposition, $\mathbf{y} = \bar{\mathbf{y}} + (\mathbf{y} - \bar{\mathbf{y}})$, is

$$\begin{bmatrix} 12.7 \\ 12.0 \\ 6.3 \\ 27.0 \\ 7.3 \end{bmatrix} = \begin{bmatrix} 13.06 \\ 13.06 \\ 13.06 \\ 13.06 \\ 13.06 \end{bmatrix} + \begin{bmatrix} -0.36 \\ -1.06 \\ -6.76 \\ 13.94 \\ -5.76 \end{bmatrix}$$

The resulting F value is

$$F = \frac{\|\bar{\mathbf{y}}\|^2}{\|\mathbf{y} - \bar{\mathbf{y}}\|^2/(n-1)} = \frac{852.818}{274.452/4} = 12.43$$

The corresponding reference distribution is $F_{1,4}$, which has 90, 95 and 99 percentiles of 4.54, 7.71 and 21.20, respectively (Table T.2). Hence our F

value of 12.43 is "large" when compared with the 95 percentile, but not when compared with the 99 percentile. That is, we have "reasonable" evidence to suggest a difference in mean rainfall between the two areas.

(c) The corresponding t value is

$$t = \frac{\pm\|\bar{\mathbf{y}}\|}{\|\mathbf{y}-\bar{\mathbf{y}}\|/\sqrt{n-1}} = \frac{\sqrt{852.818}}{\sqrt{274.452}/\sqrt{4}} = \sqrt{12.43} = 3.526$$

The corresponding reference distribution is t_4, which has 95, 97.5 and 99.5 percentiles of 2.132, 2.776 and 4.604, respectively (Table T.2). Hence our t value of 3.526 is again "large" when compared with the 97.5 percentile, but not when compared with the 99.5 percentile. That is, our (two-sided) test results agree with those obtained in (b). Also, $3.526^2 = 12.43$, so our test statistics are indeed linked by the formula $F = t^2$.

(d) The best estimate of the long-term mean percentage difference in rainfall between the two areas is $\bar{y} = 13.06\%$. Hence the best estimate of the long-term average for West Melton is $801 \times (100 - 13.06)/100 = 696$ mm. Similarly, the best estimate for the subsequent year is $927 \times (100 - 13.06)/100 = 806$ mm. (Note that these two estimates depend upon our initial assumption, that it is more appropriate to model *percentage* differences rather than raw differences. We have insufficient data to check this assumption; however, this could be done using long-term records for nearby Christchurch and Darfield.)

(2.7) (a) The resulting observation vector is

$$\mathbf{y} = [1.69, 0.95, 0.12, \ldots, 1.23, 1.92, 2.03]^T$$

(b) The mean change in red cell lead level is $\bar{y} = 1.2228$ (using a computer). The appropriate orthogonal decomposition, $\mathbf{y} = \bar{\mathbf{y}} + (\mathbf{y} - \bar{\mathbf{y}})$, is

$$\begin{bmatrix} 1.69 \\ 0.95 \\ 0.12 \\ \vdots \\ 1.23 \\ 1.92 \\ 2.03 \end{bmatrix} = \begin{bmatrix} 1.2228 \\ 1.2228 \\ 1.2228 \\ \vdots \\ 1.2228 \\ 1.2228 \\ 1.2228 \end{bmatrix} + \begin{bmatrix} 0.4672 \\ -0.2728 \\ -1.1028 \\ \vdots \\ 0.0072 \\ 0.6972 \\ 0.8072 \end{bmatrix}$$

The resulting Pythagorean breakup is

$$\begin{aligned} \|\mathbf{y}\|^2 &= \|\bar{\mathbf{y}}\|^2 + \|\mathbf{y}-\bar{\mathbf{y}}\|^2 \\ 66.27 &= 47.85 + 18.42 \end{aligned}$$

(c) The resulting F value is

$$F = \frac{||\bar{\mathbf{y}}||^2}{||\mathbf{y} - \bar{\mathbf{y}}||^2/(n-1)} = \frac{47.85}{18.42/31} = 80.54$$

The corresponding reference distribution is $F_{1,31}$, which has 90, 95 and 99 percentiles of 2.87, 4.16 and 7.53, respectively (by interpolation in Table T.2). Hence our F value of 80.54 is "extremely large," since it is huge even in comparison to the 99 percentile. That is, we have "extremely good" evidence to suggest, on average, a difference between END and PRE red cell lead levels.

(d) The corresponding t value is

$$t = \frac{\pm||\bar{\mathbf{y}}||}{||\mathbf{y} - \bar{\mathbf{y}}||/\sqrt{n-1}} = \frac{\sqrt{47.85}}{\sqrt{18.42}/\sqrt{31}} = \sqrt{80.54} = 8.975$$

The corresponding reference distribution is t_{31}, which has 95, 97.5 and 99.5 percentiles of 1.695, 2.039 and 2.744, respectively (from Table T.2). Hence our t value of 8.975 is again "extremely large" even when compared with the 99.5 percentile. That is, our (two-sided) test results agree with those obtained in (c).

E.2 Solutions for Chapter 3 Exercises

(3.1) (a) The revised orthogonal decomposition is

$$\begin{bmatrix} -4.75 \\ -2.75 \\ 1.25 \\ 6.25 \end{bmatrix} = \begin{bmatrix} -3.75 \\ -3.75 \\ 3.75 \\ 3.75 \end{bmatrix} + \begin{bmatrix} -1 \\ 1 \\ -2.5 \\ 2.5 \end{bmatrix}$$

$$(\mathbf{y} - \bar{\mathbf{y}}) = (\bar{\mathbf{y}}_i - \bar{\mathbf{y}}) + (\mathbf{y} - \bar{\mathbf{y}}_i)$$

The redrawn Figure 3.3 is given in Figure E.1. The small, highlighted triangle in the upper right of the figure shows the new decomposition.

(b) The associated Pythagorean breakup is

$$\begin{array}{ccccc} ||\mathbf{y} - \bar{\mathbf{y}}||^2 & = & ||\bar{\mathbf{y}}_i - \bar{\mathbf{y}}||^2 & + & ||\mathbf{y} - \bar{\mathbf{y}}_i||^2 \\ 70.75 & = & 56.25 & + & 14.5 \end{array}$$

The resulting test statistic is

$$F = \frac{||\bar{\mathbf{y}}_i - \bar{\mathbf{y}}||^2}{||\mathbf{y} - \bar{\mathbf{y}}_i||^2/2} = \frac{56.25}{14.5/2} = 7.76$$

as in §3.2.

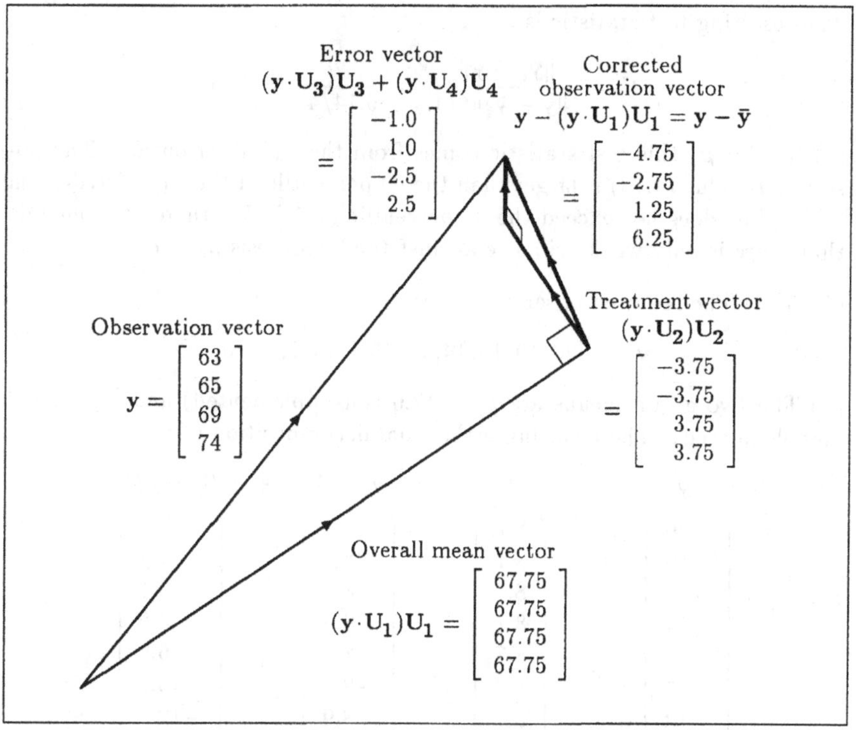

Figure E.1: Orthogonal decomposition of the observation vector for the case of independent samples of size two from two study populations, showing also the "corrected" observation vector $[\mathbf{y} - (\mathbf{y}\cdot\mathbf{U}_1)\mathbf{U}_1] = (\mathbf{y} - \bar{\mathbf{y}})$.

(3.4) (a) The observation vector **y**, and its orthogonal decomposition, are as follows (using treatment means of $\bar{y}_1 = 9.0$ and $\bar{y}_2 = 11.2$):

$$\begin{array}{ccccccc} \mathbf{y} & = & \bar{\mathbf{y}} & + & (\bar{\mathbf{y}}_i - \bar{\mathbf{y}}) & + & (\mathbf{y} - \bar{\mathbf{y}}_i) \\ \begin{bmatrix} 9.6 \\ 7.4 \\ 10.0 \\ 11.3 \\ 10.1 \\ 12.2 \end{bmatrix} & = & \begin{bmatrix} 10.1 \\ 10.1 \\ 10.1 \\ 10.1 \\ 10.1 \\ 10.1 \end{bmatrix} & + & \begin{bmatrix} -1.1 \\ -1.1 \\ -1.1 \\ 1.1 \\ 1.1 \\ 1.1 \end{bmatrix} & + & \begin{bmatrix} 0.6 \\ -1.6 \\ 1.0 \\ 0.1 \\ -1.1 \\ 1.0 \end{bmatrix} \end{array}$$

(b) The corresponding Pythagorean breakup is

$$\begin{array}{ccccccc} \|\mathbf{y}\|^2 & = & \|\bar{\mathbf{y}}\|^2 & + & \|\bar{\mathbf{y}}_i - \bar{\mathbf{y}}\|^2 & + & \|\mathbf{y} - \bar{\mathbf{y}}_i\|^2 \\ 625.46 & = & 612.06 & + & 7.26 & + & 6.14 \end{array}$$

The resulting test statistic is

$$F = \frac{||\bar{\mathbf{y}}_i - \bar{\mathbf{y}}||^2}{||\mathbf{y} - \bar{\mathbf{y}}_i||^2/4} = \frac{7.26}{6.14/4} = 4.73$$

(c) If $\mu_1 = \mu_2$ this test statistic comes from the $F_{1,4}$ distribution. The observed F value (4.73) is larger than the 90 percentile of the $F_{1,4}$ distribution (4.54), but does not exceed the 95 percentile (7.71). We therefore conclude that there is only weak evidence against the hypothesis $\mu_1 = \mu_2$.

(3.7) (a) The required observation vector is

$$\mathbf{y} = [26, 19, 18, 14, 11, 15, -13, 1, 1, -5]^T$$

(b) The two group means are $\bar{y}_1 = 17.6$ (unsupplemented) and $\bar{y}_2 = -0.2$ (supplemented). The resulting orthogonal decomposition of $\mathbf{y}$ is

$$\begin{array}{ccccccc} \mathbf{y} & = & \bar{\mathbf{y}} & + & (\bar{\mathbf{y}}_i - \bar{\mathbf{y}}) & + & (\mathbf{y} - \bar{\mathbf{y}}_i) \\ \begin{bmatrix} 26 \\ 19 \\ 18 \\ 14 \\ 11 \\ 15 \\ -13 \\ 1 \\ 1 \\ -5 \end{bmatrix} & = & \begin{bmatrix} 8.7 \\ 8.7 \\ 8.7 \\ 8.7 \\ 8.7 \\ 8.7 \\ 8.7 \\ 8.7 \\ 8.7 \\ 8.7 \end{bmatrix} & + & \begin{bmatrix} 8.9 \\ 8.9 \\ 8.9 \\ 8.9 \\ 8.9 \\ -8.9 \\ -8.9 \\ -8.9 \\ -8.9 \\ -8.9 \end{bmatrix} & + & \begin{bmatrix} 8.4 \\ 1.4 \\ 0.4 \\ -3.6 \\ -6.6 \\ 15.2 \\ -12.8 \\ 1.2 \\ 1.2 \\ -4.8 \end{bmatrix} \end{array}$$

(c) The corresponding Pythagorean breakup is

$$\begin{array}{ccccccc} ||\mathbf{y}||^2 & = & ||\bar{\mathbf{y}}||^2 & + & ||\bar{\mathbf{y}}_i - \bar{\mathbf{y}}||^2 & + & ||\mathbf{y} - \bar{\mathbf{y}}_i||^2 \\ 2099 & = & 756.9 & + & 792.1 & + & 550 \end{array}$$

The resulting test statistic is

$$F = \frac{||\bar{\mathbf{y}}_i - \bar{\mathbf{y}}||^2}{||\mathbf{y} - \bar{\mathbf{y}}_i||^2/8} = \frac{792.1}{550/8} = 11.52$$

If $\mu_1 = \mu_2$, this test statistic comes from the $F_{1,8}$ distribution. The observed F value (11.52) is larger than the 99 percentile of the $F_{1,8}$ distribution (11.26). We therefore conclude that there is strong evidence that the mean changes in selenium level between birth and 1 month of age were different between the two groups (unsupplemented and supplemented).

(d) The required observation vector is

$$\mathbf{y} = [1, -7, -3, 23, 6, 25, 2, 22, 5, 21, 33, -21, 12, 8, 23, 3, 27, 20, 25, 32]^T$$

The two group means are $\bar{y}_1 = 9.5$ (unsupplemented) and $\bar{y}_2 = 16.2$ (supplemented). The resulting orthogonal decomposition of $\mathbf{y}$ is

$$\begin{array}{ccccccc}\mathbf{y} & = & \bar{\mathbf{y}} & + & (\bar{\mathbf{y}}_i - \bar{\mathbf{y}}) & + & (\mathbf{y} - \bar{\mathbf{y}}_i) \\ \begin{bmatrix} 1 \\ -7 \\ \vdots \\ 33 \\ -21 \\ \vdots \end{bmatrix} & = & \begin{bmatrix} 12.85 \\ 12.85 \\ \vdots \\ 12.85 \\ 12.85 \\ \vdots \end{bmatrix} & + & \begin{bmatrix} -3.35 \\ -3.35 \\ \vdots \\ 3.35 \\ 3.35 \\ \vdots \end{bmatrix} & + & \begin{bmatrix} -8.5 \\ -16.5 \\ \vdots \\ 16.8 \\ -37.2 \\ \vdots \end{bmatrix}\end{array}$$

The corresponding Pythagorean breakup is

$$\begin{array}{ccccccc}||\mathbf{y}||^2 & = & ||\bar{\mathbf{y}}||^2 & + & ||\bar{\mathbf{y}}_i - \bar{\mathbf{y}}||^2 & + & ||\mathbf{y} - \bar{\mathbf{y}}_i||^2 \\ 7257 & = & 3302.45 & + & 224.45 & + & 3730.1\end{array}$$

The resulting test statistic is

$$F = \frac{||\bar{\mathbf{y}}_i - \bar{\mathbf{y}}||^2}{||\mathbf{y} - \bar{\mathbf{y}}_i||^2/8} = \frac{224.45}{3730.1/18} = 1.08$$

If $\mu_1 = \mu_2$, this test statistic comes from the $F_{1,18}$ distribution. The observed F value (1.08) is small, even in comparison with the 90 percentile of the $F_{1,18}$ distribution (3.01). We therefore conclude that we have found no real evidence that the mean changes in selenium level between 1 and 3 months of age were different between the two groups (unsupplemented and supplemented).

(3.10) (a) The projection length $\mathbf{y}\cdot\mathbf{U_1}$ works out to be $8663/\sqrt{126}$, where 8663 is the sum of the 126 heights. The projection vector $(\mathbf{y}\cdot\mathbf{U_1})\mathbf{U_1}$ is therefore the vector with $8663/126 = 68.75396825 = \bar{y}$ in all positions; that is, it is the familiar overall mean vector, $\bar{\mathbf{y}}$.

The projection length $\mathbf{y}\cdot\mathbf{U_2}$ works out to be

$$\frac{-3212/49 + 5451/77}{\sqrt{\frac{1}{49}+\frac{1}{77}}} = \frac{5.24}{\sqrt{\frac{1}{49}+\frac{1}{77}}}$$

where 3212 and 5451 are the sums of the female and male heights, respectively. The projection vector $(\mathbf{y}\cdot\mathbf{U_2})\mathbf{U_2}$ is therefore the vector

$$\frac{5.24}{\frac{1}{49}+\frac{1}{77}}\begin{bmatrix} -1/49 \\ \vdots \\ 1/77 \\ \vdots \end{bmatrix} = \begin{bmatrix} -3.20 \\ \vdots \\ 2.04 \\ \vdots \end{bmatrix}$$

Now $\bar{y}_1 = 3212/49 = 65.55$ and $\bar{y}_2 = 5451/77 = 70.79$, while $\bar{y} = 68.75$ as above, so this is the familiar vector $(\bar{\mathbf{y}}_i - \bar{\mathbf{y}})$.

(b) The orthogonal decomposition is therefore

$$\begin{bmatrix} 60 \\ 61 \\ \vdots \\ 63 \\ 65 \\ \vdots \end{bmatrix} = \begin{bmatrix} 68.75 \\ 68.75 \\ \vdots \\ 68.75 \\ 68.75 \\ \vdots \end{bmatrix} + \begin{bmatrix} -3.20 \\ -3.20 \\ \vdots \\ 2.04 \\ 2.04 \\ \vdots \end{bmatrix} + \begin{bmatrix} -5.55 \\ -4.55 \\ \vdots \\ -7.79 \\ -5.79 \\ \vdots \end{bmatrix}$$

Here the error vector is again the vector $(\mathbf{y} - \bar{\mathbf{y}}_i)$.

(c) The corresponding Pythagorean breakup is

$$\begin{aligned} ||\mathbf{y}||^2 &= \quad ||\bar{\mathbf{y}}||^2 \quad + ||\bar{\mathbf{y}}_i - \bar{\mathbf{y}}||^2 + ||\mathbf{y} - \bar{\mathbf{y}}_i||^2 \\ 597387 &= 595615.627 + \ 822.575 \ + \ 948.798 \end{aligned}$$

(d) The resulting test statistic is

$$F = \frac{822.575}{948.798/124} = 107.50$$

If $\mu_1 = \mu_2$ this test statistic comes from the $F_{1,124}$ distribution. The observed F value (107.50) is extremely large when compared with the 99 percentile of the $F_{1,124}$ distribution (6.85). We therefore conclude that we have found very strong evidence of a difference in height between female and male UCD students.

E.3 Solutions for Chapter 4 Exercises

(4.1) (a) The three relevant contrasts are:

$(\mu_1 + \mu_2)/2 - (\mu_3 + \mu_4)/2$

$\mu_1 - \mu_2$

$\mu_3 - \mu_4$

(b) The associated unit vectors are:

$$\mathbf{U_2} = \frac{\begin{bmatrix} 1 \\ 1 \\ 1 \\ 1 \\ -1 \\ -1 \\ -1 \\ -1 \end{bmatrix}}{\sqrt{8}} \qquad \mathbf{U_3} = \frac{\begin{bmatrix} 1 \\ 1 \\ -1 \\ -1 \\ 0 \\ 0 \\ 0 \\ 0 \end{bmatrix}}{\sqrt{4}} \qquad \mathbf{U_4} = \frac{\begin{bmatrix} 0 \\ 0 \\ 0 \\ 0 \\ 1 \\ 1 \\ -1 \\ -1 \end{bmatrix}}{\sqrt{4}}$$

(c) $\mathbf{U_2} \cdot \mathbf{U_3} = 0$, $\mathbf{U_2} \cdot \mathbf{U_4} = 0$ and $\mathbf{U_3} \cdot \mathbf{U_4} = 0$, which confirms their orthogonality.

(d)

$$\mathbf{U_1} = \frac{\begin{bmatrix} 1\\1\\1\\1\\1\\1\\1\\1 \end{bmatrix}}{\sqrt{8}} \quad \mathbf{U_5} = \frac{\begin{bmatrix} 1\\-1\\0\\0\\0\\0\\0\\0 \end{bmatrix}}{\sqrt{2}} \quad \mathbf{U_6} = \frac{\begin{bmatrix} 0\\0\\1\\-1\\0\\0\\0\\0 \end{bmatrix}}{\sqrt{2}} \quad \mathbf{U_7} = \frac{\begin{bmatrix} 0\\0\\0\\0\\1\\-1\\0\\0 \end{bmatrix}}{\sqrt{2}} \quad \mathbf{U_8} = \frac{\begin{bmatrix} 0\\0\\0\\0\\0\\0\\1\\-1 \end{bmatrix}}{\sqrt{2}}$$

(e)
$$\begin{aligned}
\mathbf{y} \cdot \mathbf{U_1} &= (44+40+42+50+51+55+58+52)/\sqrt{8} = 392/\sqrt{8} = 138.59,\\
\mathbf{y} \cdot \mathbf{U_2} &= (44+40+42+50-51-55-58-52)/\sqrt{8} = -40/\sqrt{8} = -14.14,\\
\mathbf{y} \cdot \mathbf{U_3} &= (44+40-42-50)/\sqrt{4} = -8/\sqrt{4} = -4,\\
\mathbf{y} \cdot \mathbf{U_4} &= (51+55-58-52)/\sqrt{4} = -4/\sqrt{4} = -2,\\
\mathbf{y} \cdot \mathbf{U_5} &= (44-40)/\sqrt{2} = 4/\sqrt{2} = 2.83,\\
\mathbf{y} \cdot \mathbf{U_6} &= (42-50)/\sqrt{2} = -8/\sqrt{2} = -5.66,\\
\mathbf{y} \cdot \mathbf{U_7} &= (51-55)/\sqrt{2} = -4/\sqrt{2} = -2.83,\\
\mathbf{y} \cdot \mathbf{U_8} &= (58-52)/\sqrt{2} = 6/\sqrt{2} = 4.24.
\end{aligned}$$

(f) The resulting Pythagorean breakup is

$$19494 = 19208 + 200 + 16 + 4 + 8 + 32 + 8 + 18$$

(g) The test statistic corresponding to question (A) is

$$F = \frac{200}{(8+32+8+18)/4} = \frac{200}{16.5} = 12.12$$

Now if the two breeds of sheep do not on average differ from the two breeds of goats, our test statistic would have come from an $F_{1,4}$ distribution. This distribution has 90, 95 and 99 percentiles of 4.54, 7.71 and 21.20, respectively. Our test statistic (12.12) exceeds the 95 percentile, so it is unusually large, and we conclude that there is some evidence that sheep and goats differ in their effectiveness as blackberry control agents.

(h) The test statistic corresponding to question (B) is

$$F = \frac{16}{(8+32+8+18)/4} = \frac{16}{16.5} = 0.97$$

The test statistic corresponding to question (C) is

$$F = \frac{4}{(8+32+8+18)/4} = \frac{4}{16.5} = 0.24$$

If the two breeds of sheep are identical in effectiveness, and similarly the two breeds of goat, these two test statistics would have both come from an $F_{1,4}$ distribution, with percentiles of 4.54, 7.71 and 21.20. Neither of the test statistics is at all large, so we conclude that there appears to be little difference between Romney and Merino sheep, and between angora and feral goats.

(4.5) The contrasts corresponding to the two questions of interest are:

$$(\mu_2 + \mu_3)/2 - \mu_1$$
$$\mu_2 - \mu_3$$

The usual equiangular direction ($\mathbf{U_1}$), and the two unit vectors associated with these contrasts ($\mathbf{U_2}$ and $\mathbf{U_3}$), are:

$$\mathbf{U_1} = \frac{\begin{bmatrix} 1 \\ 1 \\ 1 \\ 1 \\ 1 \\ 1 \\ 1 \\ 1 \\ 1 \end{bmatrix}}{\sqrt{9}} \qquad \mathbf{U_2} = \frac{\begin{bmatrix} -2 \\ -2 \\ -2 \\ 1 \\ 1 \\ 1 \\ 1 \\ 1 \\ 1 \end{bmatrix}}{\sqrt{18}} \qquad \mathbf{U_3} = \frac{\begin{bmatrix} 0 \\ 0 \\ 0 \\ 1 \\ 1 \\ 1 \\ -1 \\ -1 \\ -1 \end{bmatrix}}{\sqrt{6}}$$

The resulting orthogonal decomposition of $\mathbf{y}$ is

$$\begin{bmatrix} 45.1 \\ 46.7 \\ 47.4 \\ 56.7 \\ 57.3 \\ 54.6 \\ 53.3 \\ 55.0 \\ 54.7 \end{bmatrix} = \begin{bmatrix} 52.311 \\ 52.311 \\ 52.311 \\ 52.311 \\ 52.311 \\ 52.311 \\ 52.311 \\ 52.311 \\ 52.311 \end{bmatrix} + \begin{bmatrix} -5.911 \\ -5.911 \\ -5.911 \\ 2.956 \\ 2.956 \\ 2.956 \\ 2.956 \\ 2.956 \\ 2.956 \end{bmatrix} + \begin{bmatrix} 0 \\ 0 \\ 0 \\ 0.933 \\ 0.933 \\ 0.933 \\ -0.933 \\ -0.933 \\ -0.933 \end{bmatrix} + \begin{bmatrix} -1.30 \\ 0.30 \\ 1.00 \\ 0.50 \\ 1.10 \\ -1.60 \\ -1.03 \\ 0.67 \\ 0.37 \end{bmatrix}$$

$$\mathbf{y} = \bar{\mathbf{y}} + (\mathbf{y}\cdot\mathbf{U_2})\mathbf{U_2} + (\mathbf{y}\cdot\mathbf{U_3})\mathbf{U_3} + (\mathbf{y} - \bar{\mathbf{y}}_i)$$

When computing this expression we calculated $\bar{y} = 52.31111111$, $\mathbf{y}\cdot\mathbf{U_2} = 53.2/\sqrt{18}$ and $\mathbf{y}\cdot\mathbf{U_3} = 5.6/\sqrt{6}$. The resulting Pythagorean breakup is

$$24798.98 = 24628.072 + 157.236 + 5.227 + 8.447$$

The first test hypothesis is "fertiliser does not increase corn yield." The corresponding test statistic is

$$F = \frac{157.236}{8.447/6} = \frac{157.236}{1.408} = 111.68$$

This is extremely "large" in comparison to the percentiles of the $F_{1,6}$ distribution. We therefore reject the hypothesis, concluding that there is very strong evidence that fertilizer increased corn yield.

The second test hypothesis is "there is no difference in corn yield due to method of application." The corresponding test statistic is

$$F = \frac{5.227}{8.447/6} = \frac{5.227}{1.408} = 3.71$$

This is "small" even when compared with the 90 percentile of the $F_{1,6}$ distribution (3.78). We therefore accept the hypothesis, concluding that there is little evidence of a difference between the two methods of fertilizer application.

(4.7) (a) Appropriate orthogonal contrasts are:

$$(\mu_1 + \mu_2)/2 - \mu_3$$
$$\mu_1 - \mu_2$$

The first contrast is a comparison of "superphosphate versus no superphosphate," disregarding the time of its application. The second contrast is a comparison of "autumn versus spring" times of application.

(b) Appropriate orthogonal contrasts are:

$$(\mu_1 + \mu_3)/2 - (\mu_2 + \mu_4)/2$$
$$\mu_1 - \mu_3$$
$$\mu_2 - \mu_4$$

The first contrast is a comparison of "protein versus starch" based diets. The second contrast is a comparison of one brand of protein based diet versus the other. The third contrast is a comparison of one brand of starch based diet versus the other.

(c) Appropriate orthogonal contrasts are:

$$(\mu_1 + \mu_2 + \mu_5)/3 - (\mu_3 + \mu_4)/2$$
$$(\mu_1 + \mu_5)/2 - \mu_2$$
$$\mu_1 - \mu_5$$
$$\mu_3 - \mu_4$$

The first contrast is a comparison of "legumes versus nonlegumes" (plants which fix nitrogen versus those which do not). The second contrast is a comparison of one leguminous plant versus the other ("peas versus beans"). The third contrast is a comparison of one type of peas versus the other ("field versus garden"). The fourth contrast is a comparison of one nonleguminous plant versus the other ("oats versus mustard").

E.4 Solutions for Chapter 5 Exercises

(5.1) (a) The rewritten orthogonal decomposition is:

$$\begin{bmatrix} -139 \\ -185 \\ 334 \\ -179 \\ 169 \end{bmatrix} = \begin{bmatrix} -135.1 \\ -176.1 \\ 315.3 \\ -196.6 \\ 192.5 \end{bmatrix} + \begin{bmatrix} -3.9 \\ -8.9 \\ 18.7 \\ 17.6 \\ -23.5 \end{bmatrix}$$

$$(\mathbf{y} - \bar{\mathbf{y}}) = b(\mathbf{x} - \bar{\mathbf{x}}) + [\mathbf{y} - \bar{\mathbf{y}} - b(\mathbf{x} - \bar{\mathbf{x}})]$$

The appropriate triangle is highlighted in Figure E.2.

(b) The associated Pythagorean breakup is

$$\begin{aligned} \|\mathbf{y} - \bar{\mathbf{y}}\|^2 &= b^2\|\mathbf{x} - \bar{\mathbf{x}}\|^2 + \|\mathbf{y} - \bar{\mathbf{y}} - b(\mathbf{x} - \bar{\mathbf{x}})\|^2 \\ 225704 &= 224402 + 1302 \end{aligned}$$

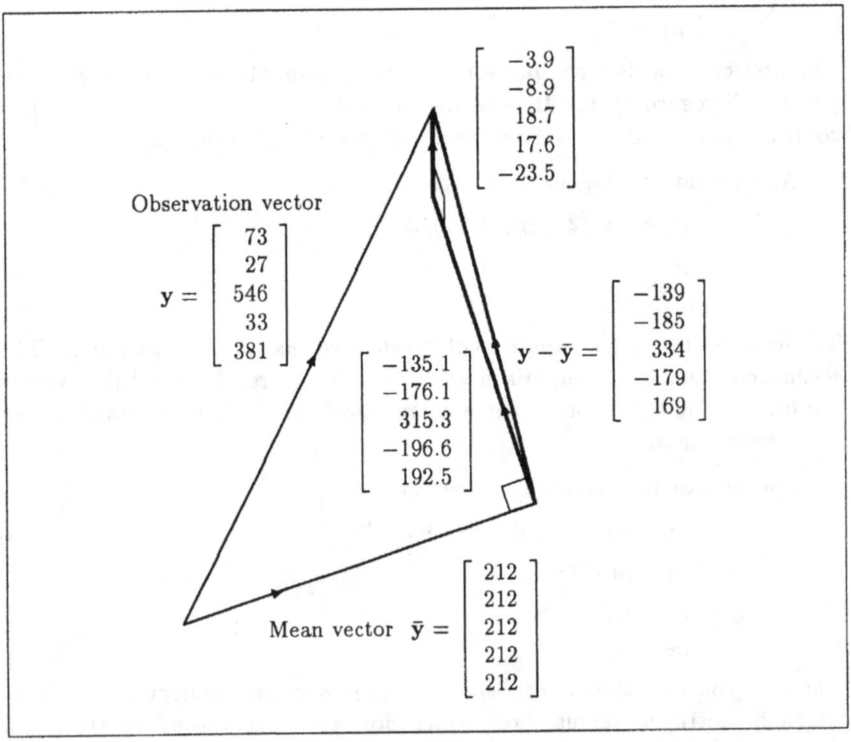

Figure E.2: Orthogonal decomposition of the observation vector for the simple regression case, showing also the "corrected" observation vector $(\mathbf{y} - \bar{\mathbf{y}})$ and highlighting the triangle corresponding to the rewritten decomposition.

The resulting test statistic is

$$F = \frac{b^2\|\mathbf{x}-\bar{\mathbf{x}}\|^2}{\|\mathbf{y}-\bar{\mathbf{y}}-b(\mathbf{x}-\bar{\mathbf{x}})\|^2/3} = \frac{224402}{1302/3} = \frac{224402}{434} = 517$$

as in §5.2.

(5.4) (a) We leave the scattergram to the reader.

(b) The unit vector is

$$\mathbf{U_2} = \frac{1}{\sqrt{64}}\begin{bmatrix} 8-12 \\ 8-12 \\ 12-12 \\ 12-12 \\ 16-12 \\ 16-12 \end{bmatrix} = \frac{1}{8}\begin{bmatrix} -4 \\ -4 \\ 0 \\ 0 \\ 4 \\ 4 \end{bmatrix} = \frac{1}{2}\begin{bmatrix} -1 \\ -1 \\ 0 \\ 0 \\ 1 \\ 1 \end{bmatrix}$$

(c) The fitted model vector, $(\mathbf{y}\cdot\mathbf{U_1})\mathbf{U_1} + (\mathbf{y}\cdot\mathbf{U_2})\mathbf{U_2}$, is

$$\begin{bmatrix} 46.667 \\ 46.667 \\ 46.667 \\ 46.667 \\ 46.667 \\ 46.667 \end{bmatrix} + \begin{bmatrix} -15.25 \\ -15.25 \\ 0 \\ 0 \\ 15.25 \\ 15.25 \end{bmatrix} = \begin{bmatrix} 46.667 \\ 46.667 \\ 46.667 \\ 46.667 \\ 46.667 \\ 46.667 \end{bmatrix} + \begin{bmatrix} 3.8125\times(-4) \\ 3.8125\times(-4) \\ 0 \\ 0 \\ 3.8125\times 4 \\ 3.8125\times 4 \end{bmatrix}$$

Hence the equation of the fitted line is

$$y = 46.7 + 3.81(x-12)$$

We shall leave it to the reader to add this line to the scattergram. Does it "look right?"

(d) The Pythagorean breakup is

$$\begin{aligned} \|\mathbf{y}\|^2 &= (\mathbf{y}\cdot\mathbf{U_1})^2 + (\mathbf{y}\cdot\mathbf{U_2})^2 + \|\text{error vector}\|^2 \\ 14009.08 &= 13066.67 + 930.25 + 12.16 \end{aligned}$$

The hypothesis is tested by calculating the test statistic

$$\frac{(\mathbf{y}\cdot\mathbf{U_2})^2}{\|\text{error vector}\|^2/4} = \frac{930.25}{12.16/4} = \frac{930.25}{3.04} = 306$$

This test statistic is extremely "large" in comparison with the percentiles of the $F_{1,4}$ distribution. We therefore conclude that the number of kiwifruit set per meter *is* related to the number of hives per hectare.

(e)

$$\mathbf{y}\cdot\mathbf{U_3} = \frac{33.4+30.6-2\times 44.7-2\times 46.3+63.7+61.3}{\sqrt{12}} = \frac{7}{\sqrt{12}} = 2.02$$

$$\mathbf{y}\cdot\mathbf{U}_4 = \frac{-33.4+30.6}{\sqrt{2}} = \frac{-2.8}{\sqrt{2}} = -1.98$$
$$\mathbf{y}\cdot\mathbf{U}_5 = \frac{-44.7+46.3}{\sqrt{2}} = \frac{1.6}{\sqrt{2}} = 1.13$$
$$\mathbf{y}\cdot\mathbf{U}_6 = \frac{-63.7+61.3}{\sqrt{2}} = \frac{-2.4}{\sqrt{2}} = -1.70$$

The average of these squared projection lengths is

$$\frac{2.02^2+(-1.98)^2+1.13^2+(-1.70)^2}{4} = \frac{12.17}{4} = 3.04$$

which is the denominator, $s^2 = 3.04$, used in the calculation of the test statistic in (d).

(5.7) (a) We leave the reader to do the scattergram.

(b) The equation of the line is $y = -0.008 + 0.018773x$. The line does seem to be a reasonably good fit (for the range to which it applies, which is 55–248 days of age).

(c) Paw Paw's rate of increase was 0.018773 kg per day (the slope of the regression line). This converts to 18.8 g per day.

(5.10) (a) We leave the reader to do the plot.

(b) The equation of the line is $y = 48.70 + 1.817x$.

(c) The hypothesis is tested by calculating the test statistic

$$\frac{(\mathbf{y}\cdot\mathbf{U}_2)^2}{\|\text{error vector}\|^2/22} = \frac{3798.3}{6135/22} = \frac{3798.3}{278.9} = 13.62$$

This test statistic is larger than the 99 percentile of the $F_{1,22}$ distribution (7.95). This means there is "strong" evidence to support the notion that Christchurch adult selenium intakes were rising during the period of the study.

(5.12) (a) We leave the reader to do the scattergram.

(b) The equation of the line is $y = 2.490 - 0.00407x$.

(c) The hypothesis is tested by calculating the test statistic

$$\frac{(\mathbf{y}\cdot\mathbf{U}_2)^2}{\|\text{error vector}\|^2/35} = \frac{7.8202}{28.75/35} = \frac{7.8202}{0.8214} = 9.52$$

This test statistic is larger than the 99 percentile of the $F_{1,35}$ distribution (7.42). This means there is "strong" evidence that summer drawdown is associated with summer rainfall.

(d) We again leave the reader to do the scattergram.

(e) The equation of the line is $y = 7.19 - 0.4722x$.

(f) The hypothesis is tested by calculating the test statistic

$$\frac{(\mathbf{y}\cdot\mathbf{U_2})^2}{\|\text{error vector}\|^2/35} = \frac{15.4356}{21.13/35} = \frac{15.4356}{0.6038} = 25.56$$

This test statistic is much larger than the 99 percentile of the $F_{1,35}$ distribution (7.42). This means there is "very strong" evidence that summer drawdown is associated with the October 1 level.

(g) The line crosses the x axis when $y = 0$, which occurs when $x = 7.19/0.4722 = 15.2$ m below ground level. When the October 1 ground water level is this low (which is unusual), our regression equation predicts the level will not fall any further during the summer period.

E.5 Solutions for Chapter 6 Exercises

(6.1) (a) See Figure E.3(a).

(b) See Figure E.3(b).

(c)

$$F = \frac{A^2}{B^2/2} = \frac{6.5^2 \times 3}{19.5/2} = \frac{126.75}{9.75} = 13.0$$

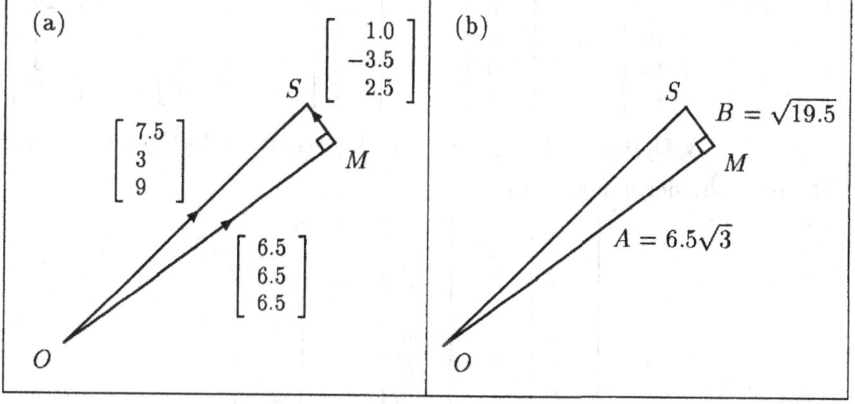

Figure E.3: (a) The statistical triangle for the 3-space example used in §2.3. (b) The triangle redrawn with lengths displayed.

$$t = \frac{A}{B/\sqrt{2}} = \frac{6.5 \times \sqrt{3}}{\sqrt{19.5}/\sqrt{2}} = \frac{11.2583}{4.4159/\sqrt{2}} = 3.606$$

These values are the same as those calculated in §2.3.

(d) We leave the two pictures to the reader. The lengths are $A = \sqrt{456.53}$ and $B = \sqrt{67.38}$. These lead to test statistics of $F = A^2/(B^2/8) = 456.53/(67.38/8) = 54.20$ and $t = A/(B/\sqrt{8}) = \sqrt{456.53}/\sqrt{67.38/8} = 7.363$ as found in §2.4.

(6.6) (a) Following on from the solution to Exercise 4.1 (in §E.3), we can write the orthogonal decomposition of $\mathbf{y}$ as

$$\begin{bmatrix} 44\\40\\42\\50\\51\\55\\58\\52 \end{bmatrix} = \begin{bmatrix} 49\\49\\49\\49\\49\\49\\49\\49 \end{bmatrix} + \begin{bmatrix} -5\\-5\\-5\\-5\\5\\5\\5\\5 \end{bmatrix} + \begin{bmatrix} -2\\-2\\2\\2\\0\\0\\0\\0 \end{bmatrix} + \begin{bmatrix} 0\\0\\0\\0\\-1\\-1\\1\\1 \end{bmatrix} + \begin{bmatrix} 2\\-2\\-4\\4\\-2\\2\\3\\-3 \end{bmatrix}$$

$$\mathbf{y} = (\mathbf{y}\cdot\mathbf{U}_1)\mathbf{U}_1 + (\mathbf{y}\cdot\mathbf{U}_2)\mathbf{U}_2 + (\mathbf{y}\cdot\mathbf{U}_3)\mathbf{U}_3 + (\mathbf{y}\cdot\mathbf{U}_4)\mathbf{U}_4 + (\mathbf{y} - \bar{\mathbf{y}}_i)$$

This can be rearranged to form the decomposition associated with the required statistical triangle (associated with $\mathbf{U}_2$) by subtraction of the projections of $\mathbf{y}$ onto $\mathbf{U}_1$, $\mathbf{U}_3$ and $\mathbf{U}_4$ from both sides:

$$\begin{bmatrix} 44\\40\\42\\50\\51\\55\\58\\52 \end{bmatrix} - \begin{bmatrix} 49\\49\\49\\49\\49\\49\\49\\49 \end{bmatrix} - \begin{bmatrix} -2\\-2\\2\\2\\0\\0\\0\\0 \end{bmatrix} - \begin{bmatrix} 0\\0\\0\\0\\-1\\-1\\1\\1 \end{bmatrix} = \begin{bmatrix} -5\\-5\\-5\\-5\\5\\5\\5\\5 \end{bmatrix} + \begin{bmatrix} 2\\-2\\-4\\4\\-2\\2\\3\\-3 \end{bmatrix}$$

$$\mathbf{y} - (\mathbf{y}\cdot\mathbf{U}_1)\mathbf{U}_1 - (\mathbf{y}\cdot\mathbf{U}_3)\mathbf{U}_3 - (\mathbf{y}\cdot\mathbf{U}_4)\mathbf{U}_4 = (\mathbf{y}\cdot\mathbf{U}_2)\mathbf{U}_2 + (\mathbf{y} - \bar{\mathbf{y}}_i)$$

The resulting decomposition is:

$$\begin{bmatrix} -3\\-7\\-9\\-1\\3\\7\\8\\2 \end{bmatrix} = \begin{bmatrix} -5\\-5\\-5\\-5\\5\\5\\5\\5 \end{bmatrix} + \begin{bmatrix} 2\\-2\\-4\\4\\-2\\2\\3\\-3 \end{bmatrix}$$

These are the three vectors which should label the sides of the triangle, which we leave to the reader to draw.

(b) The lengths for the second figure are the lengths of the second and third of these three vectors, so that $A = \sqrt{8 \times 5^2} = 10\sqrt{2} = 14.142$ and $B = \sqrt{2 \times (2^2 + 4^2 + 2^2 + 3^2)} = \sqrt{66} = 8.124$.

(c)

$$F = \frac{A^2}{B^2/4} = \frac{200}{66/4} = 12.12$$

$$t = \frac{A}{B/\sqrt{4}} = \frac{14.142}{\sqrt{66/4}} = 3.482$$

Reassuringly, the F value agrees with that obtained in §E.3!

(6.7) (a) The required statistical triangle is similar in appearance to the triangle highlighted in Figure E.2. The sides of the triangle are formed by the three vectors in the following orthogonal decomposition, obtained by subtracting the mean vector from both sides of the decomposition given in §5.3.

$$\begin{bmatrix} 141.7 \\ \vdots \\ 58.7 \end{bmatrix} = \begin{bmatrix} 53.1 \\ \vdots \\ 228.4 \end{bmatrix} + \begin{bmatrix} 88.6 \\ \vdots \\ -169.7 \end{bmatrix}$$

$$\mathbf{y} - \bar{\mathbf{y}} = b(\mathbf{x} - \bar{\mathbf{x}}) + [\mathbf{y} - \bar{\mathbf{y}} - b(\mathbf{x} - \bar{\mathbf{x}})]$$

We leave it to the reader to draw the triangle.

(b) The lengths of the two sides of interest are $A = b||\mathbf{x} - \bar{\mathbf{x}}|| = \sqrt{354781}$ and $B = ||\mathbf{y} - \bar{\mathbf{y}} - b(\mathbf{x} - \bar{\mathbf{x}})|| = \sqrt{169939}$, obtained from the Pythagorean breakup in §5.3.

(c)

$$F = \frac{A^2}{B^2/24} = \frac{354781}{169939/24} = 50.10$$

$$t = \frac{A}{B/\sqrt{24}} = \frac{\sqrt{354781}}{\sqrt{169939/24}} = 7.078$$

as obtained in §5.3.

E.6 Solutions for Appendix A Exercises

(A.1) The lengths are $\sqrt{4^2 + 1^2} = \sqrt{16 + 1} = \sqrt{17}$,

$\sqrt{6^2 + 1^2 + 0^2 + (-2)^2 + 4^2} = \sqrt{36 + 1 + 4 + 16} = \sqrt{57}$,

$\sqrt{1^2 + 1^2 + (-2)^2} = \sqrt{1 + 1 + 4} = \sqrt{6}$,

and $\sqrt{1^2 + 1^2 + 1^2 + 1^2 + 1^2} = \sqrt{5}$.

(A.2) The unit vectors are

$$\frac{1}{\sqrt{5}}\begin{bmatrix}1\\1\\1\\1\\1\end{bmatrix}, \quad \frac{1}{\sqrt{2}}\begin{bmatrix}-1\\1\\0\\0\\0\end{bmatrix}, \quad \frac{1}{\sqrt{30}}\begin{bmatrix}3\\3\\-2\\-2\\-2\end{bmatrix} \quad \text{and} \quad \frac{1}{2}\begin{bmatrix}-1\\1\\-1\\1\end{bmatrix}$$

(A.3) (a) $4\begin{bmatrix}1\\1\\0\\0\\0\end{bmatrix} + 5\begin{bmatrix}0\\0\\1\\1\\1\end{bmatrix} = \begin{bmatrix}4\\4\\0\\0\\0\end{bmatrix} + \begin{bmatrix}0\\0\\5\\5\\5\end{bmatrix} = \begin{bmatrix}4\\4\\5\\5\\5\end{bmatrix}$

(b) $3\begin{bmatrix}-1\\1\\0\\0\end{bmatrix} -2\begin{bmatrix}0\\0\\-1\\1\end{bmatrix} = \begin{bmatrix}-3\\3\\0\\0\end{bmatrix} + \begin{bmatrix}0\\0\\2\\-2\end{bmatrix} = \begin{bmatrix}-3\\3\\2\\-2\end{bmatrix}$

(A.4) To calculate each projection length, we make the specified vector into a unit vector (**U**) by dividing it by its length, calculate the dot product of **y** with **U**, and take the absolute value of $\mathbf{y}\cdot\mathbf{U}$.

(a) Here the vector is already a unit vector. Also

$$\mathbf{y}\cdot\mathbf{U} = \begin{bmatrix}4\\5\\8\\9\end{bmatrix}\cdot\frac{1}{2}\begin{bmatrix}1\\1\\1\\1\end{bmatrix} = \frac{1}{2}(4+5+8+9) = \frac{26}{2} = 13$$

Hence the length of the projection is 13.

(b)

$$\mathbf{y}\cdot\mathbf{U} = \begin{bmatrix}4\\5\\8\\9\end{bmatrix}\cdot\frac{1}{\sqrt{2}}\begin{bmatrix}1\\1\\0\\0\end{bmatrix} = \frac{1}{\sqrt{2}}(4+5+0+0) = \frac{9}{\sqrt{2}} = 6.4$$

Hence the length of the projection is 6.4.

(c)

$$\mathbf{y}\cdot\mathbf{U} = \begin{bmatrix}4\\5\\8\\9\end{bmatrix}\cdot\frac{1}{\sqrt{200}}\begin{bmatrix}10\\-10\\0\\0\end{bmatrix} = \frac{1}{\sqrt{200}}(40-50) = \frac{-10}{\sqrt{200}} = -0.7$$

Hence the length of the projection is 0.7.

(d)

$$\mathbf{y}\cdot\mathbf{U} = \begin{bmatrix} 4 \\ 5 \\ 8 \\ 9 \end{bmatrix} \cdot \frac{1}{\sqrt{4}} \begin{bmatrix} 1 \\ -1 \\ -1 \\ 1 \end{bmatrix} = \frac{1}{2}(4-5-8+9) = \frac{0}{2} = 0$$

Hence the length of the projection is 0 (meaning that **y** and **U** are at right angles to each other).

(A.5) (a) $\cos\theta = \dfrac{2+3}{\sqrt{5}\sqrt{10}} = \dfrac{5}{\sqrt{50}} = \dfrac{1}{\sqrt{2}}$. Thus $\theta = 45°$.

(b) $\begin{bmatrix} 1 \\ 1 \end{bmatrix} \cdot \begin{bmatrix} -1 \\ 1 \end{bmatrix} = 0$, so $\cos\theta = 0$ and $\theta = 90°$.

(c) $\cos\theta = \dfrac{4-7}{\sqrt{6}\sqrt{65}} = -0.15$, so $\theta = 99°$.

(d) $\cos\theta = \dfrac{32}{\sqrt{92}\sqrt{41}} = 0.52$, so $\theta = 59°$.

(A.6) (a) Yes. There are ${}_5C_2 = 10$ pairs to check.

(b) No. If labeled $\mathbf{U}_1$, $\mathbf{U}_2$, $\mathbf{U}_3$ and $\mathbf{U}_4$, then $\mathbf{U}_1$ is not orthogonal to $\mathbf{U}_3$ or $\mathbf{U}_4$. All other pairs are orthogonal.

(c) Yes. There are ${}_4C_2 = 6$ pairs to check.

(A.7) We want a vector $\begin{bmatrix} a \\ b \end{bmatrix}$ such that $\begin{bmatrix} a \\ b \end{bmatrix} \cdot \begin{bmatrix} 3 \\ 4 \end{bmatrix} = 0$. That is, we require $3a + 4b = 0$, or $b = -3a/4$. Any vector of the form $\begin{bmatrix} a \\ -3a/4 \end{bmatrix}$ will suffice. This is equivalent to the form $\begin{bmatrix} 4a \\ -3a \end{bmatrix}$. Sample answers, corresponding to $a = 1/4$, $a = 1$, and $a = -1$, are $\begin{bmatrix} 1 \\ -3/4 \end{bmatrix}, \begin{bmatrix} 4 \\ -3 \end{bmatrix}, \begin{bmatrix} -4 \\ 3 \end{bmatrix}$.

(A.8) (a) This is not an orthogonal coordinate system since

$$\frac{1}{\sqrt{3}} \begin{bmatrix} 1 \\ 1 \\ 1 \end{bmatrix} \cdot \frac{1}{\sqrt{14}} \begin{bmatrix} 2 \\ 1 \\ 3 \end{bmatrix} = \frac{1}{\sqrt{42}}(2+1+3) \neq 0$$

That is, the first and third vectors are not orthogonal.

(b) This is not an orthogonal coordinate system since there are only three coordinate axes. Four are required for 4-space.

(c) This is an orthogonal coordinate system. All three vectors are of length one, each is orthogonal to the other two vectors, and there are the correct number of them (three for 3-space).

(A.9) What we need to do is to project the vector $\mathbf{y}$ onto each of the coordinate axes $\mathbf{U_1}$, $\mathbf{U_2}$, $\mathbf{U_3}$ and $\mathbf{U_4}$. The resulting projection lengths ($\mathbf{y}\cdot\mathbf{U_1}$, and so on) are the constants that we require. These are:

$$\mathbf{y}\cdot\mathbf{U_1} = \begin{bmatrix} 3 \\ 9 \\ 11 \\ 13 \end{bmatrix} \cdot \frac{1}{2}\begin{bmatrix} 1 \\ 1 \\ 1 \\ 1 \end{bmatrix} = \frac{1}{2}(3+9+11+13) = \frac{36}{2} = 18$$

$$\mathbf{y}\cdot\mathbf{U_2} = \begin{bmatrix} 3 \\ 9 \\ 11 \\ 13 \end{bmatrix} \cdot \frac{1}{2}\begin{bmatrix} -1 \\ -1 \\ 1 \\ 1 \end{bmatrix} = \frac{1}{2}(-3-9+11+13) = \frac{12}{2} = 6$$

$$\mathbf{y}\cdot\mathbf{U_3} = \begin{bmatrix} 3 \\ 9 \\ 11 \\ 13 \end{bmatrix} \cdot \frac{1}{2}\begin{bmatrix} -1 \\ 1 \\ -1 \\ 1 \end{bmatrix} = \frac{1}{2}(-3+9-11+13) = \frac{8}{2} = 4$$

$$\mathbf{y}\cdot\mathbf{U_4} = \begin{bmatrix} 3 \\ 9 \\ 11 \\ 13 \end{bmatrix} \cdot \frac{1}{2}\begin{bmatrix} 1 \\ -1 \\ -1 \\ 1 \end{bmatrix} = \frac{1}{2}(3-9-11+13) = \frac{-4}{2} = -2$$

Hence the required answer is

$$\mathbf{y} = 18\mathbf{U_1} + 6\mathbf{U_2} + 4\mathbf{U_3} - 2\mathbf{U_4}$$

(A.10) (a) With the alternative coordinate system the orthogonal decomposition can be written as

$$\mathbf{y} = \frac{128}{\sqrt{2}}\mathbf{U_1} + \frac{143}{\sqrt{2}}\mathbf{U_2} + \frac{2}{\sqrt{2}}\mathbf{U_3} + \frac{5}{\sqrt{2}}\mathbf{U_4}$$

or more fully in the form

$$\begin{bmatrix} 63 \\ 65 \\ 69 \\ 74 \end{bmatrix} = \begin{bmatrix} 64 \\ 64 \\ 0 \\ 0 \end{bmatrix} + \begin{bmatrix} 0 \\ 0 \\ 71.5 \\ 71.5 \end{bmatrix} + \begin{bmatrix} -1 \\ 1 \\ 0 \\ 0 \end{bmatrix} + \begin{bmatrix} 0 \\ 0 \\ -2.5 \\ 2.5 \end{bmatrix}$$

(b) Pythagoras' Theorem applied to these decompositions yields

$$||\mathbf{y}||^2 = \left(\frac{128}{\sqrt{2}}\right)^2 + \left(\frac{143}{\sqrt{2}}\right)^2 + \left(\frac{2}{\sqrt{2}}\right)^2 + \left(\frac{5}{\sqrt{2}}\right)^2$$

or

$$||\mathbf{y}||^2 = 2 \times 64^2 + 2 \times 71.5^2 + 2 \times 1^2 + 2 \times 2.5^2$$

Both of these expressions lead to

$$||\mathbf{y}||^2 \;=\; 8192 \;+\; 10224.5 \;+\; 2 \;+\; 12.5$$

This is the required breakup in terms of squared lengths of projections onto the unit vectors $\mathbf{U_1}$, $\mathbf{U_2}$, $\mathbf{U_3}$ and $\mathbf{U_4}$. As a check we can calculate $||\mathbf{y}||^2$ directly as $63^2 + 65^2 + 69^2 + 74^2 = 18431$, the value on the right-hand side of our equation.

(A.11) (a) The length of the projection of $\mathbf{y}$ onto $\mathbf{U_1}$ is

$$\mathbf{y}\cdot\mathbf{U_1} \;=\; \begin{bmatrix} 6 \\ 5 \\ 9 \end{bmatrix} \cdot \frac{1}{\sqrt{3}}\begin{bmatrix} 1 \\ 1 \\ 1 \end{bmatrix} \;=\; \frac{6+5+9}{\sqrt{3}} \;=\; \frac{20}{\sqrt{3}}$$

Hence the projection vector $\mathbf{P_1y}$ is

$$\mathbf{P_1y} \;=\; (\mathbf{y}\cdot\mathbf{U_1})\mathbf{U_1} \;=\; \frac{20}{\sqrt{3}}\frac{1}{\sqrt{3}}\begin{bmatrix} 1 \\ 1 \\ 1 \end{bmatrix} \;=\; \frac{20}{3}\begin{bmatrix} 1 \\ 1 \\ 1 \end{bmatrix} \;=\; \begin{bmatrix} 6.67 \\ 6.67 \\ 6.67 \end{bmatrix}$$

The length of the projection of $\mathbf{y}$ onto $\mathbf{U_2}$ is

$$\mathbf{y}\cdot\mathbf{U_2} \;=\; \begin{bmatrix} 6 \\ 5 \\ 9 \end{bmatrix} \cdot \frac{1}{\sqrt{42}}\begin{bmatrix} -4 \\ -1 \\ 5 \end{bmatrix} \;=\; \frac{6\times(-4)+5\times(-1)+9\times 5}{\sqrt{42}} \;=\; \frac{16}{\sqrt{42}}$$

Hence the projection vector $\mathbf{P_2y}$ is

$$\mathbf{P_2y} \;=\; (\mathbf{y}\cdot\mathbf{U_2})\mathbf{U_2} \;=\; \frac{16}{\sqrt{42}}\frac{1}{\sqrt{42}}\begin{bmatrix} -4 \\ -1 \\ 5 \end{bmatrix} \;=\; \frac{16}{42}\begin{bmatrix} -4 \\ -1 \\ 5 \end{bmatrix} \;=\; \begin{bmatrix} -1.52 \\ -0.38 \\ 1.90 \end{bmatrix}$$

The length of the projection of $\mathbf{y}$ onto $\mathbf{U_3}$ is

$$\mathbf{y}\cdot\mathbf{U_3} \;=\; \begin{bmatrix} 6 \\ 5 \\ 9 \end{bmatrix} \cdot \frac{1}{\sqrt{14}}\begin{bmatrix} 2 \\ -3 \\ 1 \end{bmatrix} \;=\; \frac{6\times 2+5\times(-3)+9\times 1}{\sqrt{14}} \;=\; \frac{6}{\sqrt{14}}$$

Hence the projection vector $\mathbf{P_3y}$ is

$$\mathbf{P_3y} \;=\; (\mathbf{y}\cdot\mathbf{U_3})\mathbf{U_3} \;=\; \frac{6}{\sqrt{14}}\frac{1}{\sqrt{14}}\begin{bmatrix} 2 \\ -3 \\ 1 \end{bmatrix} \;=\; \frac{6}{14}\begin{bmatrix} 2 \\ -3 \\ 1 \end{bmatrix} \;=\; \begin{bmatrix} 0.86 \\ -1.29 \\ 0.43 \end{bmatrix}$$

(b) Hence we obtain the orthogonal decomposition

$$\mathbf{y} \;=\; \begin{bmatrix} 6 \\ 5 \\ 9 \end{bmatrix} \;=\; \begin{bmatrix} 6.67 \\ 6.67 \\ 6.67 \end{bmatrix} + \begin{bmatrix} -1.52 \\ -0.38 \\ 1.90 \end{bmatrix} + \begin{bmatrix} 0.86 \\ -1.29 \\ 0.43 \end{bmatrix}$$

which checks out, at least to our level of accuracy (the first value is out by 0.01 due to rounding errors).

(c) The squared lengths are as follows:

$$\begin{aligned}
||\mathbf{y}||^2 &= 6^2 + 5^2 + 9^2 &&= 142 \\
||\mathbf{P_1 y}||^2 &= 6.67^2 + 6.67^2 + 6.67^2 &&= 133.5 \\
||\mathbf{P_2 y}||^2 &= (-1.52)^2 + (-0.38)^2 + 1.90^2 &&= 6.1 \\
||\mathbf{P_3 y}||^2 &= 0.86^2 + (-1.29)^2 + 0.43^2 &&= 2.6
\end{aligned}$$

(d) To confirm Pythagoras' Theorem we calculate the sum of the squared lengths of the projection vectors:

$$133.5 + 6.1 + 2.6 = 142.2$$

which is very close to the required value of $||\mathbf{y}||^2 = 142$ (it would be exact if we had carried enough decimal places in our calculations).

(e)

$$\begin{aligned}
(\mathbf{y}\cdot\mathbf{U_1})^2 &= (20/\sqrt{3})^2 = 400/3 = 133.33 \\
(\mathbf{y}\cdot\mathbf{U_2})^2 &= (16/\sqrt{42})^2 = 256/42 = 6.10 \\
(\mathbf{y}\cdot\mathbf{U_3})^2 &= (6/\sqrt{14})^2 = 36/14 = 2.57
\end{aligned}$$

Comparison with the values given in (c) confirm that $||\mathbf{P_1 y}||^2 = (\mathbf{y}\cdot\mathbf{U_1})^2$, and so on, again given the level of accuracy we used in our calculations. Note that with the greater accuracy we have obtained for our $(\mathbf{y}\cdot\mathbf{U_i})^2$ values, we can see that Pythagoras' Theorem is indeed exactly true:

$$142 = 133.33 + 6.10 + 2.57$$

E.7 Solutions for Appendix C Exercises

(C.2) (a) Columns 1–3 of the spreadsheet will look something like:

y	U1	yxU1
26	0.378	9.827
19	0.378	7.181
18	0.378	6.803
14	0.378	5.292
11	0.378	4.158
28	0.378	10.583
14	0.378	5.292
		y.U1
		49.135

(b) Columns 4–6 of the spreadsheet will look something like:

y	(y.U1)U1	[y − (y.U1)U1]
26	18.571	7.429
19	18.571	0.429
18	18.571	−0.571
14	18.571	−4.571
11	18.571	−7.571
28	18.571	9.429
14	18.571	−4.571

(c) Columns 7–9 of the spreadsheet will look something like:

y(sq)	[(y.U1)U1](sq)	[y − (y.U1)U1](sq)
676	344.9	55.19
361	344.9	0.18
324	344.9	0.33
196	344.9	20.89
121	344.9	57.32
784	344.9	88.91
196	344.9	20.89
sqlength(C)	sqlength(A)	sqlength(B)
2658	2414.3	243.7

Yes, the last two numbers do add up to the first!

(d)

$$F = \frac{A^2}{B^2/6} = \frac{2414.3}{243.7/6} = 59.44$$

The appropriate reference distribution is $F_{1,6}$, which has a 99 percentile of 13.75 (Table T.2). Our test statistic of 59.44 exceeds this by a considerable margin, so we conclude that there is very strong evidence of a drop in mean selenium level between birth and 1 month of age. (The drop is estimated to be 18.6 units.)

(C.7) Statistical computing packages and spreadsheets are many and varied, and constantly changing. We therefore discuss the output from the two such programs with which we are most familiar, Minitab and Excel. Both have a regression routine.

With Minitab (Minitab Inc., 1994), the reader goes into the Data window and puts the nine data values into column one (c1), then the nine ones into column two (c2). Then the reader moves to the Session window and enters the command "let c2 = c2 /3." This divides the ones in column two by 3. Then the reader clicks on the Stat menu and then on Regression, and on Regression again. Then c1 is specified as the Response variable, and c2 as the single Predictor variable. Then the reader clicks on Options and specifies that the intercept should not be fitted. Lastly, OK, and OK again. This results in the following lines being added to the Session window.

Regression Analysis

The regression equation is
C1 = 21.4 C2

Predictor	Coef	Stdev	t-ratio	p
Noconstant				
C2	21.367	2.902	7.36	0.000

s = 2.902
Analysis of variance

SOURCE	DF	SS	MS	F	p
Regression	1	456.53	456.53	54.21	0.000
Error	8	67.38	8.42		
Total	9	523.91			

* Note * All values in column are identical.

The regression coefficient of 21.367 is the required projection length $\mathbf{y} \cdot \mathbf{U}_1$. The required Pythagorean breakup is printed in the "SS" column, as $523.41 = 456.53 + 67.38$.

The required test statistic of $F = 456.53/(67.38/8) = 54.21$ is printed by Minitab. This F value greatly exceeds all three percentiles of the $F_{1,8}$ distribution which are given in Table T.2, so we conclude there is very strong evidence of a difference in height between male and female twins.

With Excel (Microsoft Corporation, 1993), the reader puts the nine data values into cells A1:A9 of the worksheet, then puts the value $1/3 = 0.3333$ into cells B1:B9. He or she then clicks on the Tools menu, on Data Analysis, Regression, and OK. The Input Y Range is specified as A1:A9, the Input X Range is specified as B1:B9, and the Constant is set to Zero. An appropriate Output Range, such as D1 (= top left corner), is also specified. Then OK is clicked.

The output is disappointing, though adequate for our purposes, and is not reproduced here. The projection length $\mathbf{y} \cdot \mathbf{U}_1$ is correctly printed as the X Variable Coefficient of 21.36667. The analysis of variance entries are incorrect except for the Residual line, which gives us the squared length of the error vector, as the Residual SS of 67.38. Luckily this is sufficient information for us to calculate the correct Pythagorean breakup:

$$\begin{aligned} \|\mathbf{y}\|^2 &= (\mathbf{y} \cdot \mathbf{U}_1)^2 + \|\text{error vector}\|^2 \\ &= 21.36667^2 + 67.38 \\ &= 456.53 + 67.38 \end{aligned}$$

This enables us to calculate $F = 456.53/(67.38/8) = 54.21$ as above.

Note that the problem in Excel is associated with setting the Constant to zero, as in Exercises C.7 and C.8. In the remaining exercises, C.9–C.12, the Constant is not set to zero, so the problem is not encountered.

E.8 Solutions for Appendix D Exercises

(D.1) (a) The vectors **y** and **U** are:

$$\mathbf{y} = \begin{bmatrix} 8 \\ 7 \\ 5 \\ 6 \\ 9 \end{bmatrix} \qquad \mathbf{U} = \frac{1}{\sqrt{5}} \begin{bmatrix} 1 \\ 1 \\ 1 \\ 1 \\ 1 \end{bmatrix}$$

(b) The angle θ between **y** and **U** is determined by

$$\cos\theta = \frac{\begin{bmatrix} 8 \\ 7 \\ 5 \\ 6 \\ 9 \end{bmatrix} \cdot \begin{bmatrix} 1 \\ 1 \\ 1 \\ 1 \\ 1 \end{bmatrix}}{\sqrt{8^2 + 7^2 + 5^2 + 6^2 + 9^2} \times \sqrt{5}} = \frac{35}{\sqrt{1275}} = 0.9802$$

That is, the angle is $\theta = 0.199$ radians, or $11.4°$.

(c) The 10%, 5% and 1% critical values of the distribution of the test statistic θ under the null hypothesis $\mu = 0$ are 0.753, 0.624 and 0.410 radians, respectively (from Table T.2 using $q = 4$ error degrees of freedom). Our angle of 0.199 radians is considerably *less* than the 1% critical value, so we conclude that we have very strong evidence of a mean decrease in production due to nitrogen fertilizer. (Remember that with θ, it is *low* values which cause us to reject our hypothesis.)

(d) The p value is

$$p = 1 - r - r(1 - r^2)/2 = 1 - 0.9802 - 0.9802(1 - 0.9802^2)/2 = 0.0006$$

since $r = \cos\theta = 0.9802$ from (b) above. Yes, the calculated p value is in accord with the conclusions in (c) since p is considerably less than 0.01.

(e) (Optional) Using a numerical integration routine, the calculated p value is 0.0006, in agreement with the value obtained in (d).

(D.3) (a) The trick is to draw Figure E.4 and observe that

$$r = \cos\theta = \frac{\sqrt{n}\bar{y}}{\sqrt{\sum_{i=1}^{n} y_i^2}} = \frac{\sqrt{n}\bar{y}}{\sqrt{(n-1)s^2 + n\bar{y}^2}}$$

where the last equality is obtained using the Pythagorean relationship $\sum_{i=1}^{n} y_i^2 = \sum_{i=1}^{n}(y_i - \bar{y})^2 + n\bar{y}^2 = (n-1)s^2 + n\bar{y}^2$.

(b) For this example $n = 2$, $\bar{y} = 7.5$ in., and $s = 3/\sqrt{2}$. Hence

$$r = \frac{\sqrt{n}\bar{y}}{\sqrt{(n-1)s^2 + n\bar{y}^2}} = \frac{\sqrt{2} \times 7.5}{\sqrt{(2-1)3^2/2 + 2 \times 7.5^2}} = 0.9806$$

This is in agreement with the value given in Table D.1.

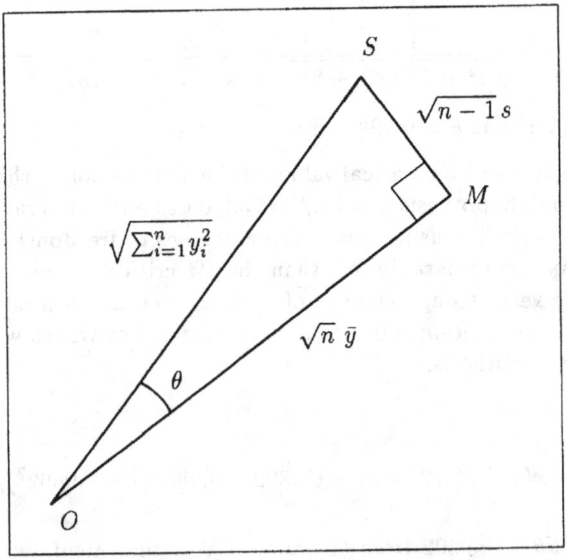

Figure E.4: The lengths of the sides of the "Statistical Triangle" for the general paired samples case. Note that the length of MS, $\sqrt{\sum_{i=1}^{n}(y_i - \bar{y})^2}$, has been rewritten as $\sqrt{(n-1)s^2} = \sqrt{n-1}\,s$.

(D.6) (a) The required "statistical triangle" is based on the following orthogonal decomposition (using treatment means of $\bar{y}_1 = 138$ and $\bar{y}_2 = 230$):

$$\mathbf{y} - \bar{\mathbf{y}} = (\bar{\mathbf{y}}_i - \bar{\mathbf{y}}) + (\mathbf{y} - \bar{\mathbf{y}}_i)$$

$$\begin{bmatrix} -64 \\ -4 \\ -44 \\ -84 \\ -34 \\ -114 \\ 226 \\ -144 \\ -94 \\ 356 \end{bmatrix} = \begin{bmatrix} -46 \\ -46 \\ -46 \\ -46 \\ -46 \\ 46 \\ 46 \\ 46 \\ 46 \\ 46 \end{bmatrix} + \begin{bmatrix} -18 \\ 42 \\ 2 \\ -38 \\ 12 \\ -160 \\ 180 \\ -190 \\ -140 \\ 310 \end{bmatrix}$$

We leave the reader to sketch the triangle. Its sides are of length $C = \|\mathbf{y} - \bar{\mathbf{y}}\| = \sqrt{234640} = 484.4$, $A = \|\bar{\mathbf{y}}_i - \bar{\mathbf{y}}\| = \sqrt{21160} = 145.5$ and $B = \|\mathbf{y} - \bar{\mathbf{y}}_i\| = \sqrt{213480} = 462.0$. The angle can be obtained using any of the trigonometric functions, such as $\cos\theta = A/C = \sqrt{21160/234640} = 0.3003$. This leads to $\theta = 1.266$ radians, or 72.5°.

(b) The F value is $A^2/(B^2/q) = 21160/(213480/8) = 0.79$. The t value is $A/(B/\sqrt{q}) = \sqrt{21160/(213480/8)} = 0.890$. The θ value has already been calculated to be 1.266 radians. The r value has also been calculated to be $\cos\theta = 0.3003$.

The corresponding 5% critical values, for eight error degrees of freedom, are 5.32, 2.306, 0.887 radians and 0.632, respectively, from Table T.2. Our calculated F, t and r values do not exceed their respective critical values, and our calculated θ value is not less than its critical value. That is, none of our test statistics yields a "statistically significant" result. This means there is insufficient evidence to conclude that there is any difference in average load weight between residential and business trailers.

(c) There are six relationships to confirm. These are as follows:

$$r = \cos\theta = \cos 1.266 = 0.3003$$

$$F = q\cot^2\theta = 8\cot^2 1.266 = 0.79$$

$$t = \sqrt{q}\cot\theta = \sqrt{8}\cot 1.266 = 0.890$$

$$F = t^2 = 0.890^2 = 0.79$$

$$F = \frac{q\,r^2}{1 - r^2} = \frac{8 \times 0.3003^2}{1 - 0.3003^2} = 0.79$$

$$t = \frac{\sqrt{q}\,r}{\sqrt{1 - r^2}} = \frac{\sqrt{8} \times 0.3003}{\sqrt{1 - 0.3003^2}} = 0.890$$

These all produce correct values, thereby confirming the relationships.

(d) (Optional) The numerical integration routine should produce a p value of 0.40. This is not less than 0.05, so agrees with our findings in (b).

Note that there is a problem with analyzing this data set as we have just done. One of the underlying assumptions has been violated. This is the assumption of equality of variances, $\sigma_1^2 = \sigma_2^2$. It seems that residential trailers, carrying mainly organic material such as garden refuse, were not as variable in load weight as business trailers, which tended to carry non-organic refuse. A more correct method of analysis is to use "Welch's test," something we do not wish to do here. (If it is of any comfort to the reader, a "nonsignificant" result is also obtained using Welch's test.)

(D.11) (a) The specified contrast is the second contrast, corresponding to the third unit vector $\mathbf{U_3}$. The required "statistical triangle" is therefore based on the following orthogonal decomposition (using treatment means of $\bar{y}_1 = 4.3$, $\bar{y}_2 = 7.5$, $\bar{y}_3 = 3.7$ and $\bar{y}_4 = 5.5$):

$$\mathbf{y} - (\mathbf{y}\cdot\mathbf{U_1})\mathbf{U_1} - (\mathbf{y}\cdot\mathbf{U_2})\mathbf{U_2} - (\mathbf{y}\cdot\mathbf{U_4})\mathbf{U_4} = (\mathbf{y}\cdot\mathbf{U_3})\mathbf{U_3} + (\mathbf{y} - \bar{\mathbf{y}}_i)$$

$$\begin{bmatrix} 4.1 \\ 5.2 \\ 3.6 \\ 6.3 \\ 7.5 \\ 8.7 \\ 3.0 \\ 4.7 \\ 3.4 \\ 5.7 \\ 4.7 \\ 6.1 \end{bmatrix} - \begin{bmatrix} 5.25 \\ 5.25 \\ 5.25 \\ 5.25 \\ 5.25 \\ 5.25 \\ 5.25 \\ 5.25 \\ 5.25 \\ 5.25 \\ 5.25 \\ 5.25 \end{bmatrix} - \begin{bmatrix} 0.65 \\ 0.65 \\ 0.65 \\ 0.65 \\ 0.65 \\ 0.65 \\ -0.65 \\ -0.65 \\ -0.65 \\ -0.65 \\ -0.65 \\ -0.65 \end{bmatrix} - \begin{bmatrix} -0.35 \\ -0.35 \\ -0.35 \\ 0.35 \\ 0.35 \\ 0.35 \\ 0.35 \\ 0.35 \\ 0.35 \\ -0.35 \\ -0.35 \\ -0.35 \end{bmatrix} = \begin{bmatrix} -1.25 \\ -1.25 \\ -1.25 \\ 1.25 \\ 1.25 \\ 1.25 \\ -1.25 \\ -1.25 \\ -1.25 \\ 1.25 \\ 1.25 \\ 1.25 \end{bmatrix} + \begin{bmatrix} -0.2 \\ 0.9 \\ -0.7 \\ -1.2 \\ 0.0 \\ 1.2 \\ -0.7 \\ 1.0 \\ -0.3 \\ 0.2 \\ -0.8 \\ 0.6 \end{bmatrix}$$

This simplifies to the following decomposition:

$$\begin{bmatrix} -1.45 \\ -0.35 \\ -1.95 \\ 0.05 \\ 1.25 \\ 2.45 \\ -1.95 \\ -0.25 \\ -1.55 \\ 1.45 \\ 0.45 \\ 1.85 \end{bmatrix} = \begin{bmatrix} -1.25 \\ -1.25 \\ -1.25 \\ 1.25 \\ 1.25 \\ 1.25 \\ -1.25 \\ -1.25 \\ -1.25 \\ 1.25 \\ 1.25 \\ 1.25 \end{bmatrix} + \begin{bmatrix} -0.2 \\ 0.9 \\ -0.7 \\ -1.2 \\ 0.0 \\ 1.2 \\ -0.7 \\ 1.0 \\ -0.3 \\ 0.2 \\ -0.8 \\ 0.6 \end{bmatrix}$$

We leave the reader to sketch the triangle. Its sides are of length $C = \|\mathbf{y} - (\mathbf{y}\cdot\mathbf{U}_1)\mathbf{U}_1 - (\mathbf{y}\cdot\mathbf{U}_2)\mathbf{U}_2 - (\mathbf{y}\cdot\mathbf{U}_4)\mathbf{U}_4\| = \sqrt{25.59} = 5.059$, $A = \mathbf{y}\cdot\mathbf{U}_3 = \sqrt{18.75} = 4.330$ and $B = \|\mathbf{y} - \bar{\mathbf{y}}_i\| = \sqrt{6.84} = 2.615$. The angle can be obtained using any of the trigonometric functions, such as $\cos\theta = A/C = \sqrt{18.75/25.59} = 0.8560$. This leads to $\theta = 0.5433$ radians, or 31.1°.

(b) The F value is $A^2/(B^2/q) = 18.75/(6.84/8) = 21.93$. The t value is $A/(B/\sqrt{q}) = \sqrt{18.75/(6.84/8)} = 4.683$. The θ value has already been calculated to be 0.5433 radians. The r value has also been calculated to be $\cos\theta = 0.8560$.

The corresponding 1% critical values, for eight error degrees of freedom, are 11.26, 3.355, 0.700 radians and 0.765, respectively, from Table T.2. Our calculated F, t and r values all exceed their respective 1% critical values, and our calculated θ value is less than its 1% critical value. That is, all of our test statistics yield "statistically significant" results ($p < 0.01$), suggesting that ordinary cats are on average heavier than Siamese cats.

(c) There are six relationships to confirm. These are as follows:

$$r = \cos\theta = \cos 0.5433 = 0.8560$$

$$F = q\cot^2\theta = 8\cot^2 0.5433 = 21.93$$

$$t = \sqrt{q}\cot\theta = \sqrt{8}\cot 0.5433 = 4.683$$

$$F = t^2 = 4.683^2 = 21.93$$

$$F = \frac{q\,r^2}{1-r^2} = \frac{8\times 0.8560^2}{1-0.8560^2} = 21.93$$

$$t = \frac{\sqrt{q}\,r}{\sqrt{1-r^2}} = \frac{\sqrt{8}\times 0.8560}{\sqrt{1-0.8560^2}} = 4.683$$

These all produce correct values, thereby confirming the relationships.

(d) (Optional) The numerical integration routine produces a p value of 0.0016. This is less than 0.01, so agrees with our findings in (b).

(D.16) The angle θ is the angle between the vectors $(\mathbf{y} - \bar{\mathbf{y}})$ and $(\mathbf{x} - \bar{\mathbf{x}})$. The required test statistic is

$$r = \cos\theta = \frac{\begin{bmatrix} 0.708 \\ 0.318 \\ -0.382 \\ -0.802 \\ 0.318 \\ 2.108 \\ -0.832 \\ -0.242 \\ -0.272 \\ -0.922 \end{bmatrix} \cdot \begin{bmatrix} 4.9 \\ 4.9 \\ 4.9 \\ -15.1 \\ 34.9 \\ 44.9 \\ -15.1 \\ -29.1 \\ 4.9 \\ -40.1 \end{bmatrix}}{\sqrt{7.61116}\times\sqrt{6240.9}} = \frac{176.258}{\sqrt{47500.48844}} = 0.809$$

The 1% critical value for the correlation coefficient (with $n - 2 = 8$ error degrees of freedom) is 0.765. This is exceeded by our test statistic. Hence we conclude there is strong evidence of a relationship between red cell lead levels and the number of rounds fired per week.

Appendix T

Statistical Tables

In this appendix we provide tables of random numbers (Table T.1) and percentiles of the distributions of $F_{1,q}$, t_q, the correlation coefficient (r value) under the null hypothesis, and the angle θ under the null hypothesis (Table T.2).

The random numbers in Table T.1 were produced using the computing package *Gauss*. The percentiles in Table T.2 were obtained as follows. Firstly, the percentiles of the distribution of the angle θ were calculated by using the computing package Maple (Char et al. 1991) to solve equations such as

$$\frac{\int_0^{\theta} \sin^{q-1} u \, du}{2\int_0^{\pi/2} \sin^{q-1} u \, du} = 0.025$$

(for the 2.5 percentile) as described in Appendix D, where θ is in radians. The percentiles of the other three distributions were then calculated using the formulas summarized in Figure D.17.

3	7	2	6	7	4	5	0	4	7	4	5	4	1	9	4	2	1	8
6	8	8	5	7	8	6	0	4	4	2	4	1	3	7	5	2	1	8
7	6	4	3	7	8	5	4	9	3	0	8	3	6	3	8	1	5	1
0	0	8	8	4	6	0	2	5	0	2	3	4	2	7	3	3	5	4
7	3	7	1	5	1	0	5	2	0	4	1	9	7	9	4	7	5	2
2	2	4	9	6	7	6	7	7	5	2	8	3	9	7	8	3	1	7
6	8	7	8	1	2	0	0	7	8	3	1	5	6	1	3	9	8	3
8	5	5	9	9	8	3	3	1	4	5	9	8	9	0	6	4	8	8
5	1	1	9	4	4	3	2	2	6	7	2	4	5	3	1	5	7	0
3	9	2	9	7	1	0	9	9	9	6	9	4	6	1	5	0	9	1
3	6	5	0	5	5	5	7	7	8	0	2	3	6	8	7	6	8	9
2	6	2	5	4	6	4	5	3	5	2	5	5	9	3	5	7	6	6
6	2	5	4	0	4	1	2	3	1	4	8	2	4	4	4	2	2	7
0	0	1	2	7	0	8	5	7	1	2	5	9	5	1	8	4	5	8
3	2	3	4	7	9	0	9	0	2	8	3	5	2	9	1	3	3	7
9	5	5	8	1	7	3	8	8	3	4	9	3	2	3	2	8	0	5
7	5	8	3	5	7	5	7	4	2	1	5	9	8	5	2	3	7	4
7	7	9	6	1	8	0	3	3	0	7	4	1	7	5	7	7	9	0
1	7	7	3	8	3	8	3	3	6	1	9	3	8	3	7	7	3	1
8	9	1	2	7	4	3	6	6	5	3	2	8	5	6	7	9	8	2
6	0	3	2	1	4	3	4	4	6	2	3	0	9	9	8	6	5	4
0	7	6	4	2	4	4	0	6	8	2	6	8	8	4	3	1	6	5
6	9	3	2	0	1	3	7	8	1	8	1	2	2	9	3	6	9	8
7	5	2	9	7	3	9	1	4	2	5	1	2	2	3	6	1	6	6
2	8	8	4	9	1	9	6	4	8	3	0	4	9	1	7	3	4	1
6	3	4	8	5	2	2	2	0	6	6	6	2	2	1	1	0	7	7
2	0	2	2	5	9	2	0	8	2	1	6	1	6	1	7	9	8	9
7	7	2	3	5	7	4	3	8	3	2	3	5	8	4	7	3	3	3
1	5	1	1	8	5	8	9	5	4	7	9	3	0	4	0	1	1	1
6	7	7	7	4	1	7	6	3	4	6	7	3	6	1	8	9	1	7
3	2	9	3	4	1	8	8	0	1	8	4	7	8	2	9	7	8	5
2	1	8	1	1	5	2	5	3	1	5	5	3	2	4	5	9	9	3
4	4	2	7	5	5	7	2	8	1	3	3	6	1	6	4	6	2	9
9	3	9	4	0	9	2	0	0	7	1	2	6	7	1	7	6	9	1
7	8	7	5	9	5	1	2	0	8	3	6	3	2	2	5	3	6	2
8	9	5	7	0	9	6	0	4	8	2	3	9	6	0	3	4	8	1

Table T.1: Table of random numbers. To use the table, begin at a random point and read down individual columns for single-digit numbers, pairs of columns for two-digit numbers, and so on. Ignore numbers which are outside the desired range. When the bottom of the table is reached, go to the top of the next column or group of columns.

Degrees of Freedom (q)	Percentiles of $F_{1,q}$		Percentiles of	t_q	r	Percentiles of θ	
	90	40	95	6.31	0.988	5	0.15708
1	95	161	97.5	12.71	0.997	2.5	0.07854
	99	4052	99.5	63.66	1.000	0.5	0.01571
	90	8.5	95	2.920	0.900	5	0.45103
2	95	18.5	97.5	4.303	0.950	2.5	0.31756
	99	98.5	99.5	9.925	0.990	0.5	0.14154
	90	5.54	95	2.353	0.805	5	0.63447
3	95	10.13	97.5	3.182	0.878	2.5	0.49842
	99	34.12	99.5	5.841	0.959	0.5	0.28828
	90	4.54	95	2.132	0.729	5	0.75350
4	95	7.71	97.5	2.776	0.811	2.5	0.62425
	99	21.20	99.5	4.604	0.917	0.5	0.40980
	90	4.06	95	2.015	0.669	5	0.83734
5	95	6.61	97.5	2.571	0.754	2.5	0.71592
	99	16.26	99.5	4.032	0.875	0.5	0.50634
	90	3.78	95	1.943	0.621	5	0.90015
6	95	5.99	97.5	2.447	0.707	2.5	0.78592
	99	13.75	99.5	3.707	0.834	0.5	0.58386
	90	3.59	95	1.895	0.582	5	0.94936
7	95	5.59	97.5	2.365	0.666	2.5	0.84145
	99	12.25	99.5	3.499	0.798	0.5	0.64736
	90	3.46	95	1.860	0.549	5	0.98920
8	95	5.32	97.5	2.306	0.632	2.5	0.88680
	99	11.26	99.5	3.355	0.765	0.5	0.70039
	90	3.36	95	1.833	0.521	5	1.02230
9	95	5.12	97.5	2.262	0.602	2.5	0.92471
	99	10.56	99.5	3.250	0.735	0.5	0.74544
	90	3.29	95	1.812	0.497	5	1.05035
10	95	4.96	97.5	2.228	0.576	2.5	0.95699
	99	10.04	99.5	3.169	0.708	0.5	0.78429

Table T.2: Table of percentiles of the distributions of $F_{1,q}$, t_q, the correlation coefficient (or r value) under the null hypothesis, and the angle θ under the null hypothesis, for varying error degrees of freedom (q). For a two-sided hypothesis test the values in this table are 10%, 5% and 1% *critical values.* The angle θ is as defined in Appendix D, and is given here in radians. The excessive accuracy for θ is to enable the reader to verify the relationships in Figure D.17 (given the provisos at the end of this table).

Table T.2 (continued).

Degrees of Freedom (q)	Percentiles of $F_{1,q}$		Percentiles of	t_q	r	Percentiles of θ	
11	90	3.23	95	1.796	0.476	5	1.07452
	95	4.84	97.5	2.201	0.553	2.5	0.98491
	99	9.65	99.5	3.106	0.684	0.5	0.81821
12	90	3.18	95	1.782	0.458	5	1.09561
	95	4.75	97.5	2.179	0.532	2.5	1.00935
	99	9.33	99.5	3.055	0.661	0.5	0.84815
13	90	3.14	95	1.771	0.441	5	1.11424
	95	4.67	97.5	2.160	0.514	2.5	1.03098
	99	9.07	99.5	3.012	0.641	0.5	0.87481
14	90	3.10	95	1.761	0.426	5	1.13084
	95	4.60	97.5	2.145	0.497	2.5	1.05030
	99	8.86	99.5	2.977	0.623	0.5	0.89875
15	90	3.07	95	1.753	0.412	5	1.14575
	95	4.54	97.5	2.131	0.482	2.5	1.06769
	99	8.68	99.5	2.947	0.606	0.5	0.92040
16	90	3.05	95	1.746	0.400	5	1.15925
	95	4.49	97.5	2.120	0.468	2.5	1.08346
	99	8.53	99.5	2.921	0.590	0.5	0.94009
17	90	3.03	95	1.740	0.389	5	1.17154
	95	4.45	97.5	2.110	0.456	2.5	1.09783
	99	8.40	99.5	2.898	0.575	0.5	0.95811
18	90	3.01	95	1.734	0.378	5	1.18279
	95	4.41	97.5	2.101	0.444	2.5	1.11100
	99	8.29	99.5	2.878	0.561	0.5	0.97468
19	90	2.99	95	1.729	0.369	5	1.19315
	95	4.38	97.5	2.093	0.433	2.5	1.12314
	99	8.18	99.5	2.861	0.549	0.5	0.98997
20	90	2.97	95	1.725	0.360	5	1.20271
	95	4.35	97.5	2.086	0.423	2.5	1.13436
	99	8.10	99.5	2.845	0.537	0.5	1.00416

Table T.2 (continued).

Degrees of Freedom (q)	Percentiles of $F_{1,q}$		Percentiles of	t_q	r	Percentiles of θ	
22	90	2.95	95	1.717	0.344	5	1.21985
	95	4.30	97.5	2.074	0.404	2.5	1.15449
	99	7.95	99.5	2.819	0.515	0.5	1.02967
24	90	2.93	95	1.711	0.330	5	1.23481
	95	4.26	97.5	2.064	0.388	2.5	1.17207
	99	7.82	99.5	2.797	0.496	0.5	1.05203
26	90	2.91	95	1.706	0.317	5	1.24800
	95	4.23	97.5	2.056	0.374	2.5	1.18760
	99	7.72	99.5	2.779	0.479	0.5	1.07184
28	90	2.89	95	1.701	0.306	5	1.25975
	95	4.20	97.5	2.048	0.361	2.5	1.20145
	99	7.64	99.5	2.763	0.463	0.5	1.08954
30	90	2.88	95	1.697	0.296	5	1.27030
	95	4.17	97.5	2.042	0.349	2.5	1.21390
	99	7.56	99.5	2.750	0.449	0.5	1.10549
35	90	2.85	95	1.690	0.275	5	1.29261
	95	4.12	97.5	2.030	0.325	2.5	1.24024
	99	7.42	99.5	2.724	0.418	0.5	1.13932
40	90	2.84	95	1.684	0.257	5	1.31059
	95	4.08	97.5	2.021	0.304	2.5	1.26149
	99	7.31	99.5	2.704	0.393	0.5	1.16672
45	90	2.82	95	1.679	0.243	5	1.32548
	95	4.06	97.5	2.014	0.288	2.5	1.27911
	99	7.23	99.5	2.690	0.372	0.5	1.18948
50	90	2.81	95	1.676	0.231	5	1.33808
	95	4.03	97.5	2.009	0.273	2.5	1.29403
	99	7.17	99.5	2.678	0.354	0.5	1.20879
60	90	2.79	95	1.671	0.211	5	1.35837
	95	4.00	97.5	2.000	0.250	2.5	1.31808
	99	7.08	99.5	2.660	0.325	0.5	1.23998

Table T.2 (continued).

Degrees of Freedom (q)	Percentiles of $F_{1,q}$		Percentiles of	t_q	r	Percentiles of θ	
70	90	2.78	95	1.667	0.195	5	1.37414
	95	3.98	97.5	1.994	0.232	2.5	1.33678
	99	7.01	99.5	2.648	0.302	0.5	1.26428
80	90	2.77	95	1.664	0.183	5	1.38684
	95	3.96	97.5	1.990	0.217	2.5	1.35187
	99	6.96	99.5	2.639	0.283	0.5	1.28392
90	90	2.76	95	1.662	0.173	5	1.39737
	95	3.95	97.5	1.987	0.205	2.5	1.36437
	99	6.93	99.5	2.632	0.267	0.5	1.30021
100	90	2.76	95	1.660	0.164	5	1.40627
	95	3.94	97.5	1.984	0.195	2.5	1.37494
	99	6.90	99.5	2.626	0.254	0.5	1.31400
150	90	2.74	95	1.655	0.134	5	1.43647
	95	3.90	97.5	1.976	0.159	2.5	1.41084
	99	6.81	99.5	2.609	0.208	0.5	1.36091
200	90	2.73	95	1.653	0.116	5	1.45447
	95	3.89	97.5	1.972	0.138	2.5	1.43226
	99	6.76	99.5	2.601	0.181	0.5	1.38894
300	90	2.72	95	1.650	0.095	5	1.47582
	95	3.87	97.5	1.968	0.113	2.5	1.45767
	99	6.72	99.5	2.592	0.148	0.5	1.42223
400	90	2.72	95	1.649	0.082	5	1.48855
	95	3.86	97.5	1.966	0.098	2.5	1.47282
	99	6.70	99.5	2.588	0.128	0.5	1.44210

In Table T.2 the percentiles for the distributions of $F_{1,q}$, t_q, and the correlation coefficient with q error degrees of freedom, agree in all cases except one with those given in the "Biometrika" tables of Pearson and Hartley (1966) for the error degrees of freedom which occur in both sets of tables. The one very minor disagreement occurs with the 97.5 percentile of the correlation coefficient for 12 degrees of freedom; here our θ percentile is 1.0956145 radians (to seven decimal places), which corresponds to $r = \cos\theta = 0.4575001$, which we rounded to 0.458, in comparison with the Pearson and Hartley value of 0.457.

In Table T.2 we have rounded our calculated percentiles for θ to five decimal places; this level of accuracy is sufficient for the accurate calculation of the other three columns except in the following cases. The 99 percentile of θ with one degree of freedom needs to be given to six decimal places (as $\pi/200 = 0.015708$ radians). Also, the 99 percentile of θ with six degrees of freedom needs to be given to six decimal places (as 0.583859 radians).

It is interesting to note that the 97.5 percentile for t_{100} in Table T.2, of 1.984, does not agree with the value of 1.982 given in Snedecor and Cochran (1967) and Little and Hills (1978). The corresponding value for $F_{1,100}$, of 3.94, is in agreement between the three sources; also, the square of 1.9825, the maximum number which rounds to 1.982, is 3.93, not 3.94, again suggesting that 1.982 is an erroneous value. This small error was presumably caused by the usage of an approximating function for t_{100}.

References

Box, J.F. (1978). *R.A. Fisher, The Life of a Scientist.* New York: Wiley.

Box, G.E.P., Hunter, W.G. and Hunter, J.S. (1978). *Statistics for Experimenters: An Introduction to Design, Data Analysis and Model Building.* New York: Wiley.

Bryant, P. (1984). Geometry, statistics, probability: Variations on a common theme. *The American Statistician* 38: 38–48.

Chance, W.A. (1986). A geometric derivation of the distribution of the correlation coefficient $|r|$ when $\rho = 0$. *American Mathematical Monthly* 93: 94–98.

Char, B.W., Geddes, K.O., Gonnet, G.H., Leong, B.L., Monogan, M.B. and Watt, S.M. (1991). *Maple V Language Reference Manual.* New York: Springer-Verlag.

Corsten, L.C.A. (1958). Vectors, a tool in statistical regression theory. *Mededelingen van de Landbouwhogeschool te Wageningen, Nederland* 58: 1–92.

Darlow, B.A., Inder, T.E., Sluis, K.B., Nuthall, G., Mogridge, N. and Winterbourn, C.C. (1995). Selenium status of New Zealand infants fed either a selenium supplemented or a standard formula. *Journal of Pediatrics and Child Health* 31: 339–344.

Dolamore, B.A., Brown, J., Darlow, B.A., George, P.M., Sluis, K.B. and Winterbourn, C.C. (1992). Selenium status of Christchurch infants and the effect of diet. *New Zealand Medical Journal* 105: 139–142.

Durbin, J. and Kendall, M.G. (1951). The geometry of estimation. *Biometrika* 38: 150–158.

Dyke, G.V. (1988). *Comparative Experiments with Field Crops.* London: Griffin.

George, P.M., Walmsley, T.A., Currie, D. and Wells, J.E. (1993). Lead exposure during recreational use of small bore rifle ranges. *New Zealand Medical Journal* 106: 422–424.

Herr, D.G. (1980). On the history of the use of geometry in the general linear model. *The American Statistician* 34: 43–47.

Little, T.M. and Hills, F.J. (1978). *Agricultural Experimentation: Design and Analysis.* New York: Wiley.

Margolis, M.S. (1979). Perpendicular projections and elementary statistics. *The American Statistician* 33: 131–135.

Microsoft Corporation (1993). *User's Guide: Microsoft Excel, Version 5.0.* Redmond: Microsoft.

Minitab Inc. (1994). *MINITAB Reference Manual, Release 10 for Windows.* State College: Minitab.

Money, D.F.L. (1970). Vitamin E and selenium deficiencies and their possible aetiological role in the sudden death in infants syndrome. *New Zealand Medical Journal* 71: 32–34.

Money, D.F.L. (1978). Vitamin E, selenium, iron and vitamin A content of livers from Sudden Infant Death Syndrome cases and control children: interrelationships and possible significance. *New Zealand Journal of Science* 21: 41–55.

Pearson, E.S. and Hartley, H.O. (1966). *Biometrika Tables for Statisticians.* Cambridge: Cambridge University Press, for Biometrika Trustees.

Saville, D.J. and Wood, G.R. (1986). A method for teaching statistics using *N*-dimensional geometry. *The American Statistician* 40: 205–214.

Saville, D.J. and Wood, G.R. (1991). *Statistical Methods: The Geometric Approach.* New York: Springer-Verlag.

Scheffé, H. (1959). *The Analysis of Variance.* New York: Wiley.

Sluis, K.B., Darlow, B.A., George, P.M., Mogridge, N., Dolamore, B.A. and Winterbourn, C.C. (1992). Selenium and glutathione peroxidase levels in premature infants in a low selenium community (Christchurch, New Zealand). *Pediatric Research* 32: 189–194.

Snedecor, G.W. and Cochran, W.G. (1967). *Statistical Methods,* 6th edition. Ames: Iowa State University Press.

Winterbourn, C.C., Saville, D.J., George, P.M. and Walmsley, T.A. (1992). Increase in selenium status of Christchurch adults associated with deregulation of the wheat market. *New Zealand Medical Journal* 105: 466–468.

Index

9 780387 947051